MOLECULAR BIOLOGY
OF THE
CARDIOVASCULAR SYSTEM

MOLECULAR BIOLOGY OF THE CARDIOVASCULAR SYSTEM

Edited by Shu Chien, M.D., Ph.D.

Professor of Bioengineering and Medicine
Department of Applied Mechanics and Engineering Sciences
and Department of Medicine
Member, Center for Molecular Genetics
University of California, San Diego

LEA & FEBIGER   PHILADELPHIA · LONDON

1990

Lea & Febiger
200 Chester Field Parkway
Malvern, PA 19355
U.S.A.
800-444-1785

Lea & Febiger (UK) Ltd.
145a Croydon Road
Beckenham, Kent BR3 3RB
U.K.

Library of Congress Cataloging-in-Publication Data

Molecular biology of the cardiovascular system / edited by Shu Chien.
    p.   cm.
  Includes bibliographical references.
  ISBN 0-8121-1312-8
  1. Cardiovascular system—Physiology. 2. Molecular biology.
3. Cardiovascular system—Diseases—Molecular aspects. I. Chien,
Shu
  [DNLM: 1. Cardiovascular System—cytology. 2. Cardiovascular
System—physiology. 3. Gene Expression Regulation. 4. Molecular
Biology. WG 102 M718]
QP102.M635 1990
612.1—dc20
DNLM/DLC
for Library of Congress   90-5465
                                   CIP

PRINTED IN THE UNITED STATES OF AMERICA

Print number: 5 4 3 2 1

Dedicated to the American Physicological Society and Its Members

FOREWORD

Molecular biology has greatly altered the nature of basic and applied biomedical research by providing investigators with a variety of powerful experimental techniques and strategies. The past 15 years have witnessed an array of impressive research advances including the identification and chromosomal localization of specific genetic sequences that are closely linked to, or responsible for, inherited diseases; the development of a new generation of sensitive and specific diagnostic tests for human diseases; the production of new and/or increased quantities of biologically active compounds; the development of transgenic animals that allow investigators to study the biologic effects of specific genes in molecular, biochemical, cellular, and physiologic terms; an enhanced understanding of the mechanisms controlling gene expression in response to developmental events and hormonal and cellular factors; and the investigation of structure–function relationships of proteins using recombinant DNA techniques such as site-directed mutagenesis and the construction of chimeric genes.

In spite of the widespread application and success of molecular biologic approaches in such areas as oncology, immunologic disorders, and neurologic and skeletal muscle diseases, the technologies of molecular biology have been underused in many important basic and clinical research areas of the circulatory system. The most significant advances employing molecular biologic approaches, outside of a few notable exceptions, have fallen in the area of hematologic research. In recent years, however, more widespread application and greater use of these technologies is leading to important advances in our understanding of cardiovascular development, organization, structure, function, and pathologic disease processes.

This volume provides selected examples of these exciting developments and, as such, represents new avenues to dissect the complex interactions and interrelationships of the various components of the cardiovascular system and to provide a detailed understanding of the molecular mechanisms underlying both normal and abnormal cardiovascular function. The subject matter includes the molecular biology of the heart, humoral factors, vascular smooth muscle, endothelium, red cell membrane, and lipoproteins. In addition, it includes an extensive and highly useful discussion of the experimental principles, techniques, and concepts of modern molecular biology.

Continued and increased use of these methodologies and conceptual approaches in cardiovascular research areas will provide broad opportunities for creative explorations that will generate penetrating insights about challenging topics at the forefront of biology, such as morphogenesis, gene expression, cellular communication, cell-cycle control, gene-replacement therapy, and gene mapping of polygenic diseases. Consequently, the application of molecular biologic technologies, in conjunction with conventional approaches, to basic and clinical cardiovascular problems, will provide important breakthroughs in our understanding of the etiology and pathogenesis of cardiovascular diseases; will lead to innovative and novel diagnostic, preventive, and therapeutic strategies; and will yield insight into key fundamental mechanisms that are important to all biology.

Stephen C. Mockrin, Ph.D.
Claude Lenfant, M.D.

National Heart, Lung, and Blood Institute

PREFACE

This book summarizes some of the recent advances in research on molecular biology of the cardiovascular system. The first chapter introduces the principles and techniques used in molecular biology; it is intended as a starting point for readers unfamiliar with this area to facilitate their reading of this book. The next nine chapters cover the current knowledge on molecular biology of the heart, blood vessels, and elements in the blood, including muscle differentiation, the hypertrophic responses of cardiac muscle to thyroid hormone and adrenergic stimulation, the atrial natriuretic factor and its relation to cardiac hypertrophy, the vascular smooth muscle, the endothelium and its interactions with leukocytes, angiogenesis, the red cell membrane skeleton, and lipoproteins. These chapters are written by scientists working at the forefront of these areas of cardiovascular molecular biology. The book concludes with a chapter summarizing the materials presented on various aspects of the heart, blood vessels, and blood, with special attention to the common principles and the application of a molecular biologic approach to physiologic and pathophysiologic investigations of the cardiovascular system.

This book is suitable for both the readers who have little or no prior knowledge of the field and those who are working actively in the area and wish to have a state-of-the-art reference book. It is hoped that the materials presented will further stimulate researchers to find new ways to elucidate the intricate mechanisms of cardiovascular regulation in health and disease, especially by applying molecular biologic information to in-vivo investigations at the organ-system level. This book should be valuable to basic scientists, clinical investigators, and students interested in the cardiovascular system and/or molecular biology.

This book was initiated when I was on sabbatical leave at the Institute of Biomedical Sciences, Academia Sinica, Taipei, Taiwan, R.O.C., prior to my move from Columbia University to the University of California, San Diego. To my colleagues at these three institutions I express my gratitude for their help and encouragement. I thank the American Physiological Society and its Cardiovascular Section, who provided the impetus for the production of this book by sponsoring a symposium on this topic on October 11, 1988 in Montreal, Canada. I wish to express my sincere appreciation to Drs. Stephen C. Mockrin and Claude Lenfant for writing the inspiring Foreword. I am most grateful to the authors for their excellent contributions and for their valuable comments and criticisms on Chapters 1 and 11.

San Diego, California

Shu Chien, M.D., Ph.D.

CONTRIBUTORS

Joseph J. Bahl, Ph.D.
Research Assistant Professor
Department of Internal Medicine
University of Arizona College of Medicine
Tucson, Arizona

Pamela S. Becker, M.D.
Postdoctoral Fellow in Hematology
Department of Internal Medicine
Yale University School of Medicine
New Haven, Connecticut

Edward J. Benz, Jr., M.D.
Professor of Internal Medicine and Human Genetics
Chief, Hematology Section
Associate Chairman, Department of Internal Medicine
Yale University School of Medicine
New Haven, Connecticut

Eric Boerwinkle, Ph.D.
Assistant Professor
Center for Demographic and Population Genetics
The University of Texas
Houston, Texas

Lawrence C.B. Chan, M.D.
Professor
Departments of Cell Biology and Medicine
Baylor College of Medicine
Houston, Texas

Shu Chien, M.D., Ph.D.
Professor of Bioengineering and Medicine
Department of Applied Mechanics and Engineering Sciences and Department of
    Medicine
Member, Center for Molecular Genetics
University of California, San Diego
La Jolla, California

Tucker Collins, M.D., Ph.D.
Assistant Professor
Department of Pathology
Harvard Medical School and Brigham and Women's Hospital
Boston, Massachusetts

Charles P. Emerson, Jr., Ph.D.
Commonwealth Professor of Biology
Department of Biology
University of Virginia
Charlottesville, Virginia

Irwin L. Flink, Ph.D.
Research Assistant Professor
Department of Internal Medicine
University of Arizona College of Medicine
Tucson, Arizona

Edward A. Fox, Ph.D.
Lecturer
Department of Microbiology and Molecular Genetics
Harvard Medical School
Boston, Massachusetts

Michael A. Gimbrone, Jr., M.D.
Elsie T. Friedman Professor of Pathology
Department of Pathology
Harvard Medical School
Director, Vascular Research Division
Brigham and Women's Hospital
Boston, Massachusetts

Larry R. Karns, Ph.D.
Postdoctoral Fellow
Department of Medicine/Cardiovascular Research Institute
University of California at San Francisco
Veterans Administration Medical Center
San Francisco, California

Wen-Hsiung Li, Ph.D.
Professor
Center for Demographic and Population Genetics
The University of Texas
Houston, Texas

Carlin S. Long, M.D.
Postdoctoral Fellow
Department of Medicine/Cardiovascular Research Institute
University of California at San Francisco
Veterans Administration Medical Center
San Francisco, California

Bruce E. Markham, Ph.D.
Assistant Professor
Department of Physiology
Medical College of Wisconsin
Milwaukee, Wisconsin

Eugene Morkin, M.D.
C. Leonard Pfeiffer Professor of Internal Medicine
Director, University Heart Center
University of Arizona College of Medicine
Tucson, Arizona

Sonia Pearson-White, Ph.D.
Research Assistant Professor of Biology
Department of Biology
University of Virginia
Charlottesville, Virginia

Deborah F. Pinney, Ph.D.
Research Associate of Biology
Department of Biology
University of Virginia
Charlottesville, Virginia

James F. Riordan, Ph.D.
Professor of Biochemistry
Center for Biochemical Biophysical Sciences and Medicine
Harvard Medical School
Boston, Massachusetts

Christine E. Seidman, M.D.
Assistant Professor of Medicine
Cardiovascular Division
Brigham and Women's Hospital and Harvard Medical School
Boston, Massachusetts

Paul C. Simpson, M.D.
Associate Professor
Department of Medicine/Cardiovascular Research Institute
University of California at San Francisco
Veterans Administration Medical Center
San Francisco, California

Mark B. Taubman, M.D.
Assistant Professor
Brookdale Center for Molecular Biology and Department of Medicine
Mount Sinai School of Medicine
New York, New York

Richard W. Tsika, Ph.D.
Research Associate
Department of Physiology and University Heart Center
University of Arizona College of Medicine
Tucson, Arizona

CONTENTS

Chapter 1

MOLECULAR BIOLOGY AND ITS APPLICATION TO THE CARDIOVASCULAR SYSTEM

Shu Chien

Recent advances in molecular biology have provided new approaches to probe the structural and functional bases of physiologic events at the molecular level.[1] Such applications of molecular biologic concepts and methodologies have been particularly notable in some areas of research, e.g., the neuromuscular and endocrine systems. In the last few years, significant advances have also been made in the use of molecular biology to elucidate the structure and function of the cardiovascular system.

The aim of this book is to summarize some of these recent advances in research on molecular biology of the cardiovascular system, with chapters on muscle differentiation (Chapter 2), the hypertrophic responses of cardiac muscle to thyroid hormone (Chapter 3) and adrenergic stimulation (Chapter 4), the atrial natriuretic factor (ANF) and its relation to cardiac hypertrophy (Chapter 5), vascular smooth muscle (Chapter 6), the endothelium and its interactions with leukocytes (Chapter 7), angiogenesis (Chapter 8), the red cell membrane skeleton (Chapter 9), and lipoproteins (Chapter 10), as well as a chapter of summary and conclusions (Chapter 11). These chapters cover some of the current knowledge on molecular biology of the heart, blood vessels, and elements in the blood.

This chapter provides an elementary introduction to molecular biology, with the aim of serving as a starting point for readers unfamiliar with the field to facilitate their reading of this book. A simpler version of this chapter has been presented by the author previously[2] and more detailed information on such background material is available in many textbooks (e.g., references 3 and 4). In presenting the basic concepts and techniques of molecular biology in this chapter, examples are given of their applications in the various chapters in this book. The aims of referring to these examples are (1) to relate the basic introduction to real systems, (2) to preview some aspects of the various chapters, and (3) to provide a correlation of the contents in different chapters, especially when common principles are involved. Therefore, reference citations are made mainly to these chapters rather than the primary sources, which can be found in the bibliographies of various chapters. Some original references are given when the authors' names are used in association with the methods.

DNA, RNA, AND PROTEIN: REPLICATION, TRANSCRIPTION, AND TRANSLATION

Specific proteins confer the structural and functional characteristics to each type of cell, and hence to the tissues, organs, and systems. The properties of these proteins are genetically determined, with the genetic information stored in the deoxyribonucleic

1

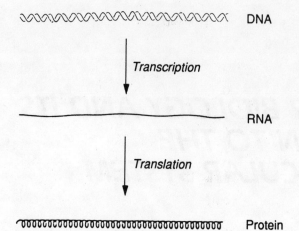

Fig. 1–1. The central dogma in molecular biology, showing the transcription of DNA to RNA and the translation of RNA to protein.

acids DNAs. The genetic code is first transcribed to ribonucleic acids (RNAs) and then translated to amino acids during protein synthesis. In this manner, the types of proteins synthesized by the cell are genetically controlled. The transcription from DNA to RNA and the translation from RNA to protein constitute the central dogma in molecular biology (Fig. 1-1). Quantitative and qualitative changes in protein synthesis during physiologic and pathophysiologic conditions are controlled by alterations in transcriptional, post-transcriptional but pretranslational, translational, or post-translational events. The exquisitely regulated genetic events form the molecular basis for the appropriate expressions of proteins and functions in specific cells and tissues, as well as their adaptations to physiologic needs and alterations in pathologic conditions.

DNA and RNA

DNAs are linear polymers of nucleotides that are composed of four types of bases (Fig. 1-2), i.e., two purines, adenine (A) and guanine (G), and two pyrimidines, cytosine (C) and thymine (T). These bases are linked together by a sugar–phosphate backbone in which the phosphate group and the deoxyribose sugar form 5′ and 3′ phosphodiester bonds, giving rise to a 5′ end and a 3′ end for each strand of the DNA molecule (Fig. 1-3).

Inside the cell, DNA molecules are packed in chromosomes and are referred to as *chromosomal DNA* or *genomic DNA*. In eukaryotes, i.e., cells with a well-defined nucleus (e.g., animal and plant cells), the negatively charged DNA molecules are packaged in the nucleus in association with the positively charged histone proteins to form chromatin (Fig. 1-4). The eukaryotic DNAs are double stranded as a result of pairing of the bases on the two strands, which are aligned in antiparallel directions (one 5′ to 3′, and the other 3′ to 5′). The opposing strands are held in precise register by hydrogen bonding between complementary base pairs (bp), i.e., adenine with thymine (A-T) and guanine with cytosine (G-C) (Fig. 1-5). The specific nucleotide sequence and the base pairing in DNA provide the mechanisms for the storage and replication of genetic information. In prokaryotes, i.e., cells without a defined nucleus (e.g., bacteria), DNA is not isolated from the cytoplasmic matrix. Prokaryotes have nonchromosomal DNA molecules, which may be single stranded and circular; these are called *plasmids*.

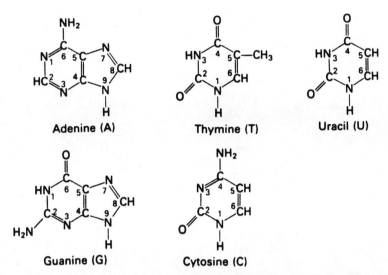

PURINES   PYRIMIDINES

Adenine (A)   Thymine (T)   Uracil (U)

Guanine (G)   Cytosine (C)

Fig. 1–2.   The purine and pyrimidine bases making up DNA and RNA. From Chien, S.: An introduction to molecular biology for physiologists. *In* Molecular Biology in Physiology. Edited by S. Chien. New York, Raven Press, 1989, p. 3.

RNAs are also linear polymers of nucleotides that contain four types of bases; three of these (A, C, G) are the same as those in DNA, but the fourth is uracil (U) instead of thymine. RNA also differs from DNA in that the sugar in the RNA backbone is ribose rather than deoxyribose (Fig. 1-2). RNAs are generally single stranded.

Replication of DNA

To preserve and transmit the genetic information from generation to generation and to keep the structural and functional characteristics of proteins in the cell lineage, it is necessary to have DNA replication, which is mediated by the enzyme DNA polymerase. DNA polymerase splits a portion of the DNA double strand into two single strands,

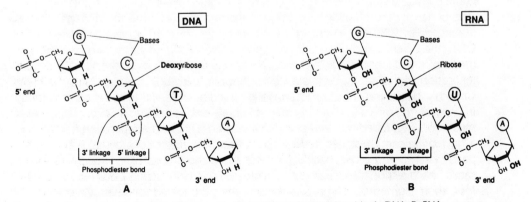

Fig. 1–3.   The sugar–phosphate backbone and attached bases for nucleic acids. A, DNA; B, RNA.

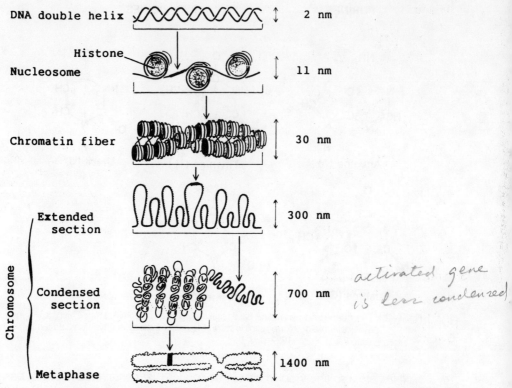

DNA double helix ⟷ 2 nm

Histone
Nucleosome ⟷ 11 nm

Chromatin fiber ⟷ 30 nm

Extended
section ⟷ 300 nm

Condensed
section ⟷ 700 nm   *activated gene is less condensed*

Chromosome

Metaphase ⟷ 1400 nm

Fig. 1–4. Schematic diagram to show the packaging of DNA with histone to form nucleosome, chromatin fiber, and chromosome. Drawings from bottom to top show serial enlargements. From Chien, S.: An introduction to molecular biology for physiologists. *In* Molecular Biology in Physiology. Edited by S. Chien. New York, Raven Press, 1989, p. 6. Modified from Molecular Biology of the Cell. Edited by B. Alberts, D. Gray, J. Lewis, M. Raff, K. Roberts, and J. Watson. New York, Garland Publishing, 1983, p. 399.

with each of these strands serving as a template for the sequential addition of complementary nucleotides. The addition of individual nucleotides to the DNA template is preceded by the attachment of a primer, which is a short nucleic acid segment complementary to a short stretch of the template. Thereafter, the daughter strand is lengthened by the addition of one nucleotide at a time, using the deoxynucleotide triphosphates (dNTPs) present in the nucleoplasm, to form a new double-stranded DNA (the parent template strand plus the complementary daughter strand). The synthesis of the daughter strand can only proceed in the 5' to 3' direction. Therefore, replication of the parent template strand that has a 3' to 5' orientation can proceed smoothly, beginning at the replication fork formed by the splitting of the parent strands. In contrast, the replication of the parent template strand that has a 5' to 3' orientation must work in a retrograde and piecemeal fashion. That is, the replication begins from a nucleotide some distance downstream of the fork (i.e., closer to the 3' end of this parent strand) and works backward (i.e., toward the 5' end of the parent strand) to satisfy the directional requirement of synthesis; this backstitching process is repeated for a short segment at a time, and the segments are joined together by the action of DNA ligase.

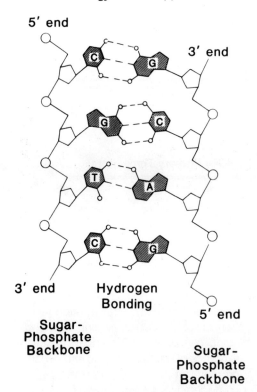

5' end

3' end

3' end **Hydrogen Bonding** **5' end**

Sugar-Phosphate Backbone

Sugar-Phosphate Backbone

Fig. 1–5. Base pairing of two DNA strands. The pentagon represents the deoxyriboses in the sugar–phosphate backbone. From Chien, S.: An introduction to molecular biology for physiologists. *In* Molecular Biology in Physiology. Edited by S. Chien. New York, Raven Press, 1989, p. 5.

In normal replication, the two newly generated double-stranded DNA molecules are identical to the two strands of the parent molecule. Errors in the replication process do occur, and this is usually corrected by the proofreading and self-correcting capability of the DNA polymerase. Such self-correction reduces the error of replication from one nucleotide in 10^4 to 10^5 to one in 10^9. Uncorrected errors lead to gene mutations, which either may result in abnormal protein expressions and functional derangements or may be silent. The mutation can be silent because (1) it does not occur in the coding region of the DNA; (2) it does not change the amino acid coding (see the section entitled Translation From RNA to Protein, p. 9, regarding degeneracy of codon); or (3) the resulting change in the amino acid sequence does not affect the function of the protein.

Transcription of DNA to RNA

The promoter is a region of the DNA molecule containing the binding site for the enzyme RNA polymerase. Eukaryotes have three types of RNA polymerases (I, II, and III). RNA polymerase II is responsible for the transcription of genomic DNA into messenger RNA (mRNA); it searches for the binding site on the promoter to initiate the transcription process and catalyzes the synthesis and elongation of the mRNA chain (Fig. 1-6). In the eukaryote, the promoter region generally has a nucleotide sequence rich in T and A (TATA or TATAA box), which controls the initiation of transcription 25 to 30 bp downstream (i.e., toward the 5' end of the DNA template strand). Another nucleotide sequence commonly found in the promoter region is the CCAAT box. In the transcription process for some of the proteins, e.g., the anion exchange membrane

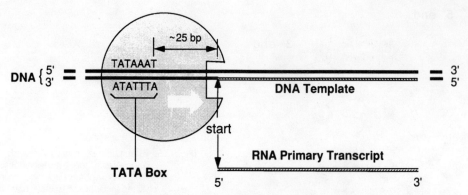

Fig. 1–6. A polymerase II (round figure with an opening on the right) bound to the TATA box of the promoter region of a double-stranded DNA to start the formation of an RNA transcript.

protein band 3 (p. 167), no such characteristic boxes (TATA or CCAAT) are in the promoter regions. The region upstream to the coding sequence of the genomic DNA contains other specific nucleotide sequences called *cis*-acting sequences, which have affinity for various transcription factors (*trans*-acting factors). The transcription factors are proteins that can regulate the transcription process via their binding to the *cis*-acting sequences, and the result of the binding can be either an enhancement or inhibition of transcription. An example is the existence of multiple positive and negative thyroid hormone response elements in the promoter of the cardiac muscle α-myosin heavy chain (α-MHC) gene; these elements provide various mechanisms for the regulation of the level of gene activity during development and in response to physiologic and pathophysiologic stimuli (p. 46). The interactions between *trans*-acting factors and *cis*-acting sequences are the basis of the regulation of gene transcription and are the subject of discussion in several chapters of this book.

Another type of *cis*-regulatory element is the enhancer, which consists of a nucleotide sequence that can enhance the transcription of genes. The enhancer can be located on the DNA at some far distance from the promoters, e.g., with a separation of several kilobases (kb), and it has no orientation requirement, i.e., the enhancer can work whether it is 5' to 3' or 3' to 5'. An example of an enhancer is the 72-bp sequence, which forms tandem repeats in the early promoter of the simian virus SV40.

In the transcription process, RNA polymerase II splits a portion of the DNA double strand into two single strands and uses only one of the strands (the one in the 3' to 5' direction) as a template to sequentially add complementary nucleotides, thus leading to the growth of an RNA chain in the 5' to 3' direction (Fig. 1-7). The elongation of the RNA chain is followed by its separation from the DNA template. The transcription process is terminated when the RNA polymerase encounters the terminator portion of the DNA molecule, which is signified by specific nucleotide sequences (Fig. 1-8). In eukaryotes, transcription often extends way beyond the end of the gene (e.g., by kilobases), and the size of the mRNA is controlled by polyadenylation. In contrast, termination sites of transcription are well defined in prokaryotes.

In prokaryotes, the RNA transcript does not undergo any modification, and transcription is directly coupled with translation. In contrast, transcription in eukaryotes occurs in the nucleus, and the primary RNA transcript molecules are modified before they can exit through nucleopores of the nuclear membrane into the cytoplasm as

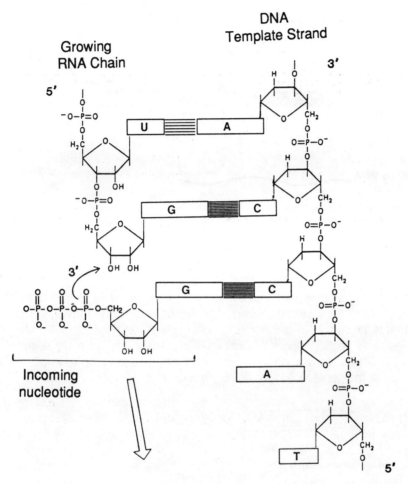

Fig. 1–7. The transcription process showing the growing of an RNA chain from the 5' to the 3' direction by adding an incoming nucleotide, which forms base pairing with the DNA template strand.

mature mRNA (Fig. 1-9). These modifications or maturation processes include the capping with 7-methyl guanosine at the 5' end (m^{7}Gppp in Fig. 1-9) and usually the addition at the 3' end of a poly (A) tail, which contains a large number of adenine nucleotides. The 5' capping most likely facilitates the ability of the mRNA to interact with protein factors needed for translation in the cytoplasm. The poly (A) tail is thought by some to enhance mRNA stability; it has also been postulated that it plays a role in the termination of transcription or in the transport of mRNA from nucleus to cytoplasm. Generally, only some regions of the primary RNA transcript become the mature mRNA found in the cytoplasm; these regions are called the *exons*. Most parts of the exons represent protein-coding sequences to be translated into amino acids, except for the untranslated regions at the 5' and 3' ends (including the cap and the tail, respectively). In contrast to the mature mRNA, which is transported through the nucleopore, the remaining parts of the primary transcript are spliced out during maturation and stay inside the nucleus; these regions are called the *introns*, or *intervening sequences*. Thus, the primary transcript, as well as the genomic DNA coding for it, is composed of exons

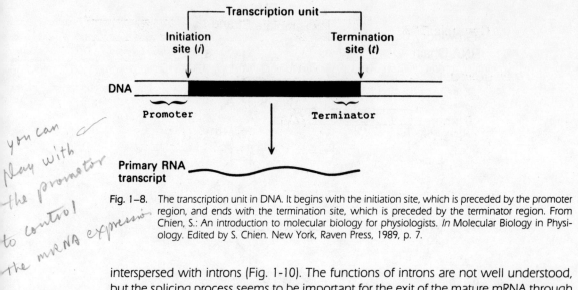

you can play with the promoter to control the mRNA expression

Fig. 1–8. The transcription unit in DNA. It begins with the initiation site, which is preceded by the promoter region, and ends with the termination site, which is preceded by the terminator region. From Chien, S.: An introduction to molecular biology for physiologists. *In* Molecular Biology in Physiology. Edited by S. Chien. New York, Raven Press, 1989, p. 7.

interspersed with introns (Fig. 1-10). The functions of introns are not well understood, but the splicing process seems to be important for the exit of the mature mRNA through the nucleopore. Because of the presence of various lengths of introns during maturation, the mRNA precursors in the nucleus are large and heterogeneous in size; they are referred to as *heterogeneous nuclear RNA*. In a few cases, an entire segment of the genomic DNA is transcribed, i.e., there is no intron splicing; an example is the transcription of the gene coding for angiogenin (p. 140).

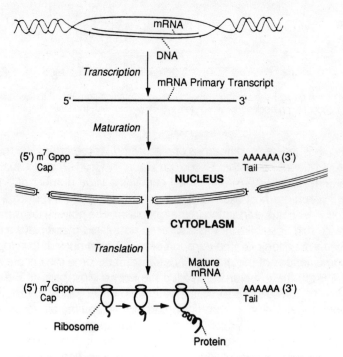

Fig. 1–9. Transcription in eukaryotes showing that RNA is modified in the nucleus before its exit into the cytoplasm.

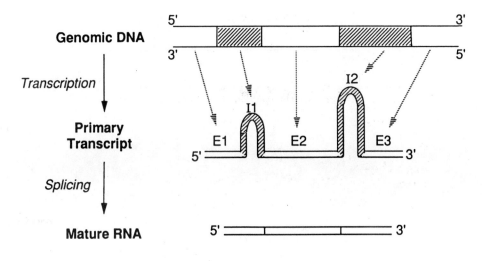

E : Exon   I: Intron

Fig. 1–10. Schematic drawing to show the removal of introns and the retention of exons when the primary RNA transcript is spliced to form mature mRNA.

In a given type of cell, a specific mode of splicing of the introns generates the mature mRNA; that is, the splicing pattern may be tissue-specifically regulated. Variations in the way the primary transcript is spliced into exons and introns (*alternative splicing*) in different types of cells and tissues is one of the ways of generating isoforms of similar proteins; e.g., the production of the erythroid and nonerythroid isoforms of protein 4.1 and β-spectrin (Chapter 9), the tissue isoforms of α-tropomyosins and myosin alkali light chains (Chapter 6), or the isoforms of the A chain of platelet-derived growth factor (PDGF) (Chapter 7).

Translation From RNA to Protein

The genetic information transcribed from DNA to mRNA is translated into a sequence of amino acids during protein synthesis in the ribosome. In the translation process, every three nucleotides form a triplet code (codon) to be read by the ribosome as encoding a specific amino acid (Table 1-1). There exists a panel of transfer RNA (tRNA) molecules, each carrying on one hand a triplet nucleotide anticodon (to make complementary pairing with the appropriate codon) and on the other hand a specific amino acid (in accordance with the coding shown in Table 1-1). Thus, each type of tRNA molecule mediates with fidelity the translation from a specific triplet of mRNA to the corresponding amino acid (Fig. 1-11). The codon AUG codes for methionine and also serves as a start codon, i.e., to begin the translation process. The codons UAA, UAG, and UGA serve as stop codons, and translation will terminate when one of these is encountered. During the translation process in the ribosome, the triplets of the mRNA are sequentially read, and the corresponding amino acids are added in turn to form a growing polypeptide chain.

The nucleotide sequence of an mRNA (or its corresponding DNA) can be divided into triplet units to provide a reading frame (Fig. 1-12). In the absence of other infor-

Table 1–1. The Genetic Code for Translation of mRNA to Protein

First Position (5' End)	Second Position				Third Position (3' End)
	U	C	A	G	
U	Phe	Ser	Tyr	Cys	U
	Phe	Ser	Tyr	Cys	C
	Leu	Ser	stop	stop	A
	Leu	Ser	stop	Trp	G
C	Leu	Pro	His	Arg	U
	Leu	Pro	His	Arg	C
	Leu	Pro	Gln	Arg	A
	Leu	Ser	Gln	Arg	G
A	Ile	Thr	Asn	Ser	U
	Ile	Thr	Asn	Ser	C
	Ile	Thr	Lys	Arg	A
	Met*	Thr	Lys	Arg	G
G	Val	Ala	Asp	Gly	U
	Val	Ala	Asp	Gly	C
	Val	Ala	Glu	Gly	A
	Val	Ala	Glu	Gly	G

*AUG is part of the initiation signal, and it codes for internal methionines as well.

mation, such as an established position for the initiation codon, a segment of mRNA with a given orientation (5' to 3') can have three possible reading frames, which are shifted from one another by one nucleotide. These reading frames would result in peptides with different lengths because each reading frame will end (i.e., become closed) when a termination codon is reached. The portion of the mRNA that is translatable without encountering a termination codon is called *an open reading frame.* Under physiologic conditions, the mature mRNA has one open reading frame, which is translated to generate the specific protein in the particular cell. The deletion (or addition) of a single nucleotide in the mRNA would shift the reading frame (frame shift) and lead to the generation of a protein with altered amino acid sequence beyond the site of modification (Fig. 1-12) and also a change in molecular size. Such frame shifts can cause gene mutation in hereditary disorders; e.g., in the abnormal apolipoprotein C-II$_{Toronto}$, a single nucleotide deletion impairs the ability of the protein to activate lipoprotein lipase (p. 203). A single nucleotide change and the consequent early closure of reading frame is used physiologically as a mechanism of generating organ-specific expressions of apolipoprotein B (apoB) of markedly different sizes (Chapter 10). In the midsection of the 14-kb apoB mRNA in the liver, there is a codon CAA (encoding glutamine), but in the intestine this is changed to UAA, which is a stop codon. As a result, the apoB-48 in the intestine is composed of only the first 264 kD of the 550-kD apoB-100 in the liver, thus making possible the differential functions served by the two forms of apoBs.

As indicated in Table 1-1, the third position of the triplet nucleotides is usually not as critical as the other two positions in amino acid coding. Therefore, variations in the third position often do not cause a change in the amino acid translated. For example, as long as the first two nucleotides in the triplet are CG, the codon always codes for arginine, regardless of what the third nucleotide is; AGA and AGG also code for arginine

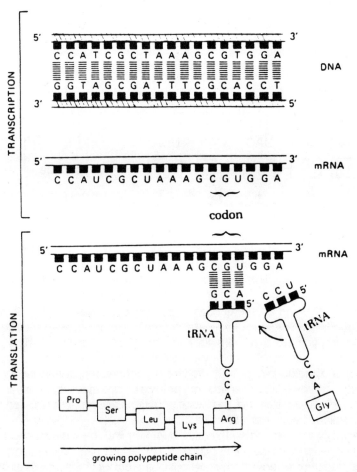

Fig. 1–11. Following transcription of DNA to mRNA, the triplet codon in the mRNA is used for translation mediated by tRNA, which carries the anticodon for base pairing and the amino acid for protein synthesis. From Chien, S.: An introduction to molecular biology for physiologists. *In* Molecular Biology in Physiology. Edited by S. Chien. New York, Raven Press, 1989, p. 11.

(Table 1-1). Such degeneracy of the codons (i.e., different codons coding for the same amino acid) allows some nucleotide substitutions to remain synonymous in terms of the amino acid translated. DNA mutations involving synonymous substitutions would be silent because the protein expressed does not change. The codon degeneracy also explains the greater degree of homology of the amino acid sequences between isoforms of a protein than the corresponding nucleotide sequences; e.g., human and chicken α-spectrins have 95% amino acid homology but only 82% nucleotide homology (p. 162). The degeneracy of the codon provides nonunique answers when one tries to back-translate from amino acid sequence to nucleotide sequence, e.g., in attempting to synthesize an oligonucleotide from the known amino acid sequence of a peptide (see the section entitled Screening of Library, p. 19).

The mRNA is composed of the coding region, which is translated into amino acids, and of the untranslated regions at the 5' and 3' ends (the flanking regions). The 3' untranslated region includes the polyA tail. The 5' untranslated region includes nu-

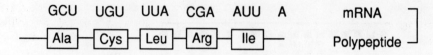

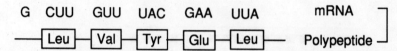

Fig. 1–12. An example of reading frame shift in translation. The mRNA segments on the top and the bottom have identical 16 nucleotides, but the shift of reading frame by one nucleotide results in the translation of polypeptides with completely different amino acid composition in this region.

cleotide sequences that bind with ribosomes and with proteins called *initiation factors,* which can regulate the initiation of translation. The 5' end of the coding region of mRNA corresponds to the amino terminus of the protein, and the 3' end to the carboxyl terminus.

The protein synthesized on the ribosomes can undergo co-translational and post-translational modifications. For example, the nascent secretory protein precursors often have a signal peptide (15 to 60 amino acids long), which provides the signal for intracellular sorting. This signal peptide is usually removed in the finished protein, e.g., in ANF (p. 88) and angiogenin (p. 140). Proteins may undergo post-translational modifications such as glycosylation, deamidation, proteolytic cleavage, acylation, and phosphorylation. These co-translational and post-translational events are discussed for ANF (p. 88) and apolipoproteins (pp. 189–192).

In summary, DNA directs the synthesis of RNA, and RNA then directs the synthesis of proteins. Translation from the genetic code into the appropriate amino acid sequence results in the proper expression of the functions of the protein following its appropriate modifications. Alterations in transcription, translation, and/or post-translational events may modify the expression of a given protein in a cell type.

DNA RECOMBINANT TECHNOLOGY: DNA CLONING

Base pairing is a key element in the coding of genetic information in DNA and RNA, the replication of DNA, the transcription of DNA to RNA, and the translation of RNA to protein. Molecular biologists have taken advantage of these and other characteristics of nucleotides to probe and to manipulate the molecular structures of DNA and RNA. Genetic engineering techniques have been developed to ligate the mammalian DNA of interest to a prokaryotic DNA (the vector), thus generating recombinant DNA molecules (hybrid or chimeric DNA), which are introduced into a host cell for replication. The recombinant DNA molecule can then be recovered from the bacteria colonies by appropriate selection procedures to obtain the desired DNA clones in high purity and large quantity. The methodology for these procedures can be found in standard manuals for recombinant DNA technology (e.g., references 5 and 6); only a brief introduction of the principles is given here.

Some Enzymes Used in DNA Cloning

Recombinant DNA technology involves the use of several enzymes that have made possible the cloning of DNA. Two types of enzymes present in some viruses and bacteria, viz., reverse transcriptase and restriction endonuclease, are particularly important in the cloning of mammalian DNA. These are described briefly in the following paragraphs.

REVERSE TRANSCRIPTASE: REVERSE TRANSCRIPTION

In eukaryotes, RNA can be synthesized from the DNA template, but DNA cannot be synthesized by using RNA as a template. Some prokaryotes (the retroviruses, e.g., the acquired immunodeficiency syndrome [AIDS] virus), however, contain the enzyme reverse transcriptase, which can cause the synthesis of a complementary DNA (cDNA) from an mRNA template. Therefore, the normal flow of genetic processes can be reversed by using the following approach. First, mRNA can be purified from the total RNA extracted from a tissue by taking advantage of its poly(A) tail. Because of the affinity of poly(A) for oligonucleotides composed of deoxythymidine (dT), the oligo dT column is used to allow specific retention and subsequent elution of mRNA molecules. By using a retrovirus reverse transcriptase, one can obtain a cDNA that is complementary to the purified mRNA. The availability of cDNA provides important advantages in molecular biologic studies, because DNA molecules are much more stable than RNA and easier to handle and because they can be replicated and manipulated. In contrast to the genomic DNA, the cDNA does not contain introns (Fig. 1-13).

RESTRICTION ENDONUCLEASES: GENERATION OF DNA FRAGMENTS WITH SPECIFIC ENDS

To create a recombinant DNA molecule, it is necessary to obtain segments of mammalian DNA and vector DNA that will anneal to each other. This has been made possible by the use of restriction enzymes, which can cleave DNAs with restricted specificities at sites of defined base sequence. The restriction enzymes acting on the inner parts of a DNA are called *restriction endonucleases*. Cleavage of a specific DNA molecule by using a particular restriction endonuclease generates a unique family of fragments. Some of the restriction endonucleases make a blunt cut, whereas others

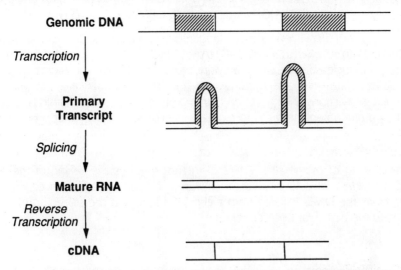

Fig. 1-13.   The transcription of genomic DNA to primary transcript of mRNA, the splicing of introns during mRNA maturation, and the reverse transcription of mRNA to cDNA. Note the presence of introns in genomic DNA and their absence in cDNA.

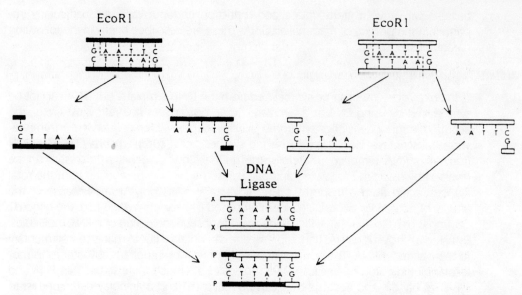

Fig. 1–14. The restriction enzyme *Eco*RI makes staggered, symmetric cuts in DNA at specific recognition sites, leaving "sticky" ends, which can anneal to any other sticky ends produced by *Eco*RI cleavage.

create a staggered cut across the two strands of DNA to result in cleaved molecules with "sticky" ends, each of which contains a short single-stranded region (an overhang) capable of base pairing with the complementary end on another DNA molecule. Such paired ends of the two DNA molecules can then be covalently ligated by the action of DNA ligase, thus producing a recombinant DNA. An example of restriction endonuclease is *Eco*RI, which has its enzymatic action restricted to a 6-bp sequence:

$$\begin{array}{ccc}
\ldots \text{G}|\text{AATTC}\ldots & \xrightarrow{\;\;\textit{EcoRI}\;\;} & \ldots \text{G} \qquad + \text{AATTC}\ldots \\
\ldots \text{CTTAA}|\text{G}\ldots & & \ldots \text{CTTAA} \qquad\quad \text{G}\ldots
\end{array}$$

The action of *Eco*RI results in a staggered cut, leading to the formation of DNA fragments with sticky ends (Fig. 1-14). To protect internal restriction sites from the action of the endonuclease, one can methylate these sites before the digestion procedure and demethylate them afterward. Methylation–demethylation alters the response of nucleotides to actions of enzymes and other proteins, and it is a physiologic mechanism used by a pluripotent stem cell to regulate its expression of myogenic genes (p. 36).

Vectors and Libraries

The cloning of mammalian DNAs of interest involves the linking of the cDNA or genomic DNA molecule to a vector DNA molecule, the production of a library that contains all the DNAs isolated from a cellular origin, and the screening of the library to obtain the particular DNA of interest.

Vectors

When the mammalian DNA and the vector DNA have the same restriction endonuclease cleavage site, they can be cut to create complementary sticky ends for making the hybrid molecule. Such common restriction sites can also be experimentally added to the mammalian DNA by the attachment of a linker, which is a short segment of synthetic double-stranded DNA containing the sequence of a desired restriction site.

Many vectors multiply in the host but the inserted DNA does not express protein products. In contrast, the genes or cDNA in expression vectors can be transcribed and translated into proteins in the host cell. The plasmid, bacteriophage, and cosmid vectors have the following common properties: (1) They contain restriction endonuclease sites for the insertion of DNA to be cloned, (2) the resulting recombinant DNA molecules are capable of autonomous replication following their introduction into a particular bacterial host, and (3) they contain selectable markers. Each of these types of vectors, however, has particular features that are advantageous for certain applications, as described following.

Plasmids are small, circular, double-stranded DNA molecules found in several species of prokaryotes, which generally encode proteins that are required for antibiotic resistance, e.g., the plasmid pBR322 has ampicillin- and tetracycline-resistance genes; these genes provide selectable markers for the recombinant plasmid clones when the bacteria are treated with the appropriate antibiotic. Plasmids have a DNA replication origin to give high copy numbers and suitable restriction endonuclease sites for the insertion of desired DNA. The major features of cloning in plasmids are shown in Figure 1-15. An appropriate restriction endonuclease (e.g., *Eco*RI) is used to digest both the circular plasmid DNA and the mammalian DNA, and DNA ligase is used to produce the recombinant closed circular DNA composed of a single vector and a single insert molecule. The bacteria host is made competent by exposure to one or more divalent cations, so that it will take up the extracellular recombinant plasmid DNA for intracellular replication. The transfected bacteria population is spread on an agar plate containing one or more of the antibiotics for which the plasmid has the resistance genes. On this bacteria lawn, the clones that survive and grow are the ones containing the recombinant plasmid DNA. The recombinant molecules can be isolated from the selected clones, and the desired DNA fragment collected with the use of the appropriate restriction endonuclease. An example of the use of an expression plasmid system is found in the cloning of the angiogenin gene (p. 141).

Bacteriophages (or phages) are viruses that infect bacteria, multiply in them, and can cause their lysis. Phages can be either single stranded (e.g., M13) or double stranded (e.g., λ). The major features of cloning in bacteriophage λ are shown in Figure 1-16. The replication of λ involves a lysogenic phase (insertion of λ-genome into the bacteria without killing it) and a lytic phase (bacteria killing and phage release). In the lytic phase, the λ-genome (50 kb) contains at its two ends complementary 12-nucleotide single-stranded tails (sticky ends). Following the infection of a bacteria, these ends of an initially linear phage ligate to form a circular λ-molecule. The circular λ-monomer DNA replicates by a "rolling circle" mechanism to form a long, linear concatemer resulting from the pairing of the complementary 12-nucleotide sticky ends on the monomers (Fig. 1-16). The paired sticky ends and adjacent sequences are known together as the *cos sites*. When the λ-genome is packaged into a newly synthesized phage head, the cos site serves as a recognition site for cleaving the long concatemer into a unit-length monomer in each λ-phage head. The recombinant λ-phage DNA created by the uses of restriction endonuclease and ligase can be packaged into a viable phage head in vitro.

Cosmids are cloning vectors that can incorporate longer segments of DNA (up to 45 kb) than those possible with the other two types of vectors. Like plasmids, cosmid

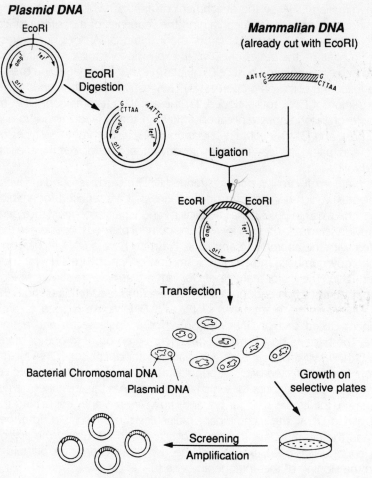

Fig. 1–15. Cloning of DNA in a plasmid vector. The mammalian DNA to be cloned and the plasmid DNA are both digested with *Eco*RI and ligated with DNA ligase. The circular hybrid DNA is transfected into bacteria that is grown on selective plates for screening and amplification. Modified from Fritsch, E.F., and Maniatis, T.: Methods of molecular genetics. In The Molecular Basis of Blood Diseases. Edited by G. Stamatoyannopoulos, A.W. Nienhaus, P. Leder and P.W. Majerus. Philadelphia, W.B. Saunders, 1987, p. 5.

vectors have a replication origin, a selectable marker, and cloning sites. In addition, they have the cos sites, which are used by the bacteriophage λ in cleaving the concatemer and packaging the DNA monomer into the phage head. In the cosmid concatemer, the cos sites are separated by 38 to 52 kb. After infection of a bacterium, the monomer DNA circularizes via the sticky ends and forms a large plasmid. Cosmid genomic DNA transfection has been used to study the myogenic regulatory genes (p. 36).

Thus, with the use of any of these vectors, recombinant DNA molecules can be produced and introduced into bacteria, wherein the hybrid vector DNA will replicate autonomously. This makes possible the cloning of multiple copies of the DNA, the construction of DNA libraries, and the selection of specific clones that contain the sequence of interest, as discussed in the subsequent sections.

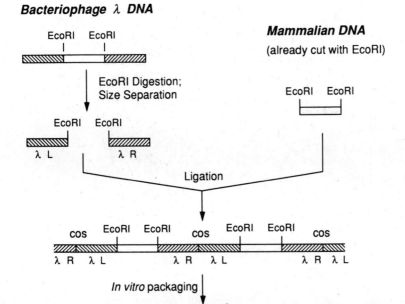

Fig. 1–16. Cloning of DNA in a bacteriophage λ-vector. The complementary *Eco*RI sites are ligated to form a linear hybrid for in vitro packaging into a phage head for bacterial transfection. Note the joining of the complementary ends of bacteriophage λ to form a cos site, which is recognized during packaging. Modified from Fritsch, E.F., and Maniatis, T.: Methods of molecular genetics. In The Molecular Basis of Blood Diseases. Edited by G. Stamatoyannopoulos, A.W. Nienhaus, P. Leder and P.W. Majerus. Philadelphia, W.B. Saunders, 1987, p. 8.

DNA LIBRARIES

A statistical representation of all genomic DNAs or all cDNAs isolated from a cellular origin is called a *library* (Fig. 1-17). From the nuclei of a given type of cell, one can extract the genomic DNAs to produce a genomic DNA library, which is useful in studying gene transcription and its regulation. From the cytoplasm of a given type of cell, one can extract all types of mRNAs contained within these cells, which represent a specific subset of the total genetic information of the organism, and generate a cDNA library by reverse transcription. cDNA libraries are especially useful in studying gene expression in tissues. In constructing a cDNA library, one begins with a tissue known to contain the mRNA of interest. For example, to study the molecular biology of erythrocyte membrane proteins, cDNA libraries are obtained from erythroid or hematopoietic cells or tissues, e.g., reticulocytes, erythroleukemic cell line, fetal liver, anemic spleen, etc. (Chapter 9).

Libraries are usually produced with the mammalian DNA ligated to a vector DNA molecule, using the procedures described previously (in the section entitled Vectors).

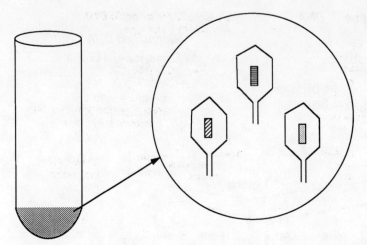

Fig. 1–17. A schematic drawing of a DNA library in a bacteriophage vector showing the presence of different DNA inserts in the three phages shown out of millions of such phages in the library.

In constructing a cDNA library from the mRNA population obtained from a given type of cells, a primer is used to anneal to a particular sequence on the mRNA. Most commonly, the primer is an oligo dT, 12 to 18 nucleotides in length, which binds to the poly(A) tail of the mRNA. The reverse transcriptase then acts to copy the entire mRNA into a cDNA strand. The cDNA clones created can be amplified into many copies and stored. The construction, amplification, and storage of a high-quality cDNA library provides continuing access to a set of cDNA clones for all the genes being expressed in a cell at a particular stage of development or differentiation. Once constructed, a cDNA library must be screened to identify the clone of interest, as described in the following section (p. 19).

Cloning of genomic DNA is best accomplished through the construction of a genomic DNA library. The procedures for the construction of a genomic library are shown in Figure 1-18. The high molecular weight genomic DNA is first digested with a restriction endonuclease that has frequent recognition sites, e.g., Sau3AI, which recognizes specifically the four-base sequence 5' . . . GATC . . . 3'. By controlling the condition of digestion, the average size of the DNA products can be made to be 18 to 20 kb, and DNA in this size range is selected by agarose gel electrophoresis to generate a collection of relatively uniformly sized DNA fragments from the genome. In the example shown in Figure 1-18, a bacteriophage λ vector containing cloning sites for *Bam*HI is chosen because this endonuclease produces a single-stranded end CTAG, which is complementary to the GATC sequence produced by Sau3AI at the ends of the genomic DNA fragment:

$$5' \ldots \text{G}|\text{GATCC} \ldots 3' \xrightarrow{\textit{Bam}\text{HI}} 5' \ldots \text{G}._{\text{OH}}$$
$$3' \ldots \text{CCTAG}|\text{G} \ldots 5' \quad\quad\quad 3' \ldots \text{CCTAG}$$

The ligation of these complementary ends under proper conditions results in long, concatemeric molecules containing the vector and the insert fragments, which can be used for in vitro packaging into a phage head to form an infectious phage particle. As a result, a large collection of recombinant phage is obtained with nearly all sequences

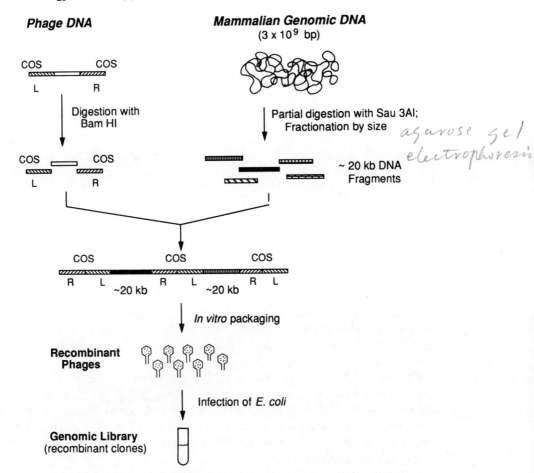

Fig. 1–18. Schematic drawing to show the procedure for the construction of a genomic library in a bacteriophage vector. The phage DNA is digested with *Bam*HI, and the mammalian DNA is digested with *Sau*3AI and size fractionated. The ~20-kb mammalian DNA is ligated to the phage DNA. The linear hybrid with cos sites are packaged into the phage head for infection of *Escherichia coli* and the construction of the genomic library. Modified from Fritsch, E.F., and Maniatis, T.: Methods of molecular genetics. In The Molecular Basis of Blood Diseases. Edited by G. Stamatoyannopoulos, A.W. Nienhaus, P. Leder and P.W. Majerus. Philadelphia, W.B. Saunders, 1987, p. 13.

in the genome represented in the library. The library can then be amplified and stored for repeated usage to isolate different genes of interest.

SCREENING OF LIBRARY

A library consists of a very large number of different DNA inserts in a vector. These clones must be screened to identify the recombinant clone of interest. One way of identifying the cloned gene that codes for the desired protein is to make a cDNA library in an expression vector (either plasmid or bacteriophage λ), such that the bacterial host expresses the recombinant fusion proteins. The aim is to synthesize in the bacterium one or more antigenic determinants of the coded protein of interest, so that the bacterial colonies containing such an antigen can be identified by the specific binding of a polyclonal or monoclonal antibody.

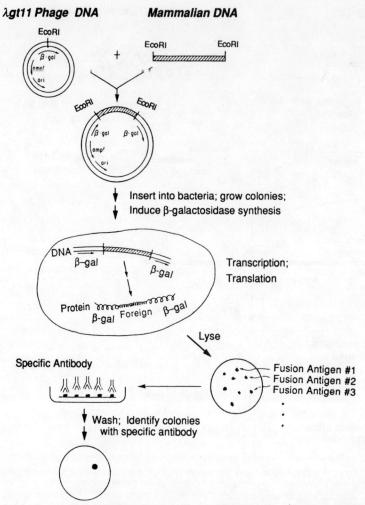

λgt11 Phage DNA   **Mammalian DNA**

Fig. 1–19. λgt11 cloning by expression in *Escherichia coli* and antibody screening. The mammalian DNA is ligated to the λgt11 phage DNA after *Eco*RI cut and inserted into bacteria for colony growth. The β-galactosidase gene in λgt11 is induced for transcription and translation. The fusion proteins generated can be released after bacterial lysis and blotted onto nitrocellulose paper for identification with specific antibody. Modified from Fritsch, E.F., and Maniatis, T.: Methods in molecular genetics. In The Molecular Basis of Blood Diseases. Edited by G. Stamatoyannopoulos, A.W. Nienhaus, P. Leder and P.W. Majerus. Philadelphia, W.B. Saunders, 1987, p. 17.

Figure 1-19 shows an example of cloning with λgt11 by expression in *Escherichia coli* and antibody screening. λgt11 has a single *Eco*RI site located within its gene coding for β-galactosidase, which is an enzyme capable of digesting the chromogenic sugar 5-bromo-4-chloro-3-indolyl-β-D-galactoside (X-gal) to yield a product with blue color. With the use of *Eco*RI, the cDNA clone can be ligated to the β-galactosidase gene of λgt11. Following transcription and translation in the bacteria, the fusion protein would have the protein expressed from the mammalian cDNA, which separates the β-galactosidase molecule into two sections. When the bacteria lawn is treated with X-gal in the presence of isopropylthiogalactoside (IPTG), an inducer for the β-galactosidase gene, the clones without the inserts will show blue color, reflecting the presence of the intact β-galactosidase. In contrast, the clones with the positive inserts will have lost the β-galactosidase activity and hence will be colorless under this condition. This facilitates the identification of the insert-containing clones for subcloning. Specific identification of the clones containing the desired cDNA insert is made by transferring (blotting) parts of each colony from the bacteria lawn to a nitrocellulose filter paper (NC paper, which has an affinity for nucleic acids), lysing the transferred bacteria, and reacting the filter with a specific antibody. Once the desired clone(s) are identified on the filter, one can return to the bacteria lawn to subclone the corresponding colonies situated at the same geometric locations. This antibody screening method has been used in the molecular cloning of several adhesion receptors on the endothelial cell (p. 130) and several erythrocyte membrane proteins (Chapter 9). A more detailed description of the method of using λgt11 as a cloning vector is given in Chapter 7 (p. 118).

Instead of using a bacterial expression vector, as described for λgt11, a eukaryotic expression system can also be used. For example, DNA cloning has been performed by transfections of myogenic regulatory genes into the mouse embryo cell line 10T1/2 with a cosmid shuttle vector (p. 36), the human α-MHC gene into cultured rat fetal myocytes with the expression vector pSVOCAT (p. 47), genes of several endothelial cell adhesive molecules into monkey kidney COS cells with pCDM8 (p. 130), and the angiogenin gene into baby hamster kidney cells with pMTAG (p. 141), etc. These vectors contain appropriate restriction sites and several replication origins, with the inclusion of SV40, to allow high-level expression. In applying this method to cDNA cloning of endothelial cell surface molecules, the selection of COS cells expressing these antigens is facilitated by the use of specific monoclonal antibodies and the adhesion of the cells to antibody-coated dishes (p. 130). From the adherent cells, the hybrid vector is extracted and transformed back into *E. coli* for further transfection and selection.

If the expressed protein is biologically active, e.g., polypeptide hormones and growth factors, identification can also be achieved by a biologic assay, such as the determination of responses of cultured cells or the correction of the genetic defect in a mutant strain. If the expressed protein is secreted by the transfected cells into the medium, as in the case of ANF (Chapter 5), PDGF (Chapter 7), or angiogenin (Chapter 8), the biologic activity can be assayed using the medium.

Another way to detect specific recombinant clones in bacterial colonies is colony hybridization, which requires the availability of a ^{32}P-labeled probe for the DNA of interest. The bacterial colonies containing the clones are blotted onto NC paper. After bacterial lysis and DNA denaturation with NaOH, the single-stranded DNA is fixed to the NC paper by baking at 80°C and is hybridized with the ^{32}P-labeled probe. After washing, the NC paper is subjected to autoradiography by being put in contact with a photographic film for a period of time in lightproof condition. The presence of the desired recombinant and its location will be detected on the developed autoradiographic film.

The cDNA probe used for ^{32}P labeling can be an existing cloned DNA or a synthetic oligonucleotide with a sequence deduced from the partial amino acid sequence of the purified protein. The use of synthetic oligonucleotides as probes has proven valuable in the isolation of both cDNA and genomic clones for many proteins, and this is an important development in DNA technology. The synthetic oligonucleotide probes are prepared in accordance with a known sequence of amino acids (usually 30 or so) of the purified protein. In selecting regions of the known amino acid sequence to make the synthetic oligonucleotide probe, one takes into account the degeneracy of the genetic code, i.e., several different nucleotide sequences give the same amino acid sequence (see the preceding section entitled Translation from RNA to Protein, p. 10). This nonuniqueness of the nucleotide decoding makes it necessary to test more than one combination of nucleotide sequences. To increase the probability of choosing the correct nucleotide sequence and to minimize the testing of a large number of combinations, one generally chooses the stretch of amino acids with the least degeneracy. Usually, two sets of DNA oligonucleotides (each about 20 nucleotides long) are so selected from two regions of the known amino acid sequence for chemical synthesis and radioactive labeling. Bacterial colonies on NC paper that hybridize with both sets of DNA probes are strong candidates for containing the desired gene, and they can then be subjected to further characterization.

The technique of oligonucleotide synthesis is also used for template-independent polymerization of nucleic acids of defined sequence. By this method, one can readily obtain any desired single-stranded or double-stranded DNA sequence up to 100 to 150 bases long. This makes possible the custom design of short DNA stretches for almost any experimental purpose.

Examples for the use of synthetic oligonucleotides in DNA manipulation are found in the cloning of the ANF gene (p. 83), PDGF gene (p. 126), angiogenin gene (p. 143), and α-spectrin, actin, and glycophorin genes (Chapter 9).

Methods for the Study of DNA Structure and Gene Expression

SOUTHERN BLOTTING

An important technique for the characterization of cloned DNA is the hybridization of its restriction-digested fragments with specific probes. This procedure, developed by Edward Southern[7] is known as Southern blotting. The principle is the same as that described for colony hybridization, except that in Southern blotting, the hybridization is performed after electrophoretic separation of DNA fragments by size. The DNA fragments obtained after digestion with one or more restriction endonucleases are separated by agarose gel electrophoresis and denatured on the gel to form single strands. An NC paper (or a nylon membrane) is placed under a stack of absorbent paper and above the gel, which is positioned above a sheet of paper towel or filter paper with its ends soaked in a buffer solution. The fluid flow, which is due to capillarity from this paper towel sheet through the gel toward the NC paper, leads to the transfer of the DNA fragments from the gel onto NC, thus preserving with high resolution the spatial pattern created by electrophoresis (Fig. 1-20). The transfer can also be achieved by applying an electrophoretic field perpendicular to the NC paper and the gel, instead of using fluid mechanical flow. After having been baked in vacuo to make the DNA binding firmer, the NC paper is incubated with a radiolabeled DNA probe, which adheres by base pairing only to the specific DNA fragments carrying the complementary sequences. The position of these fragments can then be visualized by autoradiography. This provides important size and structural information on the DNA of interest, e.g., the localization of a small gene in a large fragment of cloned genomic DNA, the identification of gene defects in abnormal chromosomes, and the construction of a coarse map of the structure of the gene by using a number of restriction endonucleases. Figure 1-21 shows the principle of aligning different DNA fragments from such restriction mapping. This approach helps to line up different clones obtained from a given

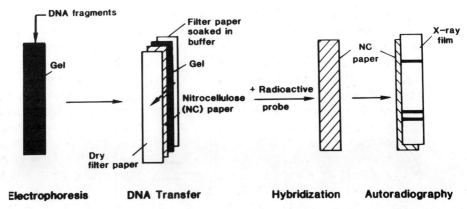

Electrophoresis   **DNA Transfer**   **Hybridization**   **Autoradiography**

Fig. 1–20. Steps involved in Southern blotting. From Chien, S.: An introduction to molecular biology for physiologists. *In* Molecular Biology in Physiology. Edited by S. Chien. New York, Raven Press, 1989, p. 13.

DNA cloning study and facilitates the final construction of the detailed nucleotide sequence map.

Each chromosome contains a large number of restriction sites. With the use of appropriate restriction enzymes, the genomic DNA can be cut into a number of restriction fragments. Following gel electrophoresis to separate the fragments by length and Southern blotting with a radioactive DNA probe, a restriction map is obtained showing the distribution of the fragment lengths. Alterations in the genomic DNA, e.g., the absence of one of the sites for the enzyme used, would result in length changes in the restriction fragments, referred to as restriction fragment-length polymorphism (RFLP). RFLP can be used to assay for specific variations in the human genome that are linked to a defective gene. For example, RFLP has been used to identify an abnormal ankyrin gene in hereditary spherocytosis patients and an abnormal 4.1 gene in hereditary elliptocytosis patients (Chapter 9) and to establish the role of apolipoprotein polymorphism in affecting serum cholesterol levels in epidemiologic genetic studies (p. 183).

The availability of the cDNA probe coding for a protein allows the detection of the location of the genomic DNA on the chromosomes by in situ hybridization on chromosome spreads. These and other methods have made possible the chromosomal assignment of various genes. Examples are the chromosomal mapping of various erythroid membrane protein genes (Chapter 9) and apolipoprotein genes (Chapter 10). This method is based on the same principle of DNA base pairing as Southern blotting, but no electrophoresis is involved.

NORTHERN BLOTTING

Northern blotting is a technique similar to Southern blotting, but RNAs, rather than DNAs, are separated by gel electrophoresis and then hybridized with the radiolabeled DNA probe. This procedure can be applied to determine the presence or absence of the expression of specific mRNAs in RNA samples prepared from various tissues or cells, as well as the sizes of isoforms of mRNA transcripts. Northern blotting has been used to quantify the actin iso-mRNAs (p. 65) and to study the expression of iso-mRNAs of erythroid membrane proteins such as band 4.1 in various nonerythroid tissues (p.

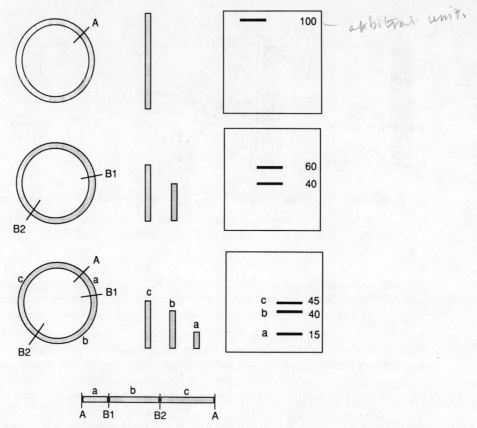

arbitrar units.

Fig. 1–21. A schematic drawing to show the principle of restriction mapping. A circular DNA is first cut with restriction enzyme A with a single site. The schematic drawing of the linear molecule in the center column and the result of gel electrophoresis show that the resulting linear DNA is uniform in length (100 in arbitrary units). When the same circular DNA is cut with restriction enzyme B with two sites (B_1 and B_2), two DNA fragments are generated with lengths of 60 and 40. When both A and B are used at the same time, three DNA fragments are obtained (a, b, and c, with lengths of 15, 40, and 45, respectively). The length relations allow the deduction of the relative positions of the restriction sites shown both on the circular DNA and the horizontal drawing at the bottom.

164). As mentioned in conjunction with the discussion on colony hybridization (see Screening of Library, p. 19), labeled DNA probe can be used to identify and select mRNA on a filter without electrophoretic separation. Unlike the Northern blotting, which identifies the specific mRNA by size, such a "dot" or "slot"-blotting method does not allow a size identification of the mRNA. Radiolabeled DNA probes can also be used to perform in situ hybridization of cell and tissue specimens for the autoradiographic detection and localization of mRNA transcripts in relation to histologic features.

Western blotting is a procedure that refers to the detection of proteins after electrophoretic separation with labeled antibodies (Fig. 1-22).

DNA SEQUENCING

Two methods are used to determine the base sequence of DNA, each using a panel of reagents that are specific for the four bases. The Maxam and Gilbert method[8] involves

Northern Blotting

RNA in gel;
hybridized with DNA

Western Blotting

Protein in gel;
hybridized with antibody

Southern Blotting

DNA in gel;
hybridized with DNA

Fig. 1–22. The definitions of Southern blotting, Northern blotting, and Western blotting. From Chien, S.: *An introduction to molecular biology for physiologists. In* Molecular Biology in Physiology. Edited by S. Chien. New York, Raven Press, 1989, p. 14.

the use of a panel of four chemicals that can break the DNA strand in a base-specific manner. The DNA is end-labeled with ^{32}P, made into single strands, and divided into four samples. Each sample is treated with a separate chemical that specifically breaks the DNA at a particular base (Fig. 1-23). The conditions of the reaction are controlled such that most DNA strands would have only one breakage site, thus resulting in a statistical distribution of DNA fragment lengths in accordance with the locations of this particular base. These fragments are separated precisely according to size by gel electrophoresis. By simultaneously running four lanes on the gel using the specimens of DNA treated with the four chemicals with specificities for the four bases, one obtains a distribution of the fragment lengths (indicated by the vertical positions of the bands) as a function of their terminal bases (indicated by the lane in which the band lies). Thus, by reading the gel from bottom to top while noting the lane location of successive bands, one obtains the nucleotide sequence of the DNA from the 5' end to the 3' end (Fig. 1-23).

Another DNA-sequencing procedure is that developed by Sanger[9] using a panel of 2', 3'-dideoxy nucleoside triphosphates. In the normal process of DNA chain elongation in vivo, the four types of dNTPs are sequentially added to synthesize a DNA complementary to the full length of the template. The incorporation of one of the 2', 3'-dideoxy nucleoside triphosphates, however, will stop further chain elongation beyond that point. By using a low concentration of the 2', 3'-dideoxy adenosine triphosphate in the presence of all four types of dNTPs, the DNA chain elongation will stop randomly at positions where there is an adenine nucleotide to be incorporated, resulting in a statistical distribution of DNAs of various lengths, all terminating with an A. By carrying out such reactions for each of the four bases and by separating each of the reaction mixtures on one of the four lanes by gel electrophoresis, one obtains the nucleotide sequence of the DNA in a manner similar to that described for the Maxam and Gilbert method.[8]

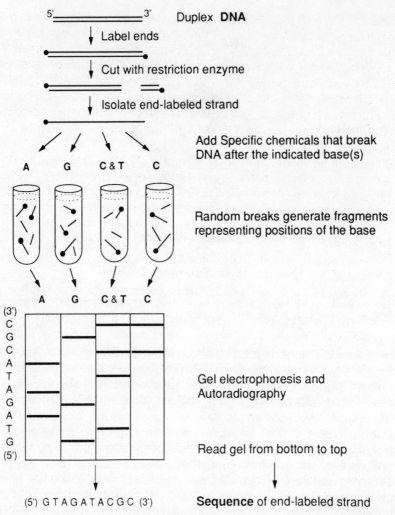

5'_____3' Duplex **DNA**

↓ Label ends

↓ Cut with restriction enzyme

↓ Isolate end-labeled strand

A   G   C & T   C

Add Specific chemicals that break
DNA after the indicated base(s)

Random breaks generate fragments
representing positions of the base

A   G   C & T   C

(3')
C
G
C
A
T
A
G
A
T
G
(5')

Gel electrophoresis and
Autoradiography

Read gel from bottom to top

(5') G T A G A T A C G C (3')   **Sequence** of end-labeled strand

Fig. 1–23. The principle of DNA sequencing by the Maxam and Gilbert method, showing four lanes of nucleotide separations after treatments of 4 aliquots of the sample with chemicals having different nucleotide specificities.

Once the base sequence has been determined, this can be used to derive the amino acid sequence of the encoded protein, i.e., its primary structure, by making use of the universal triplet codons (Table 1-1). The primary sequence of amino acids derived from molecular biologic studies can be subjected to computer analysis for the modeling of the secondary structure of the protein, i.e., the spatial arrangement of amino acids that are close to each other in the peptide chain. Various forms of secondary structures such as α-helix (coiling of the peptide chain around an imaginary cylinder), β-sheet (amino acids having the conformation of a sheet of pleated paper), random coils, etc., can be deduced from such analysis. Sometimes, one can even derive some information on the tertiary structure of the protein in terms of the folding of the helical and random coil regions with respect to each other, i.e., the three-dimensional relationship of amino acid segments that may be far apart in the linear sequence. Such structural analysis can yield information on the interaction sites of a given protein with enzymes, receptors, other neighboring proteins, etc. For membrane-spanning proteins, e.g., the erythrocyte membrane protein band 3 (p. 167), one can deduce the extracellular, transmembranous, and cytoplasmic domains of the protein and can derive the number of transmembrane loops of the protein from the hydropathy index of the amino acids by using the method of Kyte and Doolittle.[10] Knowledge of the nucleotide and amino acid sequences allows the computer search of genes and proteins with sequence homology using available data bases, thus making possible the identification of proteins with similar structures and membranes of a gene family.

In some of the DNA molecules, sequencing has demonstrated the presence of repeating units of nucleotides with a high degree of sequence homology. These repeats allow the production of large proteins, e.g., spectrins and apolipoprotein B, by using relatively few variations of the fundamental structure. The α- and β-spectrins have many 318-nucleotide (106-amino acid) repeats, which constitute the major parts of these long proteins, with molecular weights of 240 and 220 kD, respectively (p. 161). The 550-kD apoB-100 contains numerous internal repeats of different lengths, which generally have a significant degree of degeneracy (p. 195).

TRANSCRIPT MAPPING

To have a detailed knowledge of gene structure and function, one needs to analyze with precision the corresponding RNA in terms of sites for transcription initiation, termination, and processing. Together with sequencing information on the cDNA and genomic clones, the results of transcript mapping allow a detailed description of the synthesis and processing of a given mRNA. One of the methods used for transcript mapping is the Northern blotting described previously, which provides a sensitive method for detecting the presence or absence of specific mRNAs in a tissue or cell sample and for sizing and analyzing cellular distribution of related mRNA transcripts. A brief description is given here for two other methods of transcript mapping: primer extension analysis, and S_1 nuclease protection assay.

In the primer extension analysis, a small DNA primer is synthesized with its oligonucleotide sequence complementary to the mRNA of interest and with ^{32}P labeling on its 5' end. This primer is annealed in excess to the RNA sample containing the desired mRNA, and reverse transcriptase is used to extend the annealed primer until the 5' end of the mRNA is reached. The newly synthesized DNA–RNA hybrid is denatured and electrophoresed on a DNA-sequencing gel. Appropriate design of the procedure allows the identification of the site of annealing of the primer on the mRNA and the determination of the nucleotide number and sequence of the mRNA from the primer annealing site to its 5' end. In addition to the mapping of the 5' end of an mRNA, the primer extension technique can also be used to quantify mRNAs by radioactivity counting of the extension product on the gel band identified by autoradiography. This approach has been used to quantify the skeletal and cardiac iso-mRNAs of α-actin in the same myocyte RNA sample (p. 66); because of the similarity in size of these two iso-mRNAs, it would be more difficult to quantify them by Northern blotting.

S_1 nuclease is an endonuclease that has preference for single-stranded DNA or RNA over double-stranded molecules. A single-stranded DNA oligonucleotide complementary to the RNA of interest is either synthesized or prepared from a fragment of genomic DNA and end-labeled. The hybrid formed between the RNA and the labeled DNA fragment is then subjected to S_1 nuclease digestion, which results in the removal of the unhybridized portion(s) of the DNA and RNA. The protected probe fragments are separated by gel electrophoresis and detected by autoradiography to reveal the number and sizes of the protected fragments. When a piece of a genomic DNA is used as a probe, the results allow the determination of the number and sizes of exons contained within the DNA probe. By performing the digestion and separation on a sequencing gel, one can determine the nucleotides from the labeling site to the end of the next exon or to the 5' or 3' end of the mRNA. This procedure of S_1 nuclease protection assay can also be used to quantify mRNA in an RNA sample by performing radioactivity counting. Examples of application of the S_1 nuclease protection assay are the quantitation of α- and β-iso-mRNAs of MHC in myocytes (p. 67), the study of MHC isoforms in several smooth muscle tissues (p. 102), and the identification of multiple transcriptional start sites for brand 3 message in various tissues (p. 167).

The primer extension method and the S_1 nuclease protection assay allow the identification and sequencing of the 5' end of the mRNA. Therefore, these methods provide valuable information on nucleotide sequence in a region that is often not covered by cDNA cloning and also on the location of the transcription start site, which is difficult to predict based on DNA sequencing of the gene.

POLYMERASE CHAIN REACTION

The methodology in molecular biology is continuously evolving. The recently developed technique of polymerase chain reaction (PCR) allows the specific amplification of discrete fragments of DNA initially present in very small quantities (even a single copy), thus greatly simplifying most applications involving the isolation and purification of nucleic acid fragments.[11] The PCR technique circumvents to a great extent the need for subcloning and traditional biologic (plasmid) amplification, and it can be coupled with DNA-sequencing methods to eliminate the need for cloning and purifying the DNA sample to be sequenced. Therefore, PCR makes it possible to detect and characterize nucleic acid fragments with sensitivity and specificity, while reducing the time and labor required for sample preparation.

In the PCR technique, a DNA polymerase (usually the temperature-stable *Taq*DNA polymerase) and dNTPs are used to replicate a segment in a double-stranded DNA; the procedure is carried out in the presence of a large excess of two different synthetic oligonucleotides (extension primers), which anneal to the two sites flanking the segment of interest to initiate the replication. This method is based on the cyclic repetition of a set of three steps conducted in succession; the procedure mimics in principle the natural DNA replication process, except that it is controlled by temperature changes (Fig. 1-24). First, the double-stranded DNA is denatured at a high temperature (e.g., 95°C). Second, the temperature is lowered (37 to 72°C, depending on experimental conditions and purpose) to allow annealing of the primers to the separated strands of the DNA. Third, the DNA polymerase-mediated extension of the primer–template complex is allowed to proceed for 1 to 4 min at 72°C. This three-step cycle is repeated in succession for 25 to 30 cycles. Following the denaturation step in the second cycle, the single-stranded DNAs will be composed of the original template DNA and the newly synthesized DNA (long product) marked at one end by the 5' end of either one of the primers. As the cycles are repeated, the long product marked with one primer will become the template for annealing and extension with the other primer. This results in the formation of the "short product," which is the desired segment of the

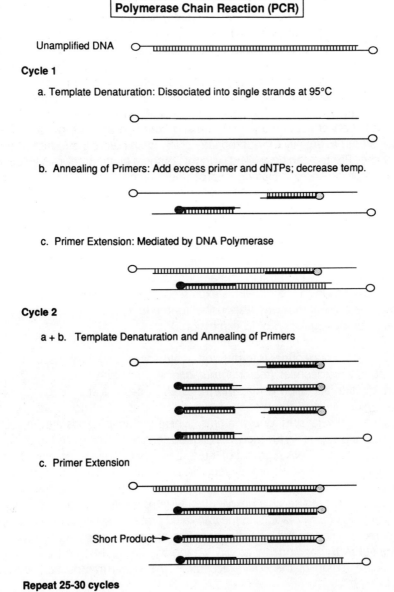

Polymerase Chain Reaction (PCR)

Unamplified DNA

Cycle 1

a. Template Denaturation: Dissociated into single strands at 95°C

b. Annealing of Primers: Add excess primer and dNTPs; decrease temp.

c. Primer Extension: Mediated by DNA Polymerase

Cycle 2

a + b. Template Denaturation and Annealing of Primers

c. Primer Extension

Short Product

Repeat 25-30 cycles

Fig. 1–24. The principle of polymerase chain reaction. See text for explanation.

template DNA demarcated by the two extension primers. Statistical considerations indicate that the short product will form with greater probability than the longer ones and that it will accumulate in an exponential fashion. Thirty cycles of PCR can lead to a 10^7-fold increase in copy number of the DNA segment of interest. The short product can be purified by size separation from the longer products, but for some experiments, purification may not be absolutely required because the short product is so overwhelmingly abundant.

The PCR technique is very useful for cloning DNA into vectors. For example, in the "linker-primer" method, nucleotides containing restriction endonuclease-recognition sites are attached to the 5' ends of the extension primers during primer synthesis, and the amplified short product obtained following PCR can be separated and digested with the appropriate restriction endonuclease to generate the desired DNA segment for subcloning. The PCR technique is also useful as the first step in DNA sequencing; one of the advantages is that the requirement for purifying the amplified product is reduced to a minimum. By adding a third (sequencing) primer, it is possible to perform the sequencing reaction directly following the amplification.

In the detection of RFLP (see section entitled Southern Blotting, p. 22), if the PCR technique is used to amplify the appropriate DNA region before the restriction enzyme treatment, the RFLP test can be applied without Southern blotting; i.e., the radioisotope-labeled DNA probe and the blotting step can be omitted.

GENE MANIPULATION

Alterations of Gene Sequence

Genetic engineering techniques are available to change the nucleotide sequence of the coding or regulatory regions of a gene, with consequent alterations in the functional properties of the protein, the amount of protein produced, and/or the particular cell type in which the protein is expressed. These procedures allow the identification of the parts of the gene responsible for the functional properties of its coded protein and for the regulation of its expression.

The coding sequence of a gene can be altered by in vitro mutagenesis involving either one or a few nucleotides (site-directed or site-specific mutagenesis) or a relatively large number of nucleotides (regional mutagenesis). In site-directed mutagenesis, a specific nucleotide alteration is made in the DNA to cause a specific change in one or a few amino acids of the protein expressed. Attempts for such mutagenesis are usually made at several sites, one at a time, where the amino acids are suspected to play a critical role in protein function. An oligonucleotide is synthesized to contain the same sequence as the DNA segment of interest, except for the nucleotide(s) that code for the amino acid(s) to be altered. By replacing that DNA segment with the synthetic oligonucleotide containing the altered coding at the desired location, a mutant DNA is generated. With the use of an expression system, the functional role of the site-specific alteration of amino acid(s) in the translated protein can be evaluated. Examples are the demonstration of the critical importance of several amino acids (aspartic-116 and lysine-40) in the angiogenic activity of angiogenin (p. 145) and the analysis of tyrosine kinase region of the PDGF receptor in the signal transduction after PDGF binding (p. 129).

In regional mutagenesis, a relatively larger number of nucleotides are replaced by a new sequence, resulting in the expression of a protein with a stretch of altered amino acids. Usually, this is carried out, when no specific amino acid is suspected, to first identify the region of interest before performing the more specific procedure of site-directed mutagenesis. An example of regional mutagenesis is the construction of a hybrid gene in which 15 nucleotides in the middle of the angiogenin gene (nucleotides no. 58 to 72) are replaced by an oligonucleotide encoding the ribonuclease A enzyme, which has a high degree of homology with angiogenin but is different in this region (p. 148). The protein expressed from the hybrid gene shows a marked reduction of angiogenic activity accompanied by an increase in ribonucleolytic activity, indicating the functional importance of the five amino acids encoded by this segment of the angiogenin gene.

To determine the parts of a eukaryotic gene responsible for the regulation of its expression, a recombinant DNA is made by joining the suspected regulatory DNA sequence of the gene to the coding sequence of a prokaryotic gene that encodes a readily detected marker not normally present in eukaryotic cells. One of the most commonly used marker enzymes is the bacterial protein chloramphenicol acetyltransferase (CAT). The recombinant DNA molecule is inserted into a eukaryotic cell to test the CAT enzyme activity after an overnight incubation, thus determining whether the suspected regulatory sequence has promoted the expression of the CAT gene and by inference the regulatory DNA sequence of the gene under study. This procedure has been used to localize the regulatory elements of the cardiac muscle α-MHC gene (p. 47), the cardiac muscle ANF gene (p. 85), and the smooth muscle α-actin gene (p. 98).

The redesigning of genes can involve not only replacement but also the deletion or addition of selected nucleotide sequences. For example, deletion analysis has been used to assess the functional significance of different regulatory elements in the α-MHC gene (p. 48).

Elimination of Specific Protein Translation With Antisense RNA

Because RNA polymerase II effects RNA synthesis only in the 5' to 3' direction (see preceding section entitled Transcription of DNA to RNA, p. 5), normally, only one of the two DNA strands in the duplex is transcribed, and it is always the same strand for a given gene. If a cloned gene is engineered so that the opposite DNA strand is transcribed, the result will be an antisense RNA with a sequence complementary to the normal RNA transcripts. Such an antisense RNA will hybridize with the "sense" RNA made by the normal gene and thereby inhibit the translational synthesis of the corresponding protein. This leads to the creation of a specific dominant mutation in which the normal protein is not expressed; hence, the functional role of this protein can be assessed by comparing the phenotypes of such mutant cells with the normal cells. An example of this approach is the use of antisense c-*erb*-A mRNA to evaluate the role of the c-*erb*-A protein in the regulation of the α-MHC gene by thyroid hormone (p. 50). c-*erb*-A is a proto-oncogene product that has been shown to be the high-affinity nuclear receptor for T_3, the intracellular form of thyroid hormone.

Radiolabeled antisense RNA can also be used as a probe for localizing RNA transcripts in cells and tissues by in situ hybridization; this approach has been used to localize ANF gene expression in the fetal heart (p. 84, Fig. 5-1).

Transgenic Animals

The ultimate test of the function of an altered gene is to study the effect of replacing the normal gene with the altered gene after its reinsertion into an organism. Although this can be done relatively easily in the yeast, it is difficult in mammalian cells because they have enzymes that rapidly ligate the inserted DNA fragments end to end to form long tandem arrays to be integrated into a chromosome at some random site. When the modified chromosome is present in the germ line cells (eggs or sperm), the animal (e.g., mouse) will pass the newly adopted genes on to its progeny to produce transgenic animals with a permanently altered gene containing the insertion. Although the presence of the normal gene complicates the interpretation of results, studies on such transgenic animals do provide useful insights into the mechanism of regulation of mammalian genes and the role of oncogenes in causing cancer.

Recently, direct gene replacements have been made possible in mice by the following more elaborate approach. A DNA fragment containing a desired mutant gene is first transfected into a special line of pluripotent embryo-derived stem cells in culture. After cell proliferation, the rare clones with the desired gene replacement are identified by Southern blotting. The selected cells are implanted into an early mouse embryo, wherein these embryo-derived stem cells often proliferate to produce major portions of the mouse. Such mice are then bred to produce both males and females, each heterozygous for the gene replacement (i.e., having one normal and one mutant copy of the gene). Breeding of these heterozygous mice would lead to homozygous mutants in one fourth of the progeny, which can be used to study more directly the function of the altered gene. This transgenic mice method provides a powerful new tool for dissecting the function of genes in an intact organism. The application of this technology to studies on the vascular system is discussed in Chapter 7 (p. 133).

COMMENTS

It is remarkable that the encoding and transmission of the complex genetic information of life is accomplished by using such simple units, i.e., the nucleotides, which have variations only among four types of bases for either DNA or RNA. The combination and permutation of these simple units and the unique features of base pairing between them allow the formation of the complex molecules DNA and RNA and the precise regulation of the intricate processes of replication, transcription, and translation. It is also admirable that molecular biology has taken advantage of the very same principles to devise ingenious approaches for the investigation and manipulation of these complex molecules and cellular functions.

The available molecular biologic technologies allow the identification of the gene coding for a protein, the mapping of the distribution of its message transcript in various types of cells and tissues, the determination of the nucleotide sequence of the genomic DNA and cDNA, the derivation of the primary sequence and secondary structure of the protein, the understanding of the molecular mechanisms regulating the transcription and translation processes under physiologic and pathophysiologic conditions, the manipulation of the gene and the protein expressed in experimental animals, the estimation of pathways of evolution of the protein, and the elucidation of molecular alterations in disease states. These applications will become more apparent in Chapters 2 to 10 and are summarized in Chapter 11. The application of molecular biology to cardiovascular research has a very recent beginning, but it has already made impressive progress, and an exciting future awaits this new area of investigation.

REFERENCES

1. Moskowitz, J., and Lenfant, C.: Physiology: is it time to cross the new frontier? J. Appl. Physiol., *61*:1609, 1986.
2. Chien, S.: An introduction to molecular biology for physiologists. *In* Molecular Biology in Physiology. Edited by S. Chien. New York, Raven Press, 1989. pp. 1–17.
3. Alberts, B., Bray, D., Lewis, J., Raff, M., Roberts K., and Watson, J.: Molecular Biology of the Cell. 2nd Ed. New York, Garland Publishing, 1989.
4. Darnell, J., Lodish, H., and Baltimore, D.: Molecular Cell Biology. New York, Scientific American Books, 1986.
5. Ausubel, F.M., Brent, R., Kingston, R.E., Moore, D.D., Seidman, J.G., Smith, J.A., and Struhl, K.: Current Protocols in Molecular Biology. New York, Greene Publishing Associates and Wiley-Interscience, 1987.

6. Maniatis, T., Fritsch, E.F., and Sambrook, J.: Molecular Cloning: A Laboratory Manual. 2nd Ed. Cold Spring Harbor, NY, Cold Spring Harbor Lab., 1988.
7. Southern, E.M.: Detection of specific sequences among DNA fragments separated by gel electrophoresis. J. Mol. Biol., *98:*503, 1975.
8. Maxam, A.M., and Gilbert, W.: A new method for sequencing DNA. Proc. Natl. Acad. Sci. USA, *74:*560, 1977.
9. Sanger, F., and Coulson, A.R.: A rapid method for determining sequences in DNA by primed synthesis with DNA polymerase. J. Mol. Biol., *94:*441, 1975.
10. Kyte, J., and Doolittle, R.F.: A simple method for displaying the hydropathic character of a protein. J. Mol. biol., *157:*105–132, 1982.
11. Mullis, K.B., and Faloona, F.A.: Specific synthesis of DNA in vitro via a polymerase catalysed chain reaction. Meth. Enzymol., *155:*335, 1987.

Chapter 2

REGULATORY GENES THAT CONTROL THE DETERMINATION AND DIFFERENTIATION OF SKELETAL MYOBLAST STEM CELL LINEAGES

Charles P. Emerson, Jr. • Sonia Pearson-White • Deborah F. Pinney

Stem cells arise during very early stages of embryonic development by regulatory processes that generally have been assumed to be complex. In vertebrates, stem cell lineage determination has been particularly difficult to investigate because of the cellular complexity of the vertebrate embryo and a lack of developmental genetic approaches. However, in-vitro approaches involving clonal cell culture and gene transfer by DNA transfection recently have provided a novel means to identify and characterize regulatory genes that control the determination of skeletal muscle stem cell lineages in mammalian embryos. In this chapter, we review the discovery of these myogenic regulatory genes using cell culture and DNA transfection approaches. We discuss current information on the functions of these genes in the determination and differentiation of skeletal myogenic cell lineages and the general application of these approaches to the study of regulatory genes in other cell types, including cardiac and smooth muscle.

MULTIPOTENTIAL 10T1/2 CELLS CAN UNDERGO MYOBLAST LINEAGE DETERMINATION

Myogenic regulatory genes have been identified by transfection of recombinant DNAs into the mouse embryo cell line, C3H10T1/2C18(10T1/2).[1] The 10T1/2 cell is a host cell uniquely suited for transfection analysis of the genetic basis of myogenic lineage determination. This clonal cell line is multipotential,[2] and 5-azacytidine, which causes DNA hypomethylation,[3] converts 10T1/2 cells at high frequencies into stable, clonal lineages of myoblast, adipoblast, and chondroblast stem cells.[4] 10T1/2-derived myoblast lineages are similar to embryonic myoblasts derived from the muscle-forming regions of normal embryos in that these myogenic cells can proliferate extensively and yet stably inherit the potential to differentiate into muscle[5,6] and to activate coordinately muscle protein genes[7,8] in response to growth-factor depletion and entry into a G_1/G_0 differentiated state.[9,10] The 10T1/2 cell, therefore, is a model for a multipotential embryonic mesodermal cell that can undergo lineage determinations to form myoblasts, as well as adipoblasts and chondroblasts.

To explain the observed high frequencies of 5-azacytidine-induced clonal conversion of 10T1/2 cells into these new lineages, we hypothesized that the 10T1/2 cell is genetically blocked in a multipotential state because of methylation at one or a few critical sites required for transcription of different lineage-specific regulatory gene loci that control the determination of myogenic, adipogenic, and chondrogenic lineages.[4] Because 5-azacytidine treatment generates apparently random demethylation of methylated cytosine residues in DNA, we reasoned that cells that have one of their regulatory gene loci demethylated and activated then convert into a new lineage. This model for 5-azacytidine conversion of 10T1/2 cells correctly predicted that these lineage-specific regulatory genes, particularly myogenic regulatory genes, could be identified by transfection of DNA molecules in an unmethylated state.[4] The significant conclusion of these DNA transfection studies, discussed in the following section, is that the determination of skeletal muscle myoblast lineages has a simple genetic basis and, in 10T1/2 cells, is controlled by activation of only a few muscle-specific regulatory genes.

IDENTIFICATION OF MYOGENIC REGULATORY GENES BY COSMID GENOMIC DNA TRANSFECTION

DNA transfection of the 10T1/2 cell has led to the identification of four different myogenic regulatory genes that function in myogenic lineage determination, *myd, myoD1, mdf1,* and *myf5.* These myogenic regulatory genes have been identified by transfection of 10T1/2 cells with genomic DNA prepared from myoblasts,[11,12] with recombinant cosmid library DNA,[13] and with muscle-specific, *myc*-related cDNAs cloned into expression vectors with constitutive promoters.[14-17] The initial experiments that established the feasibility of this approach, however, involved transfection of 10T1/2 cells with genomic DNA prepared from myoblasts, which was assumed, based on the 5-azacytidine experiments, to have an unmethylated myogenic regulatory gene.[4] These experiments showed that genomic DNA isolated from myoblasts, but not from parental 10T1/2 cells, induces a low frequency of myogenic colony formation when transfected into 10T1/2 cells.[11,12] This genomic DNA transfection approach, however, did not provide a means to trace and identify specific DNA molecules transfected into 10T1/2 cells. Subsequent cosmid genomic DNA transfection studies have provided clear evidence that transfection of single genomic DNA segments can convert 10T1/2 cells into myogenic colony-forming cells.[13] These cosmid genomic DNA experiments involved transfection of 10T1/2 cells with human leukocyte DNA cloned into a cosmid shuttle vector.[18] Because recombinant cosmids replicate in bacterial hosts, cytosine methylation patterns in human leukocyte DNA are erased during cloning. Thus, a putative myogenic lineage determination gene in the cosmid library would be transfected as an unmethylated, active gene. The advantage of this approach is that cosmid vector DNA sequences, including the cos site and associated *neo* and *amp* drug-resistance genes, can be traced by hybridization and G418 neomycin-resistance selection to directly demonstrate gene transfer into mammalian cells. Integrated cosmid segments also can be recloned from mammalian cells.[18]

Cosmid genomic DNA transfection experiments have led to the identification of a human regulatory gene locus, referred to as *myd.*[13] The *myd* gene converts 10T1/2 cells into myogenic colony-forming cell lines at frequencies of about $\frac{1}{5000}$ G418 neomycin-resistant cells when transfected with recombinant cosmid genomic library DNA. Primary transfected cell lines have 10 to 20 copies of library DNA integrated into their chromosomes, including recombinant human segments linked to cosmid vector sequences and G418 neomycin-resistance genes used for selection of transfected cells. Proof that the *myd* gene locus is a transfectable segment of DNA comes from secondary transfection experiments in which 10T1/2 cells were transfected with genomic DNA from primary cosmid-transfected myogenic cells. These 10T1/2 cells converted at high frequency both to G418 neomycin resistance and to the myogenic cell type. This result establishes linkage of transfected *myd* and vector neomycin resistance genes in the primary transfected myogenic cell line. *Myd*-transfected cells express at least two other myogenic regulatory genes, myoD1 and mdf1, identified independently by transfection of muscle-specific cDNAs, as discussed in the following section.

Myd-transfected cells are myogenic, as established by their capacity to differentiate into myofibers in response to mitogen-deficient media, activating the expression of the set of muscle protein genes, such as myosin heavy chain, characteristic of differentiated muscle. Molecular hybridization studies show that DNA from the myogenic cell lines that result from *myd* transfection do not have integrated human genes for myoD1,[14] myf5,[17] or mdfl,[15,16] indicating that the *myd* gene is distinct from these three more recently identified *myc*-related myogenic genes. Molecular experiments are now in progress to characterize the structure of the *myd* gene and its encoded protein.

IDENTIFICATION OF MYOGENIC REGULATORY GENES BY cDNA CLONING

A significant aspect of the cosmid genomic DNA transfection experiments is that the biologic activity of the transfected *myd* DNA is assumed to be derived from the transcriptional activity of the promoter of the *myd* gene. Alternative cDNA cloning approaches have identified cDNA clones of three different muscle-specific regulatory genes, myoD1, myf5, and mdf1, which are expressed in cultured myogenic cells and in embryonic and adult muscles.[14,17] These muscle-specific cDNAs cause conversion into myogenic cells when transfected into 10T1/2 cells under the control of constitutively transcribed promoters. In contrast to the genomic DNA transfection approach, the cDNA transfection approach assays regulatory functions of proteins encoded by transfected cDNAs independent of the transcriptional activities of the natural promoters of their genes. The promoters of such regulatory genes, for instance, may require specific transcription factors not present in 10T1/2 cells before conversion to myogenic stem cells. The cDNA cloning approach, therefore, provides an assay for identifying regulatory genes that encode proteins that function downsteam in a regulatory gene pathway.

The myoD1 cDNA was identified by screening a cDNA library derived from 5-aza-cytidine-converted myoblast RNA with subtracted myoblast-specific cDNA probes.[13,14] Of the three cDNA clones recovered in this screen, one clone, myoD1, induces a high frequency of myogenesis when transfected into 10T1/2 cells and other tissue culture cell lines using plasmid and retroviral vectors with a Maloney sarcoma virus long terminal repeat (LTR) constitutive promoter.[14,19] The decoded protein sequence of the myoD1 cDNA has localized homology to the *myc* oncogenes, as well as to other developmentally significant proteins (Fig. 2-1). The localized *myc*-homology domain and an adjacent domain of basic amino acids in an expression vector are by themselves sufficient to convert 10T1/2 cells to a myogenic phenotype.[20]

MyoD1 transcripts are detected only in mouse skeletal muscle myoblasts and embryonic and adult skeletal muscle. Avian homologues of myoD1 cDNA have also been cloned from embryonic muscle,[21,22] and these avian mRNAs are also expressed in myoblasts and embryonic and adult muscles. The levels of expression of myoD1 vary greatly in different myogenic cell lines. 10T1/2-derived, 5-azacytidine myoblasts express very high levels of myoD1 mRNA, whereas primary embryonic myoblasts in culture, differentiated muscle, and *myd*-transfected myogenic cells express much lower levels

Helix/Loop/Helix Motifs in Regulatory Proteins

```
            BASIC DOMAIN   AMPHIPATHIC HELIX I                          AMPHIPATHIC HELIX II
myoD     DRRKAATMRERRRLSKVNEAFETLKRCTSSNP---------------NQRLPKVEILRNAIRYIEGLQALLR
myf5     DRRKAATMRERRRLKKVNQAFETLKRCTTTNP---------------NQRLPKVEILRNAIRYIESLQELLR
mdf1     DRRRAATLREKRRLKKVNEAFEALKRSTLLNP---------------NQRLPKVEILRSAIQYIERLQALLS
E12      DRRVANRILADRLNVRD INDAFKELGNIOQLIILIIODKR           QTKLLILIIQAVOVILILDQQVR
E47      ERRMANNARERVRVRDINEAFRELGRMCQMHLKSDKA-------------QTKLLILQQAVQVILGLEQQVR
da       ERRQANNARERIRIRDINEALKELGRMCMTHLKSDKP-------------QTKLGILNMAVEVIMTLEQQVR
twist    NQRVMANVRERQRTQSLNDAFKSLQQIIPTL---------------PSDKLSKIQTLKLATRYIDFLCRMLS
AS-C T4  VQRR--NARERNRVKQVNNSFARLRQHIPQSIITDLTKG--G-GRGPHKKISKVDTLRIAVEYIRSLQDLVD
AS-C T5  VIRR--NARERNRVKQVNNGFSQLRQHIPAAVIADLSNGRRGIGPGANKKLSKVSTLKMAVEYIRRLQKVLH

                *   *   *  *   *                         *  **  ** **  *  **
CONSENSUS   R    N RER R     N  F   L                     K   IL  A  YI  L
            T    K         L                             T      V
```

Fig. 2–1. Comparison of sequence similarities of the myogenic regulatory proteins and other developmental regulatory proteins. Amino acid sequences have been aligned in the *myc* homology regions of the myogenic genes, myoD1 (aa108-164),[14] myf5 (aa83-139),[17] and mdf1 (aa81-137),[15] and in the E12 and E47 enhancer-binding proteins, and the *daughterless* (*da*),[25] *twist*,[27] and T4 and T5 *achaete-ascute*[26] *Drosophila* developmental regulatory genes, as previously described.[28] The asterisk (*) symbol indicates positions of hydrophobic amino acid residues at positions of a four-three repeat that defines the hydrophobic surfaces of two hypothetical amphipathic helices, delineated by the lines shown above the sequences. These helical regions are potential domains involved in dimer formation.[28] The central sequence region between helix I and helix II is postulated to be an unfolded loop of the helix/loop/helix structure, and the basic region adjacent to helix I may be involved in interactions of these proteins with DNA.

of myoD1 transcripts,[13] and some cultured myogenic cell lines, rat L6,[15] and mouse BC3H1,[16] do not express detectable levels of myoD1. These findings suggest that myoD1 is not an essential gene for myogenesis, although the L6 and BC3H1 lines may express a functional homologue of myoD1 that has not yet been isolated.

A third myogenic regulatory gene, mdf1 (also called *myogenin*), has been identified by screening a rat L6 cell line cDNA library with subtracted, early differentiation-specific cDNA probes.[15] The mdf1 cDNA also encodes a *myc*-related protein homologous to myoD1 in its basic and *myc* domains (Fig. 2-1). A nearly identical mouse mdf1 cDNA has been isolated based on its sequence homology to myoD1 by screening a cDNA library derived from BC3H1 cells.[16] The rat and mouse mdf1 cDNAs convert 10T1/2 cells to a myogenic phenotype, although the myogenic conversion frequencies may be lower than obtained with myoD1 transfection. As discussed previously, the rat L6 cell line and the BC3H1 cell line do not express detectable levels of myoD1. Mdf1 transcripts are differentiation specific and accumulate to high levels in L6 and BC3H1 cells, as well as in 10T1/2-derived 5-azacytidine myoblasts, primary human myoblasts, and *myd*-transfected myoblasts following transfer into mitogen-deficient differentiation medium (Pearson-White, Pinney and Emerson, unpublished observations). In L6 and BC3H1 cells, mdf1 accumulation immediately precedes the accumulation of mRNAs encoding muscle proteins. The differentiation-specific expression of mdf1 contrasts with myoD1, which is expressed before differentiation in proliferating normal human myoblasts, 5-azacytidine-derived myoblasts, and continues to be present in differentiated muscle. Their different temporal regulation suggests that myoD1 and mdf1 have different functions in myogenesis. In any case, myoD1 and mdf1 transcripts are both expressed in normal myogenic cells, and in the newly formed somites of mouse and quail embryos.[15,22] Somites are a major site of origin of myogenic stem cells in vertebrates.[23]

A fourth muscle-specific regulatory gene, myf5, has been isolated by screening a cDNA library of human fetal muscle with myoD1 probes.[17] Myf5 has extensive protein-coding homology to myoD1 and to mdf1 in their *myc* and basic domains (Fig. 2-1). Myf5 cDNA also converts 10T1/2 cells to a myogenic phenotype, although the frequency of myf5-directed conversion relative to myoD1 conversion is unclear. Detailed studies of the expression of myf5 in myoblasts and differentiated muscle cells of primary muscle cultures and the various clonal myogenic lines are not yet reported, but the available data indicate that myf5 is abundant in fetal muscle and some, but not all, myoblast cell lines.

MYOGENIC cDNAs ENCODE PROTEINS HOMOLOGOUS TO THE *myc*-ONCOGENE AND TO OTHER DEVELOPMENTAL REGULATORY PROTEINS

MyoD1, mdf1, and myf5 are members of a large *myc* gene family and share remarkable protein sequence and structural homology, particularly in their basic/*myc* domains (Fig. 2-1). In addition, these myogenic proteins also share structural and sequence homology with proteins encoded by developmentally important genes involved in sex determination and neurogenesis,[24-26] mesoderm formation,[27] and immunoglobulin gene-enhancer binding and transcription[28] (Fig. 2-1). The basic/*myc* domains of myoD1, mdf1, and myf5 encode nearly identical sequences, which, as discussed earlier, are sufficient to convert 10T1/2 cells to the myogenic phenotype.[20] The specific myogenic function of this myoD1 basic/*myc* domain and the functions of these different myogenic proteins, however, are unclear because transfected myoD1 apparently activates expression of the endogenous chromosomal myoD1 genes, which may then produce the biologically active, intact myoD1 protein or other myogenic proteins that direct the myogenic conversion of 10T1/2 cells.[20]

The current data suggest that myoD1 and other related myogenic proteins are localized in the nucleus during the process of differentiation. Polyclonal antibodies made to myoD1/trpE fusion protein apparently react avidly with the nuclei of differentiated muscle cells of 5-azacytidine-derived myoblast lines but react only infrequently and weakly with nuclei of apparently undifferentiated myoblasts.[20] Because the mdf1 and myf5 myogenic proteins share extensive homology with myoD1, these myoD1 antibodies likely also cross-react with these other myogenic proteins in differentiated muscle. These observations are consistent with the idea that these *myc*-related myogenic proteins function in the differentiation process,[13] which would explain why 10T1/2 cells often respond to myoD1 cDNA transfection by rapidly differentiating and expressing muscle-specific proteins.[20]

The specific nuclear functions of myogenic proteins in differentiation processes remain to be discovered, although a reasonable hypothesis is that these are DNA binding proteins and function as coordinate transcription activators of muscle protein genes during differentiation. In support of this idea, the basic/*myc* sequence domains of the myogenic proteins share sequence homology with enhancer binding proteins E12 and E47 that regulate immunoglobulin gene expression[28] (Fig. 2-1). The E12 and E47 enhancer binding proteins, the myoD1, myf5, and mdf1 myogenic genes, and the *myc*-related *Drosophila* developmental genes have similar deduced helix/loop/helix structures in their basic/*myc* domains of protein sequence (Fig. 2-1). These structurally similar domains likely are involved in sequence-specific DNA binding and in formation of protein homo- or heterodimers through interactions along the amphipathic surfaces of these deduced alpha helical domains.[28,29] Further analyses of the DNA-binding reactivity of the myogenic proteins with muscle gene enhancer sequences and the formation of protein dimers will be of considerable importance in revealing the specific functions of these proteins in myogenic differentiation.

PERSPECTIVES

Cosmid genomic DNA transfection experiments and myogenic cDNA cloning and expression vector transfection experiments have provided a definitive test of the regulatory gene model of myogenic lineage determination in 10T1/2 cells.[4] Four different but functionally related myogenic regulatory genes, *myd*, myoD1, mdf1, and myf5, have been identified to date, as demonstrated by their activity in the transfection of 10T1/2 cells to a myogenic phenotype. These myogenic genes and their species-related homologues are summarized in Table 2-1.

Some of these myogenic regulatory genes appear to be expressed in a temporally ordered sequence during myogenesis of cultured cells, suggesting that these genes are part of a transcriptional regulatory gene pathway and that each of these genes likely has specific functions in a myogenic regulatory gene pathway leading to muscle differentiation.[13] A descriptive model of a proposed pathway for myogenic gene activation is shown in Figure 2-2. However, the activities of these regulatory genes also influence each other's expression. The *myd* gene is hypothesized to be repressed by specific DNA methylation in 10T1/2 cells. *Myd* expression leads to myoD1 and mdf1 gene expression[6]; myoD1 can regulate its own expression, at least in 10T1/2 cells[20]; and transfection of the later stage differentiation-specific gene, mdf1, apparently converts 10T1/2 cells into an earlier stage proliferative myoblast phenotype.[15] Such regulatory interactions among these myogenic genes provide a mechanism for stable maintenance of a myogenic phenotype. It will be of interest to determine whether there are additional muscle-specific regulatory genes that act upstream, downstream, or intermediate in the pathway of action of the *myd*, myoD1, myf5, and mdf1 regulatory genes. The available data also do not distinguish whether the *myd*, myoD1, myf5, and mdf1 genes function as an interdependent regulatory gene pathway or, alternatively, whether these genes have redundant functions in myogenic determination. Gene knockout experiments using antisense RNA[30] or gene-specific recombination[31] approaches could distinguish between these possibilities. Characterization of the *cis* control elements and *trans*-acting regulatory factors that regulate the transcription of the *myd* gene, as well as the myoD1, myf5, and mdf1 genes, also is of particular interest. The myogenic activities of the myoD1, myf5, and mdf1 genes have been defined only by cDNA transfection analysis. The transcriptional activities of their natural promoters may reveal the genetic complexity of regulatory interactions among these and other myogenic regulatory genes.

Table 2–1. Myogenic Regulatory Genes

Gene	Species (Name)	Source and Reference
myd	Human (*myd*1)	Lymphocyte cosmid genomic DNA[13]
myoD	Mouse (myoD1)	10T1/2-derived myoblast cDNA[13,14]
	Human (myf3)	Fetal muscle cDNA[17]
	Chicken (cmd1)	Embryonic breast myofiber cDNA[21]
	Quail (qmf1)	Embryonic breast myofiber cDNA[22]
myf	Human (myf5)	Fetal muscle cDNA[17]
mdf	Rat (myogenin)	L6 rat myofiber cDNA[15]
	Mouse (myo8)	BC3H1 myofiber cDNA[16]

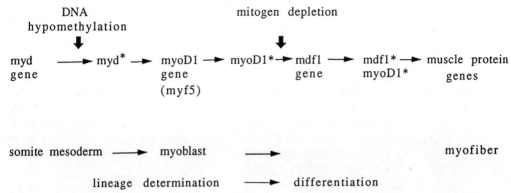

Fig. 2–2. A regulatory gene pathway model of myogenesis. This model proposes that myoblast lineage determination and skeletal muscle differentiation are controlled through the sequential activation of muscle-specific regulatory genes.[13] *Myd*, a myoblast determination gene, is activated in somite mesoderm, perhaps by hypomethylation of a *myd* gene promotor control sequence.[4] The *myd* protein, myd*, activates the myoD1 gene (or a functionally homologous gene such as myf5) and other genes characteristic of myoblasts.[4] The functional activity of the myoD1 protein in myoblasts is modulated through a mitogen-mediated signal transduction pathway, perhaps by myoD1 protein phosphorylation.[14] When extracellular mitogens are depleted, myoD1 protein, myoD1*, functions as a transcription factor that activates the mdf1 (myogenin) gene. The active mdf1* protein, perhaps in combination with myoD1* protein, then acts as a transcription factor that interacts with muscle protein gene transcription enhancers and coordinates the activation of the large set of muscle protein genes expressed in differentiated myofibers.[8]

The *myd*, myoD1, myf5, and mdf1 have remarkable regulatory power to coordinate the genetically complex process of myogenic cell lineage determination and differentiation in 10T1/2 cells. Genetic studies of *Drosophila* and *Caenorhabditis elegans* development also reveal that individual regulatory genes, acting as master regulatory genes or through regulatory gene pathways, control other complex developmental processes, including determination of body plan,[32] determination of sex,[24] and determination of neuronal cell lineages.[33–36] These findings are consistent with the notion that, at the regulatory level, there is a genetic simplicity to the developmental processes of determination and differentiation.

Finally, as discussed previously, the homology of myoD1, myf5, and mdf1 proteins to the *myc* gene family (Fig. 2-1) emphasizes the possibility that the myogenic genes and the other *myc*-related proteins are part of a large family of regulatory genes that control the determination and differentiation of cell types. The general applicability of the cell culture and DNA transfection approach is suggested by a recent report on the identification of a fat cell regulatory gene by genomic DNA transfection of cultured cell lines.[36] In this regard, it will be of particular interest to investigate whether developmental regulatory genes that control the determination and differentiation of other closely related vertebrate cell types such as heart and smooth muscle, and chondrocytes and fat also can be identified and studied by DNA transfection and tissue culture approaches or by direct cloning of genes and cDNA clones based on their structural and sequence homology to the myogenic genes.

ACKNOWLEDGMENTS

We thank the Muscular Dystrophy Association and the National Institutes of Health for research support. We also thank Margaret Ober for her expert assistance in the computer analysis of the myogenic protein sequences.

REFERENCES

1. Reznikoff, C.A., Brankow, D.W., and Heidelberger, C.: Establishment and characterization of a cloned line of C3H mouse embryo cells sensitive to postconfluence inhibition of division. Cancer Res., *33*:3231–3238, 1978.

2. Taylor, S.M., and Jones, P.A.: Multiple new phenotypes induced in 10T1/2 and 3T3 cells treated with 5-azacytidine. Cell, *17*:771–779, 1989.

3. Jones, P.A., and Taylor, S.M.: Cellular differentiation, cytidine analogs and DNA methylation. Cell, *20*:85–93, 1980.

4. Konieczny, S.F., and Emerson, C.P., Jr.: 5-Azacytidine induction of stable mesodermal stem cell lineages from 10T1/2 cells: Evidence for regulatory genes controlling determination. Cell, *38*:791–800, 1984.

5. Konigsberg, I.R.: Clonal analysis of myogenesis. Science, *140*:1273–1284, 1963.

6. Yaffe, D., and Saxel, O.: Serial passaging and differentiation of myogenic cells isolated from dystrophic mouse muscle. Nature (Lond), *270*:725–727, 1977.

7. Konieczny, S.F., and Emerson, C.P., Jr.: Differentiation, not determination, regulates muscle gene activation: Transfection of troponin I genes into multipotential and muscle lineages of 10T1/2 cells. Mol. Cell. Biol. *5*:2423–2432, 1985.

8. Bucher, E.A., Maisonpierre, P.C., Konieczny, S.F., and Emerson, C.P. Jr.: Expression of the troponin complex genes: transcriptional coactivation during myoblast differentiation and independent control in heart and skeletal muscles. Mol. Cell. Biol., *8*:4134–4142, 1988.

9. Konigsberg, I.R.: Diffusion-mediated control of myoblast fusion. Devel. Biol., *26*:133–152, 1971.

10. Konigsberg, I.R.: The role of the environment in the control of myogenesis in vitro. In Pathogenesis of Human Muscular Dystrophies. 5th International Scientific Conference, Muscular Dystrophy Association. Edited by L.P. Rowland, Amsterdam, Elsevier Press, 1977. pp. 779–798.

11. Konieczny, S.F., Baldwin, A.S., and Emerson, C.P., Jr.: Myogenic determination and differentiation of 10T1/2 cell lineages: Evidence for a simple genetic regulatory system. In Molecular Biology of Muscle. Edited by C. Emerson, D.A. Fischman, B. Nadal-Ginard, and M.A.Q. Siddiqui. New York, Alan R. Liss, Inc., 1986. New Series Vol. 29, pp. 21–34.

12. Lassar, A.B., Paterson, B.M., and Weintraub, H.: Transfection of a DNA locus that mediates the conversion of 10T1/2 fibroblasts to myoblasts. Cell, *47*:649–656, 1986.

13. Pinney, D.F., Pearson-White, S.H., Konieczny, S.F., Latham, K.E., and Emerson, C.P., Jr.: Myogenic lineage determination and differentiation: evidence for a regulatory gene pathway. Cell, *53*:781–793, 1988.

14. Davis, R.L., Weintraub, H., and Lassar, A.B.: Expression of a single transfected cDNA converts fibroblasts to myoblasts. Cell, *51*:987–1000, 1987.

15. Wright, W.E., Sassoon, D.A., and Lin, V.K.: Myogenin, a factor regulating myogenesis, has a domain homologous to MyoD. Cell, *56*:607–617, 1989.

16. Edmondson, D.G., and Olson, E.N.: A gene with homology to the *myc* similarity region of MyoD1 is expressed during myogenesis and is sufficient to activate the muscle differentiation program. Genes Devel., *3*:628–640, 1989.

17. Braun, T., Buschhausen-Denker, G., Bober, E., Tannich, E., and Arnold, H.H.: A novel human muscle factor related to but distinct from MyoD1 induces myogenic conversion in 10T1/2 fibroblasts. EMBO J., *8*:701–709, 1989.

18. Lau, Y-F., and Kan, Y.W.: Versatile cosmid vectors for the isolation, expression and rescue of gene sequences: studies with the human alpha-globin gene cluster. Proc. Natl. Acad. Sci. USA, *80*:5225–5229, 1983.

19. Weintraub, H., Tapscott, S.J., Davis, R.L., Thayer, M.J., Adam, M.A., Lassar, A.B., and Miller, D. Activation of muscle-specific genes in pigment, nerve, fat, liver and fibroblast cell lines by forced expression of MyoD. Proc. Nat. Acad. Sci. USA, *86*:5434–5438, 1989.

20. Tapscott, S.J., Davis, R.L., Thayer, M.J., Cheng, P-F., Weintraub, H., and Lassar, A.B.: MyoD1: a nuclear phosphoprotein requiring a *myc* homology region to convert fibroblasts to myoblasts. Science, *242*:405–411, 1988.

21. Lin, Z., Dechesne, C.A., Eldridge, J., and Paterson, B.M.: An avian muscle factor related to MyoD1 activates muscle-specific promoters in non-muscle cells of different germ layer origin and in BrdU-treated myoblasts. Genes Devl., *3*:986–996, 1989.

22. de la Brousse, C.F., and Emerson, C.P., Jr.: Localized expression of a myogenic regulatory gene, qmf1, in the somite dermatome of avian embryos. Genes Dev., in press, 1990.

23. Konigsberg, I.R.: The embryonic origin of muscle. In Myology. Edited by A.G. Engle and B.W. Banker. McGraw-Hill, NY, 1986. Vol. 1, pp. 39–71.

24. Cronmiller, C., Schedl, P., and Cline, T.W.: Molecular characterization of *daughterless*, a Drosophila sex determination gene with multiple roles in development. Genes Devel, *2*:1666–1676, 1988.

25. Caudy, M., Vassin, H., Brand, M., Tuma, R., Jan, Y.L., and Jan, Y.N.: *daughterless*, a Drosophila gene essential for both neurogenesis and sex determination has sequence similarities to *myc* and the *achaete-scute* complex. Cell, *55*:1061–1067, 1988.

26. Villares, R., and Cabrera, C.V.: The *achaete-ascute* gene complex of *D. melanogaster:* conserved domains in a subset of genes required for neurogenesis and their homology in *myc.* Cell, *50:*415–424, 1987.
27. Thisse, B., Stoetzel, C., Gorostiza-Thisse, C., and Perrin-Schmitt, F.: Sequence of the *twist* gene and nuclear localization of its protein in endomesodermal cells of early *Drosophila* embryos. EMBO J., *7:*2175–2183, 1988.
28. Murre, C., McCaw, P.S., and Baltimore, D.: A new DNA binding and dimerization motif in immunoglobulin enhancer binding, *daughterless* myoD, and *myc* proteins. Cell, *56:*777–783, 1989.
29. Landschulz, W.H., Johnson, P.F., and McKnight, S.L.: The leucine zipper: a hypothetical structure common to a new class of DNA binding proteins. Science, *240:*1759–1764, 1988.
30. Giebelhaus, D.H., Elb, D.W., and Moon, R.T.: Antisense RNA inhibits expression of membrane skeleton protein 4.1 during embryonic development of *Xenopus.* Cell, *53:*601–615, 1988.
31. Mansour, S.L., Kirk, R., and Cepecchi, M.R.: Disruption of the proto-oncogene *int-2* in mouse embryo stem cells: a general strategy for targeting mutations to non-selectable genes. Nature, *336:*348–352, 1988.
32. Levine, M., and Hoey, T.: Homeobox proteins as sequence-specific transcription factors. Cell, *55:*537–540, 1988.
33. de la Concha, A., Dietrich, U., Weigfel, D., and Compos-Ortega, J.A.: Functional interactions of the neurogenic genes of *Drosophila melanogaster.* Genetics, *118:*499–508, 1988.
34. Desai, C., Garriga, G., McIntire, S.L., and Horvitz, H.R.: A genetic pathway for the development of the *Caenorhabditis elegans* HSN motor neurons. Nature, *336:*638–646, 1988.
35. Way, J.C., and Chalfie, M.: *mec-3*, a homeobox-containing gene that specifies differentiation of the touch receptor neurons in *C. elegans.* Cell, *54:*5–16, 1988.
36. Chien, S., Teicher, L.C., Kazim, D., Pollack, R.E., and Wise, L.S.: Commitment of mouse fibroblasts to adipocyte differentiation by DNA transfection. Science, *244:*582–585, 1989.

Chapter 3

MOLECULAR ANALYSIS OF HUMAN AND RAT α-MYOSIN HEAVY CHAIN GENES

Eugene Morkin • Joseph J. Bahl • Irwin L. Flink • Bruce E. Markham • Richard W. Tsika

Cardiac myosin heavy chain (MHC) isoforms in human and rat are the products of two highly homologous genes, α and β, located in a head-to-tail arrangement about 4.2 kilobases (kb) apart.[1,2] In rat ventricle, the cardiac MHC types interact to produce three MHC isoforms, designated as V_{1-3}, in order of decreasing electrophoretic mobility on pyrophosphate-containing gels. The V_1 and V_3 forms contain two α- and β-MHCs, respectively, while the V_2 form is thought to contain one heavy chain of each type. The cardiac MHC genes have distinct patterns of tissue and developmental regulation and also are affected by cardiac work and hormone status.[3-5] Among these multiple regulatory influences, the effects of thyroid hormone are the strongest and best studied. In the ventricular myocardium of rats and rabbits, thyroid hormone has been shown to cause accumulation of α-MHC messenger RNA (mRNA) and inhibition of β-MHC mRNA expression.[6-8]

FUNCTIONAL SIGNIFICANCE OF MYOSIN ISOFORM CHANGES

Variations in myosin composition of heart and skeletal muscles may be important physiologically because there is a direct relationship between the predominant isoform type and the intrinsic speed of muscle contraction.[4,6] Thyroid hormone increases the proportion of the V_1 isoform and the maximal velocity of shortening in intact ventricle and isolated ventricular muscle. Recent evidence suggests this may be related to increased cross-bridge cycling rate.[9,10]

Identification of Nuclear T_3 Receptors

The actions of the intracellular form of thyroid hormone, 3,5,3'-triiodo-L-thyroxine (T_3), on the expression of myosin and other proteins under its control are thought to be mediated by binding to high-affinity nuclear receptors, which have been identified as the products of the proto-oncogene, c-erb-A.[11,12] These products have been shown to bind directly to the genes under control.[13] Before describing these experiments, we will briefly review a few general points related to regulation of mammalian gene transcription.

This investigation was supported by research grants from the National Institutes of Health (5 PO1 HL20984-13, T32 HL07249-13), the Arizona Disease Research Commission (No. 033), American Heart Association/Flinn Foundation, and the Gustavus and Louise Pfeiffer Research Foundation.

Regulation of Transcription

Initiation and regulation of mRNA transcription in mammalian cells depends on several proteins in addition to RNA polymerase II. These proteins, called *transcription factors,* generally are believed to interact with sequences within the promoter region, DNA sequences 5' upstream to those that encode the protein. Most mammalian promoters contain TATAA and CAAT sequences that are necessary for the regulation and accuracy of transcription initiation. Typically, TATAA sequences are located 25 to 30 base pairs (bp) away from the transcriptional initiation site. Because they are present in many different kinds of promoters, TATAA and CAAT elements are presumed to have a general role in the transcription process, such as binding general transcription initiation factors. in addition to these basal promoter elements and their associated transcription factors, other upstream elements are required for gene activity; they usually determine the particular regulatory properties of a given promoter.

REGULATION OF α-MHC GENE EXPRESSION

Localization of T₃ Response Elements

Sequences in the 5' flanking regions of several thyroid hormone-sensitive genes have been analyzed and found to be important for hormone regulation. Thyroid hormone-dependent expression of a chimeric gene constructed by fusion of 5' flanking sequences of the rat growth hormone gene to the bacterial chloramphenicol acetyltransferase (CAT) gene has been demonstrated by transfection into rat pituitary GC2 cells.[13,14] Furthermore, a DNA fragment containing this thyroid hormone-responsive element (TRE) was shown to have a binding site for the putative thyroid hormone receptor, the product of c-*erb*-A proto-oncogene. Similar types of experiments have demonstrated the presence of a thyroid hormone-sensitive element in the 5' flanking region of the rat α-MHC gene.[15,16]

Three-Element Model of α-MHC Promoter

The results of experiments on the rat growth hormone and α-MHC genes suggest that a minimum of three elements are necessary for T₃ control: a basal promoter element containing TATAA and CAAT sequences, a TRE, and additional upstream sequences that act to amplify the activity of the basal promoter element (Fig. 3-1). As discussed in the following section, recent experiments indicate that a more complex series of regulatory elements is probably present in both the human and rat α-MHC genes.

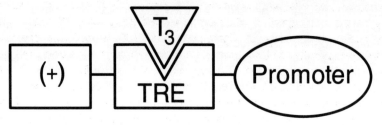

Fig. 3–1.   Schematic representation of a minimal three-element model of the human α-myosin heavy chain promoter consisting of a basal promoter, thyroid hormone-responsive element (TRE), and adjacent sequences that amplify basal activity when thyroid hormone is present.

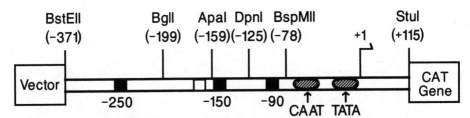

■ Putative TRE
□ Putative Muscle-Specific Element

Fig. 3–2. Schematic representation of the proximal human α-myosin heavy chain promoter region. A strategy for deletion analysis was based on identification by sequence homology of possible thyroid hormone-responsive elements (TREs) and a muscle-specific element (see text). (CAT, chloramphenicol acetyltransferase.)

Structure of Human α-MHC Promoter

A schematic representation of the 5′ flanking region of the human α-MHC gene is shown in Figure 3-2. The rat and human cardiac α-MHC genes show remarkable homology in this region for several hundred nucleotides upstream of the putative transcription initiation site,[1,17] while there is no homology for the 1000 nucleotides 3′ downstream. Conservation of sequences in the 5′ flanking region of these genes may indicate similar regulatory mechanisms. By homology with the TRE sequence described in the growth hormone gene,[12] three possible TREs were identified in the promoter of the gene at positions -54 to -100 (TRE_1), -116 to -163 (TRE_2), -216 to -247 (TRE_3). TRE_2 was found to be identical to the TRE identified in the rat α-MHC gene.[16] The degree of homology between the other two possible TREs and the rat growth hormone sequence is much less.

Examination of the DNA sequence in the proximal promoter region of the human α-MHC gene also revealed a 17-bp element identical to that found in genes encoding the muscle isoform of creatine kinase and α-actins. This sequence has been referred to as a "muscle specific" element and is thought possibly to be involved in the coordinate control of a subset of muscle genes.[18]

Expression of α-MHC/CAT Genes in Heart Cells

We have evaluated the role of T_3 in the control of human α-MHC expression by transfection of chimeric plasmids containing 5′ flanking sequences of the human gene into primary cultures of fetal rat heart cells.[19] A genomic DNA fragment containing 754 bp of 5′ flanking sequences of the human α-MHC gene and 115 nucleotides 3′ to the transcription initiation site was ligated into pSVOCAT, a eukaryotic expression vector in which the expression of CAT serves as a marker of gene activation. This construct, pSVHαMHC-754, had very little CAT activity in the absence of T_3. When 10 nM T_3 was added to the system, an approximately 50-fold induction of CAT activity was observed. These results indicate that the DNA sequences required for induction by T_3 are contained within the 5′ flanking sequences of the human α-MHC gene. Moreover, the transcription factors necessary for expression of the human gene are present in rat heart cells.

Deletion Analysis

A beginning has been made in localizing functionally significant regulatory elements by constructing a series of recombinant plasmids containing deletions in the 5′ flanking sequences. These were designed to permit evaluation of the putative TREs identified by sequence analysis. The most proximal deletion mutant, pSVHαMHC-78, contains the canonical TATAA sequence found in most eukaryotic promoters and terminates about 10 nucleotides 5′ upstream to the CAAT element. Interestingly, pSVHαMHC-78 exhibited low but significant activity in the absence of T_3 (constitutive expression)—activity that could be approximately doubled by addition of the hormone. Deletion constructs terminating upstream from −78, showed two regions of strong constitutive expression activity, which were flanked by regions of lower activity, presumably containing negative elements. In addition, there were two broad regions of T_3-inducible activity, a proximal area from −78 to −159 and a more distal region from −191 to −572. Negative regulatory elements were located at positions −159 to −199 and −572 to −754. The first of these negative elements abolished almost all T_3-inducible activity; the second abolished constitutive expression, while retaining hormone-inducible activity.

T_3-Receptor Binding Studies

The existence of at least two TREs in the proximal 5′ flanking sequences of the human α-MHC gene also was confirmed by binding of T_3 receptors to DNA sequences from these regions.[20] Double-stranded biotinylated oligonucleotides were synthesized corresponding to the rat growth hormone gene TRE (TRE_{GH}) and to the three putative TREs in the human α-MHC promoter. Rat liver T_3 receptors bound most tightly to TRE_{GH} and TRE_2. TRE_1 bound less strongly, and TRE_3 bound insignificantly. A restriction fragment containing the first 124 bp of 5′ flanking sequence competed about as well for T_3 receptors as TRE_1. These data are consistent with a second proximally located TRE in the α-MHC promoter, which is functionally active but exhibits lower affinity for the T_3 receptor. The existence of two TREs in the same promoter region is unusual, but has been reported recently in the rat growth hormone gene (21).

A synthetic oligonucleotide containing the putative TRE at about −250 did not compete in this assay, and no other candidate TRE sequences could be identified in this area. Presumably, thyroid-hormone sensitivity in the more distal areas, e.g., between positions −371 to −503, is conferred by an interaction between positive elements in this area and the more proximally located TREs. This concept is supported by experiments in which proximal α-MHC sequences were shown to confer thyroid hormone sensitivity on a heterologous promoter ($pA_{10}CAT$), but more distal sequences were inactive.[22]

Expression of α-MHC/CAT Fusion Genes in Skeletal Muscle Cells

The human α-MHC/CAT constructs also were expressed but were not T_3 regulated in L_6E_9, a permanent skeletal muscle cell line. No expression of any of the constructs was observed in 3T3 or HeLa cells, indicating that the so-called "muscle-specific" element does not restrict the expression of the human α-MHC gene to muscle cell types.[19]

A deletion analysis of the rat α-MHC gene also was carried out. A similar overall pattern of regulatory elements was observed, but the rat gene seemed to lack the additional proximal TRE found in the human gene 5′ flanking sequences.[19]

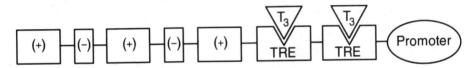

Fig. 3–3. A multicomponent model of human α-myosin heavy chain 5′ flanking sequences containing a basal promoter, two thyroid hormone-responsive elements (TREs), and multiple upstream positive and negative elements.

Assessment of Multiple Promoter Elements

A more realistic multi-element model of the human α-MHC promoter is shown in Figure 3-3. The functional significance of the repetitive positive and negative elements in the α-MHC promoter is unclear, however. Regulatory sequences in mammalian genes often are located some distance from their basal promoter elements and are thought to exert their influence by causing the intervening nucleotides to "loop out." The existence of multiple upstream positive elements suggests the possibility that selective interactions may occur with the sequences responsible for initiation of transcription. Possibly, this may provide alternative means of regulating the level of gene activity during development and in response to various stimuli.

Viewed from a functional perspective, the proximal area of the gene immediately upstream from the basal promoter containing constitutive elements and TREs has characteristics associated with expression of the α-MHC gene in atrial tissue, that is, a high level of constitutive activity that is enhanced by T_3. The functional activity of the region of the gene after the first negative element, and particularly after the second negative element, more closely resembles the expression of the gene in ventricular tissue, i.e., low constitutive activity and a high level of T_3 inducibility. It is tempting to speculate that packaging of the gene by nucleosomes or other proteins may permit selective expression of different regions or combinations of elements in atrial and ventricular tissue.

EFFECTS OF INHIBITORS OF PROTEIN SYNTHESIS

We also have investigated whether α-MHC expression is regulated as a primary event of thyroid hormone receptor action on gene expression or as a secondary effect mediated by T_3 stimulation of a gene product necessary for expression. Operationally, primary events are characterized as those that occur rapidly and without the requirement for new protein synthesis. In the case of the rat growth hormone gene, it has been reported that binding of T_3 to its receptor in rat pituitary tumor cells may be sufficient to cause gene induction even when synthesis of new protein is blocked.[23] Other reports, however, indicate that ongoing protein synthesis is required both for induction of the growth hormone gene and its continued expression.[24]

Similar experiments with inhibitors of protein synthesis in cultured fetal heart cells seems to indicate that expression of the α-MHC gene requires participation by one or more labile factors in addition to the T_3 receptor.[22] When cycloheximide was added at concentrations that inhibited protein synthesis by 95%, simultaneous addition of T_3 no longer induced α-MHC mRNA synthesis. If protein synthesis was permitted to proceed normally for 4 or 12 hours, however, α-MHC mRNA levels were about 26 and 55%, respectively, of values obtained in cultures treated with T_3 alone. Similar results were obtained with 2 mM of emitine. These observations provide further support for the importance of multiple *trans*-acting factors (proteins) in regulation of α-MHC transcription.

ANTI-SENSE c-*erb*-A EXPERIMENTS

The importance of the c-*erb*-A protein in the regulation of the α-MHC gene by T_3 was confirmed in experiments using anti-sense c-*erb*-A mRNA. For this purpose, the Rous sarcoma virus (RSV) long terminal repeat (LTR) was used to direct the transcription of human placental c-*erb*-A cDNA in the sense and anti-sense orientation in primary cultures of rat cardiomyocytes.[22] Anti-sense DNA or RNA fragments have been used to block the expression of specific genes and are postulated to act by annealing to complementary mRNA molecules and disrupting normal processing or translation of the message.[25]

In a typical experiment, only about 16 to 20% of fetal cardiomyoctyes were transfected; hence it was difficult to detect the effect of an anti-sense c-*erb*-A on expression of endogenous α-MHC mRNA. For this reason, an α-MHC/CAT reporter gene was used. Co-transfection into primary cultures of fetal rat heart cells was carried out with the sense and anti-sense c-*erb*-A constructs and pSVOMCAT containing 2.4 kilobase pairs (kbp) of 5' flanking sequences of the rat α-MHC gene fused to the CAT gene. When transfected as a control into primary cultures of rat fetal cardiomyocytes, the anti-sense construct (pRSVhEACAT$^-$) gave only a slight background of CAT activity, which was not significantly affected by T_3. Co-transfection of pSVOMCAT with increasing amounts of pRSVhEACAT (2.5 to 20 μg) produced a progressive decrease in T_3-inducible CAT activity, which at the highest dose approached background levels. Co-transfection of pSVOMCAT with similar amounts of the sense clone (pRSVhEACAT$^+$) had no effect on expression of pSVOMCAT.

These results suggest that the c-*erb*-A proto-oncogene product is important in transduction of the effects of thyroid hormone on α-MHC gene expression. They also are encouraging in suggesting that it may be possible to use anti-sense mRNAs to explore other intracellular processes in primary heart cell cultures.

COMMENTS

Molecular analysis of the human and rat α-MHC genes has revealed a surprising degree of complexity. Both genes seem to contain at least four types of elements: a basal promoter and one or more TREs followed by repetitive upstream positive and negative regulatory elements. The α-MHC/CAT fusion constructs also were expressed, but were not T_3 regulated, in a skeletal muscle cell line. No expression was observed in nonmuscle cells. The importance of the products of the c-*erb*-A proto-oncogene in mediating the effects of T_3 on MHCs and other genes under its control has been shown in several ways, including demonstration of binding to specific DNA sequences and the use of an anti-sense mRNA. Additional protein factors, however, are required for hormone-dependent regulation. Interactions among the sequence elements we have identified may be sufficient to explain many of the features that characterize the expression of the gene in atrial and ventricular tissues.

REFERENCES

1. Saez, L.J. et al.: Human cardiac myosin heavy chain genes and their linkage in the genome. Nucleic Acids Res., *15*:5443, 1987.
2. Mahdavi, V., Chambers, A.O., and Nadal-Ginard, B.: Cardiac alpha- and beta-myosin heavy chain genes are organized in tandem. Proc. Natl. Acad. Sci. USA, *81*:2626, 1984.

3. Morkin, E. et al.: Biochemical and physiologic effects of thyroid hormone on cardiac performance. Prog. Cardiovasc. Dis., *25:*435, 1983.

4. Lompre, A.-M. et al.: Expression of the cardiac ventricular α- and β-myosin heavy chain genes is developmentally and hormonally regulated. J. Biol. Chem., *259:*6437, 1984.

5. Swynghedauw, B.; Developmental and functional adaptation of contractile proteins in cardiac and skeletal muscles. Physiol. Rev., *66:*710, 1986.

6. Gustafson, T.A. et al.: Hormonal regulation of myosin heavy chain gene expression in cultured rat myocytes: analysis of the effect of thyroid hormone and other cardioactive agents on mRNA expression. J. Biol. Chem., *262:*13316, 1987.

7. Gustafson, T.A., Markham, B.E., and Morkin, E.: Effects of thyroid hormone on α-actin and myosin heavy chain expression in cardiac and skeletal muscles of the rat: measurement of mRNA content using synthetic oligonucleotide probes. Circ. Res., *59:*194, 1986.

8. Izumo, S., Nadal-Ginard, B., and Mahdavi, V.: All members of the MHC multigene family respond to thyroid hormone in a highly tissue specific manner. Science, *231:*597, 1986.

9. Rossmanith, G.H.: Influence of V1 and V3 isomyosins on the mechanical behavior of rat papillary muscle as studied by pseudo-random binary noise modulated length perturbations. J. Muscle Res. Cell Motil., *7:*307, 1986.

10. Holubarsch, C. et al.: The economy of isometric force development, myosin isoenzyme pattern and myofibrillar ATPase activity in normal and hypothyroid rat myocardium. Circ. Res., *56:*78, 1985.

11. Weinberger, C. et al.: The c-*erb*-A gene encodes a thyroid hormone receptor. Nature, *324:*641, 1986.

12. Sap, J. et al.: The c-*erb*-A protein is a high-affinity receptor for thyroid hormone. Nature, *324:*635, 1986.

13. Glass, C.K. et al.: A c-*erb*-A binding site in rat growth hormone gene mediates trans-activation by thyroid hormone. Nature, *329:*738, 1987.

14. Koenig, R.J. et al.: Thyroid hormone receptor binds to a site in the rat growth hormone promoter required for induction by thyroid hormone. Proc. Natl. Acad. Sci. USA, *84:*5670, 1987.

15. Gustafson, T.A. et al.: Thyroid hormone regulates the expression of a transfected α-myosin heavy chain fusion gene in fetal heart cells. Proc. Natl. Acad. Sci. USA, *84:*3122, 1987.

16. Izumo, S., and Mahdavi, V.: Thyroid hormone receptor α isoforms generated by alternative splicing differentially activate myosin HC gene transcription. Nature, *334:*539, 1988.

17. Yamauchi-Takihara, K. et al.: Characterization of human cardiac myosin heavy chain genes. Proc. Natl. Acad. Sci. USA, *86:*3504, 1989.

18. Jaynes, J.B. et al.: Transcriptional regulation of the muscle creatine kinase gene and regulated expression in transfected mouse myoblasts. Mol. Cell. Biol., *6:*2855, 1986.

19. Tsika, R.W. et al.: Thyroid hormone regulates the expression of a transfected human α-myosin heavy chain fusion gene. Proc. Natl. Acad. Sci. USA, *87:*379, 1990.

20. Flink, I.L. et al.: Binding of thyroid hormone receptors to specific DNA sequences in the 5′ flanking sequences of the α-myosin heavy chain gene. J. Cell. Biochem. [Suppl.], *13E:*178, 1989.

21. Norman, M.F. et al.: The rat growth hormone gene contains multiple thyroid response elements. J. Biol. Chem., *264:*12603, 1989.

22. Morkin, E. et al.: Control of cardiac myosin heavy chain gene expression by thyroid hormone. *In* Cellular and Molecular Biology of Muscle Development, U.C.L.A. Symposium on Molecular and Cellular Biology. Edited by L.H. Kedes and F.E. Stockdale. New York, Alan R. Liss, New Series Vol. 93, p. 381, 1989.

23. Spindler, S.R. et al.: Growth hormone gene transcription is regulated by thyroid and glucocorticoid hormones in cultured rat pituitary tumor cells. J. Biol. Chem., *257:*11627, 1982.

24. Santos, A. et al.: Labile proteins are necessary for T_3 induction of growth hormone mRNA in normal rat pituitary and rat pituitary tumor cells. J. Biol. Chem., *262:*16880, 1987.

25. Izant, J.G., and Weintraub, H.: Constitutive and conditional suppression of exogenous and endogenous genes by anti-sense RNA. Science, *229:*345, 1985.

Chapter 4

AN APPROACH TO THE MOLECULAR REGULATION OF CARDIAC MYOCYTE HYPERTROPHY

Paul C. Simpson • Larry R. Karns • Carlin S. Long

This chapter describes an approach to molecular regulation in myocardial hypertrophy, an important problem in clinical cardiology. The focus is on hypertrophy of the cardiac myocyte, a growth process that varies quantitatively and qualitatively and is a major component of enlargement of the intact organ. The goal is to identify the molecular signals that produce the different types of hypertrophy. Identification of the signals might permit manipulation of hypertrophy.

The proto-oncogene concept suggests that the critical signals in growth regulation are proteins in four classes: growth factors, growth-factor receptors, intracellular transducers, and transcription factors. A five-step approach to these signals in cardiac myocyte hypertrophy is outlined. The steps include (1) establishing a cell culture model; (2) demonstrating induction of hypertrophy by a single growth factor; (3) identifying specific genes that are up-regulated by the growth factor; (4) defining the level of regulation of gene induction; and (5) investigating the intracellular connection from the growth factor to the level of regulation.

This approach has identified catecholamines as growth factors for cardiac myocyte hypertrophy, and the α_1-adrenergic receptor as a growth-factor receptor. Specific genes are up-regulated at the transcriptional level by α_1-adrenergic stimulation. An intracellular pathway connecting the α_1-adrenergic receptor at the cell surface to the transcription of specific genes in the myocyte nucleus can be proposed and investigated. Protein kinase C may be an important transducer in this pathway. The approach outlined can be used to identify and study other growth factors for cardiac myocytes.

DEFINITION OF THE PROBLEM

Clinical Problems in Myocardial Hypertrophy

An increase in the mass of the heart (myocardial hypertrophy) occurs in most types of cardiac disease. Focal or generalized hypertrophy is seen in hypertension, valve disease, and coronary artery disease, and is a central feature of the dilated and hypertrophic cardiomyopathies. Hypertrophy occurs also as a response to vigorous exercise training and as an integral part of normal development. An increase in heart mass can be produced by an increase in ventricular wall thickness, dilation of the ventricular cavity, or a combination of these processes. Although hypertrophy would seem to be a beneficial adaptation to alterations in hemodynamic loading, it is also clear that this growth process is frequently maladaptive.

Maladaptive hypertrophy can take several forms, including inadequate, excessive, and pathologic growth. If ventricular wall thickness is not increased sufficiently to normalize wall stress, hypertrophy is described as "inadequate."[1] Conversely, hypertrophy in the absence of a hemodynamic stimulus is clearly "excessive" (idiopathic hypertrophy).[2] Finally, the hypertrophied myocardium in cardiac disease appears to exhibit frequently, and perhaps invariably, abnormal systolic and/or diastolic function and electrical instability.[1,3–5] Thus, the hypertrophy of cardiac disease can be called "pathologic." Evidence suggests, however, that the functional abnormalities seen in the hypertrophy associated with cardiac disease are not invariant accompaniments of hypertrophic growth. For example, hypertrophy associated with exercise training appears not to be accompanied by abnormal function, even when the absolute magnitude of hypertrophy is equal to that produced by hypertension or valve disease.[6–11] The hypertrophy of development may also be considered a "physiologic" hypertrophy.

In summary, myocardial hypertrophy is extremely common but is often maladaptive. The magnitude of growth can be inadequate or excessive relative to hemodynamic loading conditions. Myocardial function is frequently abnormal in the hypertrophy of cardiac disease, in contrast with the hypertrophy of exercise or development. Considerable clinical benefit might result if adequate hypertrophy could be stimulated, excessive growth could be prevented, or growth resulting in abnormal function could be manipulated in a more normal direction. For these reasons, elucidating molecular mechanisms in hypertrophy is of major importance.

Cardiac Myocyte Hypertrophy in Myocardial Hypertrophy

Myocardial hypertrophy is produced by diverse patterns of growth of the multiple cell types in the heart.[12,13] Growth at the cellular level can be dissected into four consecutive stages: hypertrophy, DNA synthesis, mitosis, and cytokinesis[14,15] (Fig. 4-1). Completion of this sequence results in a doubling of cell number (hyperplasia or proliferation).

About 75% of the cells in the heart are not cardiac myocytes.[12,16] The cells that are not cardiac myocytes in the intact heart include fibroblasts, endothelial cells (cardiac and vascular), smooth muscle cells, pericytes, neurons, and blood-borne cells (macrophages and others). Many of these other types of cells grow by hyperplasia during myocardial hypertrophy. The mechanisms that control growth of these cells are very important to the understanding of myocardial hypertrophy, pertinent, for example, to angiogenesis, collagen deposition, and neural regulation. This chapter however, focuses on the cardiac myocyte.

Cardiac myocytes comprise only about 25% of the total cells in the heart, although they occupy the greatest volume fraction of the organ.[17] The cardiac myocytes proliferate during early heart development but appear to lose the capacity to undergo cytokinesis, mitosis, and DNA synthesis during the early postnatal period. The appearance of binucleate myocytes in the postnatal myocardium apparently marks a failure of cytokinesis after a final round of mitosis.[12,18,19] The cardiac myocyte growth process in myocardial hypertrophy after the postnatal period is hypertrophy or enlargement, not hyperplasia. Therefore, two questions are critical: (1) What molecular mechanisms regulate cardiac myocyte hypertrophy; and (2) what explains the loss of DNA synthesis, mitosis, and cytokinesis?

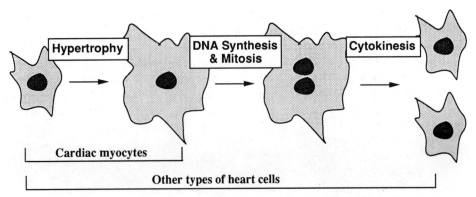

Fig. 4–1. Cell hyperplasia or proliferation can be dissected into stages. Completion of these stages results in a doubling of cell number. After early postnatal life, cardiac myocytes appear to be largely incapable of DNA synthesis, mitosis, and cytokinesis. Thus, cardiac myocyte growth in postnatal myocardial hypertrophy is by cell enlargement or hypertrophy alone. In contrast, other cell types in the heart—fibroblasts, for example—do proliferate during myocardial hypertrophy. A goal of research on molecular mechanisms in hypertrophy is to identify the extracellular signals and intracellular pathways that regulate progression through these stages. With respect to the cardiac myocyte, it is important to learn the mechanisms regulating hypertrophy, as well as those responsible for the loss of DNA synthesis, mitosis, and cytokinesis. The heterogeneity of the cardiac myocyte phenotype seen in hypertrophy in vivo, as reflected in differential isogene expression, suggests that there may be multiple molecular mechanisms in hypertrophy.

Stimulation of cardiac myocyte proliferation may be the best approach to inadequate hypertrophy. Unfortunately, little progress has been made in this area. However, new model systems, which have been reviewed briefly elsewhere,[20,21] give exciting prospects for the future. The remainder of this chapter will summarize an approach to studying the molecular mechanisms that regulate hypertrophy of the cardiac myocyte.

Cardiac Myocyte Hypertrophy Is a Heterogeneous Process

The fact that several types of myocardial hypertrophy are seen clinically suggests that cardiac myocyte hypertrophy might be a heterogeneous process, produced by more than one molecular mechanism. Quantitative variability is indicated by the syndromes of excessive and inadequate hypertrophy and by measurements of myocyte size in different models of hypertrophy.[22]

Evidence for qualitative variability has come from analysis of gene expression in hypertrophy. Contractile proteins, such as myosin heavy chain (MHC) and sarcomeric (α) actin can exist in multiple isoforms, differing in a variable number of their constituent amino acids. Each isoform of MHC and α-actin is encoded by a separate gene.[23,24] In the case of MHC, these differences in amino acid sequence have been shown to have functional significance.[25] V_1 myosin, which contains two α-MHC molecules, has higher adenosine triphosphatase (ATPase) activity than does V_3 myosin, which contains two β-MHC molecules. Functional differences between the two α-actin isoforms, cardiac α-actin and skeletal α-actin, have not yet been demonstrated.[26]

Different types of hypertrophy of the rat heart are associated with different patterns of isoform expression, as deduced by assay of specific messenger RNAs (mRNAs) and proteins.[7,27–34] This phenomenon is sometimes referred to as *isoform switching*. In hypertrophy produced by pressure overload, up-regulation of β-MHC and skeletal α-actin occurs. These two isoforms are characteristic of the fetal heart and are down-regulated during the hypertrophy of normal development. In contrast to the hypertrophy of pressure overload, the hypertrophy produced by exercise training or thyroid hormone administration is accompanied by increased expression of α-MHC, the adult isoform.

The significance of up-regulation of early developmental isogenes in some forms of hypertrophy is uncertain. Early gene induction does not appear to be *required* for hypertrophy.[29,34] The small absolute magnitude of isoform switching in some cases[31] raises questions about the significance for myocyte contractile function. The differences in gene expression in different types of hypertrophy however, do indicate clearly that cardiac myocyte hypertrophy in vivo is a heterogeneous process. Cardiac myocyte hypertrophy varies both quantitatively and qualitatively. This heterogeneity of the myocyte phenotype implies that hypertrophy may be produced by more than one molecular mechanism.

Proto-oncogenes and the Signals for Cardiac Myocyte Hypertrophy

Extracellular signals activate intracellular growth pathways.[14,20,21,35–37] An exception to this rule is neoplastic growth, which can be produced by autonomous activity of an intracellular mechanism. Soluble molecules called *growth factors* appear to be the dominant extracellular signals for growth of most types of cells. These can be classified as endocrine, paracrine, or autocrine, depending on whether they come to the responding cell from distant organs, adjacent cells, or the responding cell itself, respectively.

In addition, physical or mechanical stimuli, such as those producing cell stretch, can activate intracellular pathways associated with growth.[38–42] The "receptor" for stretch may be a cell surface ion channel.[43]

Therefore, the search for signals for cardiac myocyte hypertrophy can be conceived as an attempt to identify growth factors, either soluble or mechanical. On the basis of analogy with other types of cells, these growth factors will bind to and/or activate cell receptors that are coupled to growth-producing pathways. Furthermore, based on the heterogeneity of cardiac myocyte hypertrophy and the example of other cell types, there may be multiple growth factors for cardiac myocytes. The growth factors, receptors, and intracellular pathways will define the molecular mechanisms in cardiac myocyte hypertrophy.

The proto-oncogene concept is a useful paradigm in the search for critical regulatory elements in cardiac myocyte hypertrophy. A central feature of the proto-oncogene concept is the focus on cell growth regulation via a limited number of proteins in four functional classes[20,21,44,45] (Fig. 4-2). These critical regulatory proteins are growth factors, growth-factor receptors, intracellular transducers, and proteins that bind to DNA and modulate RNA synthesis (transcription factors).

Virtually all aspects of cellular biochemistry and physiology are altered in some way during growth. Only some of these alterations may be regulatory, rather than permissive. The fact that mutation or overexpression of proto-oncogene proteins can produce cancer emphasizes that these proteins have critical regulatory roles in growth. It seems possible that the proto-oncogenes of cardiac myocytes will be a critical focus of studies on molecular mechanisms in hypertrophy. By this scheme, excessive and inadequate hypertrophy could be conceived to represent abnormalities of myocyte proto-oncogenes.

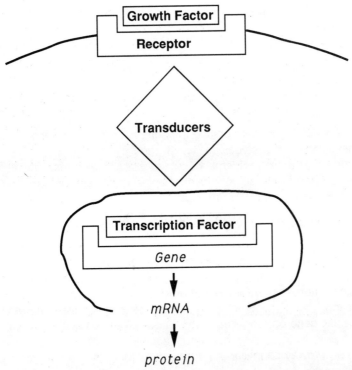

Fig. 4–2. The products of proto-oncogenes can be grouped into four functional classes: growth factors, growth-factor receptors, intracellular transducers (e.g., protein kinases and guanine nucleotide binding proteins), and nuclear proteins, including those that bind to DNA and regulate transcription by RNA polymerase (transcription factors). The critical regulatory role of proto-oncogenes is emphasized by the fact that cancer may result from mutation or overexpression of these proteins. How proto-oncogenes activate DNA synthesis is not known. However, similar proteins may regulate RNA synthesis and DNA synthesis.[117] Identification of cardiac myocyte proteins in these four functional classes is an important focus of work on molecular mechanisms in myocardial hypertrophy. (Proto-oncogenes in myocardial hypertrophy are reviewed in refs. 20 and 21.)

Problems With In Vivo Models of Myocardial Hypertrophy

Two major difficulties are encountered in the study of molecular mechanisms using in-vivo models of myocardial hypertrophy, heterogeneity of cell types, and complexity of extracellular signals.

As noted previously, most of the cells in the heart are not cardiac myocytes. This cellular heterogeneity complicates assays on the intact organ. Thus, assays of adrenergic receptors in intact myocardial tissue, for example, do not discriminate which receptors are from the myocytes. This problem is encountered with any cell product that is not myocyte specific, such as many receptors, all second messengers, total RNA and protein, and so on. Even myocyte-specific products are associated with a problem of data normalization. For example, total protein, RNA, or DNA, which is often the basis for normalization, may be altered by differential cellular responses in control and experimental hearts.

It is also very difficult to isolate potential growth factors for hypertrophy in vivo. For example, increases in sympathetic activity have been found in hypertrophy produced by several types of alterations in hemodynamic loading.[46] This observation provides a part of the basis for the long-held notion that catecholamines may be growth factors for hypertrophy in some cases.[47–49] However, it is not possible to monitor and control all potential mechanical and soluble growth factors in an experiment in vivo.

In the case of catecholamines, it is very difficult to sort out the following: whether the release of norepinephrine (NE) from sympathetic terminals is modified; whether a cardiac myocyte adrenergic receptor is actually stimulated or blocked during long-term experiments with adrenergic agonists or antagonists; and which adrenergic receptor might be responsible for a catecholamine effect on growth. Norepinephrine, the sympathetic neurotransmitter, and epinephrine, the adrenal medullary catecholamine, activate all adrenergic receptor subtypes, α_1, α_2, β_1, and β_2.

As a consequence of these problems, discussed elsewhere in more detail,[50,51] no cardiac myocyte growth factor, soluble or mechanical, had been demonstrated conclusively before the use of cell culture models.

AN APPROACH TO MOLECULAR MECHANISMS IN CARDIAC MYOCYTE HYPERTROPHY

The general steps taken in a new approach to cardiac myocyte hypertrophy follow:

1. Establish a cell culture model for hypertrophy.
2. Demonstrate induction of hypertrophy by a single, defined agonist (a growth factor).
3. Identify a specific gene up-regulated in hypertrophy induced by the growth factor.
4. Define the level at which gene induction is regulated.
5. Investigate the pathway connecting the growth factor to the level of regulation.

This approach has revealed that stimulation of the α_1-adrenergic receptor with catecholamines induces cardiac myocyte hypertrophy in culture and that regulation occurs at the level of gene transcription, at least in part. A potential intracellular pathway connecting the α_1 receptor at the myocyte surface to transcriptional regulation in the nucleus can be suggested. A similar approach should be applicable to other cardiac myocyte growth factors.

Cardiac Myocyte Culture

The first step was to establish a cell culture model for cardiac myocyte hypertrophy. This involved development of a specific culture system and assays for cell size. For studies on cardiac myocyte growth regulation, investigation of cardiac myocytes is essential. The principles of regulation of growth and gene expression may be similar in all eukaryotic cells. That is, members of the same four functional classes of proto-oncogenes may be critical regulatory elements (Fig. 4-2). Proto-oncogenes, however, may be like characters in an alphabet or code.[52] All proto-oncogene proteins may not be expressed in every cell type, and the precise role of each may vary as a function of the other proto-oncogenes expressed in a particular type of cell.[21]

A CELL CULTURE MODEL

A well-defined cardiac myocyte culture system can be used to reduce significantly the problems of heterogeneity of cell types and complexity of signals encountered with models in vivo. Use of a culture system has provided new insights into molecular mechanisms in hypertrophy.

Several years ago, we developed a model system for cardiac myocyte hypertrophy,

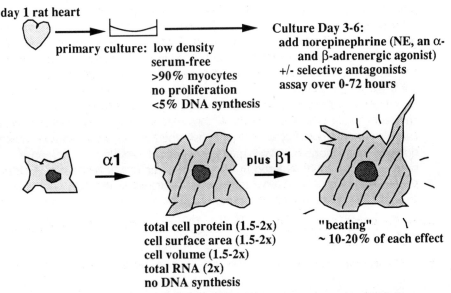

Fig. 4–3. α_1-Adrenergic-stimulated hypertrophy in culture. Myocytes are obtained from the ventricles of day-old rats. After a few days in serum-free culture, the cells are treated with adrenergic agents. Control cell size is the same as at isolation, and the myocytes do not contract spontaneously. α_1-Adrenergic-receptor stimulation induces, over 12 to 24 hours, an increase in myocyte size and changes in gene expression (Fig. 4-8). Gene-specific and total transcription is increased (Fig. 4-10), but no change occurs in DNA synthesis. If the β_1-adrenergic receptor is activated along with the α_1, the myocytes develop spontaneous contractility and each effect is augmented by ~10 to 20%. Techniques for studying hypertrophy and gene expression are discussed in this chapter. Primary data are in references 53, 55, 60, 62, 64, and 85. Additional details regarding the culture system are reviewed in references 50 and 57. The pharmacologic procedures for determining receptor specificity and other parameters of adrenergic responses are discussed in the same references and in reference 51. (NE, norepinephrine.)

using cells from the neonatal rat ventricle[50,53–55] (Fig. 4-3). In this system, over 90% of the cells are cardiac myocytes; possible signals for hypertrophy can be tested with precision; and myocyte size can be quantified. The cells are maintained at low density (low numbers of cells per dish), so that cell types and numbers can be documented unequivocally through the microscope. Contaminating fibroblasts are reduced to <10% of the total cell population by differential plating and by pulse addition of an inhibitor of DNA synthesis, bromodeoxyuridine.

Serum is not used after the initial period of cell attachment, in an attempt to eliminate nondefined growth factors. Furthermore, nonmyocyte proliferation is drastically reduced in serum-free medium, so that control experiments can be done in the absence of bromodeoxyuridine. For example, control experiments indicate that the cardiac myocytes do not proliferate in response to serum stimulation, even in the absence of an inhibitor of DNA synthesis. Insulin and transferrin are added to the medium, to promote myocyte survival in the absence of serum. Under these serum-free control conditions, neither cell size nor number change over time, and the myocytes do not display spontaneous contractile activity. Thus, changes in both myocyte size and contractility in response to defined signals can be detected with high sensitivity.

Although cell culture offers critical advantages, important cautions apply.[50,56,57] First, at high cell densities ("confluence"), the cells form multilayers;[58] and it is not possible to judge the proportion of fibroblasts, which contaminate all primary cultures. Second, even though the culture medium is defined, it may include paracrine and autocrine growth factors that depend on cell density.[54,59] Third, the time the cells have been in culture may be a determinant of the response to a potential growth factor. Finally, the trade-off for a system that can be defined is one that is very different from the highly complex situation in vivo. For example, the low cell densities required for identification of cell types and the myocyte predominance required for accurate assays on the entire culture both alter the cell–cell interactions that occur normally in vivo. However, because the cultured myocytes are nontransformed and retain the cardiac myocyte phenotype, observations in culture may be reasonably assumed to demonstrate biologic potential that may be realized in vivo. Once basic principles have been identified by studies in culture, it is possible to test their relevance to the heart in vivo.

Application of new methods in molecular biology may lead to the development of a continuously replicating cardiac myocyte cell line that would contain muscle cells only and that would be suitable for genetic studies. Findings in a transformed myocyte line, however, may need to be confirmed in primary cultures, established directly from the intact heart. Configurations of primary cultures other than those used in our studies will be useful, if the cell type producing a measured variable can be assigned unequivocally.

ASSAY OF HYPERTROPHY

Both cell number and cell size must be quantified. Cell number (and type) is determined by microscopic observation of randomly selected microscopic fields that sample the entire dish surface.[53,54] If care is taken in plating the original cell suspension, all culture dishes in a given preparation will have the same number of myocytes and the same proportion of nonmyocytes.

Cell size is measured in three ways.[53,55,60] Multiple methods to quantify cell size provide assurance that the increase in size is real. An increase in cell size defines cell hypertrophy.

The following size assays are used: cell surface area, cell volume, and cell total protein content. Surface area corresponds to what one sees looking through the microscope. It is determined by an image-analysis technique used on cells attached in the dishes. Because the measured cells can be identified as to type at low densities, the surface area method is entirely specific for myocytes. Volume is assayed using microscopy (or a cell counter), after removal of myocytes from the dishes by trypsinization. The volume assay is also specific for the myocytes, because careful adjustment of trypsinization removes the myocytes and leaves the nonmyocytes attached to the dish.

Total cell protein content is measured by a chemical assay (Bradford or Lowry) or by prolonged labeling with an amino acid that is not metabolized (e.g., phenylalanine or tyrosine). The isotopic assay is more sensitive, possibly because extracellular proteins that have accumulated before growth-factor stimulation are not measured. The chemical assay provides an absolute protein mass in picograms per cell. The result of a protein determination in a dish of cells can be divided by the number of cells in that dish to give the protein content per cell. Under our conditions, each control myocyte contains ~500 pg of total protein. The nonmyocyte contribution to total protein content can generally be ignored, for three reasons. First, nonmyocyte contamination is <10%; second, myocyte-specific size assays are used also; third, pure cultures of nonmyocytes are easily prepared and are used for control studies.

Total RNA content in picograms per cell, measured by ultraviolet spectrophotometry, can be used as an additional index of cell hypertrophy.[14] In most cases of cell hypertrophy, a coordinate increase in total RNA (which is >75% ribosomal RNA [rRNA]) and total protein occurs.

Hypertrophic growth of cardiac myocytes in culture can be observed and measured easily and is, in the broad sense, the same process of myocyte enlargement that occurs in hypertrophy of the intact heart in vivo.

ASSAY OF CONTRACTILITY

The cardiac myocytes at low density in serum-free culture do not contract spontaneously (beat). Development of contractile activity is monitored by observing through the microscope and counting the fraction of myocytes that beat regularly at a frequency of 50 to 100 per minute.[55] It is essential that temperature (37°C) and pH (7.3) remain stable during the period of observation and that, as with cell counting, representative areas covering the entire dish surface be examined.

This contractility assay is imprecise and does not provide information on the velocity and amplitude of contraction. Methods have been developed, however, that overcome both of these difficulties.[61]

A Growth Factor for Cardiac Myocyte Hypertrophy

The second step in our approach was to identify an agonist (a growth factor) that induced cardiac myocyte hypertrophy in the culture model. The work in our laboratory has focused on the catecholamine NE. This interest was based on a large body of data from experiments in vivo suggesting that catecholamines could be important for hypertrophy in some circumstances.[50] Difficulties with in-vivo models, however, reviewed previously, had made it difficult to exclude indirect effects and to identify the adrenergic receptor that might be responsible. We have found that NE is a cardiac myocyte growth factor, acting through the α_1-adrenergic receptor.

TESTING POTENTIAL GROWTH FACTORS IN THE CULTURE MODEL

Any soluble molecule and many mechanical stimuli can be tested for growth-factor activity in the culture system. (Details of pharmacologic manipulations are in ref. 57, and examples of mechanical stimuli are in refs. 38, 41, and 42.) It is important that the dose of a potential growth factor be variable, so that a dose–response relationship can be defined.

It is also crucial that the stimulus be validated. For example, a "false-negative" response may be produced by breakdown or adsorption of the potential growth factor. Ideally, the test substance is assayed at the beginning and end of the treatment period, as we have done with catecholamines. Alternately, one can use another cell type in a bioassay as a positive control. Potential false-positive considerations, although interesting and important, are production of a paracrine myocyte growth factor by the contaminating nonmyocytes in response to the test substance or an autocrine growth factor by the myocytes.

Treatment of the cultured myocytes with crude serum reproduces many of the effects of α_1-adrenergic stimulation that will be described later. Serum is enormously complex, however, containing many known and unknown hormones and growth factors. Furthermore, considerable variability occurs among different lots of serum, even from the same supplier. Thus, serum is not an ideal agonist for studies designed to identify a specific growth-transducing pathway. Treatment with serum, however, is a very useful positive control to ensure that the cultured cells are capable of a response under the conditions used and to validate the various assays.

INDUCTION OF CARDIAC MYOCYTE HYPERTROPHY BY α_1-ADRENERGIC STIMULATION

Treatment of the cultured cardiac myocytes with NE increases cell size by ~1.5- to 2-fold times control, as quanitified by measurements of cell protein content, surface area, and volume[53,55,60] (Fig. 4-3). A 2-fold increase in total RNA content also occurs.[62] The time course, dose–response relationship, and receptor specificity of the response to NE have been defined; and it has been shown that the response is not produced by metabolites or breakdown products of NE. The hypertrophic response develops over about 24 hours. The EC_{50} for NE is ~200 nM, a concentration that could be found in an active sympathetic synapse in vivo.[63] NE activates all types of adrenergic receptors, α_1, α_2, β_1, and β_2. However, studies with selective adrenergic agonists and antagonists indicate that it is the α_1 receptor specifically that transduces cardiac myocyte hypertrophy in response to NE.

The cardiac myocytes in culture do not undergo DNA synthesis, mitosis, or cytokinesis in response to NE.[55,60,64,65] Why the cultured cardiac myocyte response to α_1 stimulation is limited to hypertrophy is unknown. Other types of cultured cells complete all four stages of growth and proliferate after α_1 stimulation[66–69] (Fig. 4-1). Evidence indicates that neonatal cardiac myocytes can proliferate in vivo.[18] Some additional growth factor may be required for cardiac myocyte hyperplasia in response to α_1 stimulation in culture.[14,15] Interestingly, the cultured cardiac nonmyocytes (fibroblasts) do not grow in response to NE. Why the cardiac nonmyocytes do not respond to α_1 stimulation is unknown; they do express α_1-adrenergic receptors (unpublished results).

In summary, these studies identify the α_1-adrenergic receptor as a growth-factor receptor for cardiac myocytes and identify catecholamines as growth factors. These results provide direct support for the idea that catecholamines may be important extracellular signals for myocardial hypertrophy in vivo in some circumstances[47–49] and point to the α_1-adrenergic receptor as the mediator of the catecholamine response.

HYPERTROPHY DOES NOT REQUIRE CONTRACTILE ACTIVITY

The fact that growth occurs in culture indicates that cardiac myocyte hypertrophy does not require mechanical stimuli. Additional studies show that hypertrophy does not require myocyte contractile activity.[55,70–72] α_1-Adrenergic stimulation, however, has an important influence on the development of contractile activity[55] (Fig. 4-3).

The control myocytes are quiescent, and contractile activity does not change significantly with α_1 stimulation alone, even though hypertrophy is induced. Treatment with a β-adrenergic agonist alone does not produce a major change in either myocyte size or contractile activity. If both the α_1- and β_1-adrenergic receptors are activated, however, the proportion of myocytes beating spontaneously increases from <5% in control cultures to >95%. This response is not acute but develops over many hours, as does hypertrophy.

Therefore, α_1-stimulated cardiac myocyte hypertrophy does not *require* myocyte contractile activity. α_1-Receptor stimulation, however, has some effect that induces spontaneous contractility when the β_1-receptor is activated also. The nature of the α_1 effect(s) underlying the development of contractility in this system is (are) unknown. Protein synthesis is not required, and changes in ion fluxes might be involved.[73,74] Many growth factors and their receptors have been shown to modulate cell function, such as contractility or secretion, as well as growth, just as the α_1-adrenergic receptor does in the cardiac myocytes.[21,52] Whether the intracellular pathways mediating these multifunctional roles of growth factors differ quantitatively or qualitatively is not known.

Gene Expression in Hypertrophy in Culture

Two major reasons motivated investigation of gene expression in α_1-stimulated cardiac myocyte hypertrophy. First, different types of hypertrophy in vivo are associated with different patterns of expression of contractile protein isoforms, as described previously. That is, cardiac myocyte hypertrophy in vivo involves both qualitative and quantitative changes in gene expression. It was therefore important to ask whether α_1-adrenergic stimulation in culture induced specific genes or whether similar increases occurred in both myocyte-specific proteins and cell-nonspecific proteins involved in "housekeeping."

Second, the quantitative and qualitative changes in protein content that define hypertrophy could be produced by regulation of protein degradation, mRNA translation into protein, mRNA processing, or DNA transcription into mRNA (Fig. 4-4). To define an intracellular pathway, it was necessary to identify the level of regulation, the rate-limiting step in gene expression. In light of the proto-oncogene model, it was particularly important to know whether the α_1-receptor regulates gene transcription and thus must transmit a signal to the myocyte nucleus. For technical reasons, precise

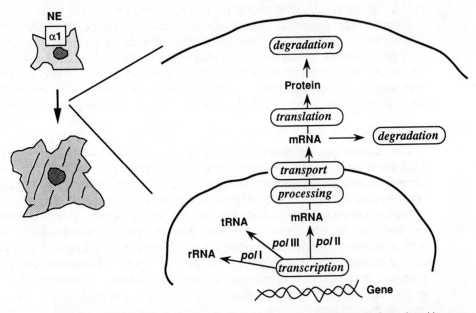

Fig. 4–4. Possible levels of regulation in hypertrophy. Cardiac myocyte hypertrophy is produced by quantitative and qualitative changes in cell protein content. These changes could be regulated at one or several steps between the gene and the protein product. The level of regulation must be defined in order to identify the intracellular pathway. Different RNA polymerase (pol) molecules catalyze synthesis of different RNA species: pol I for ribosomal RNA (rRNA), pol II for messenger RNA (mRNA), and pol III for transfer RNA (tRNA). rRNA and tRNA are required for translation of mRNA into protein. (NE, norepinephrine.)

identification and study of the level of regulation is greatly facilitated if a specific gene can be examined.

The focus here is on gene expression in the context of cell hypertrophy. Modulation of gene expression also is studied in systems that do not involve cell growth (e.g., see refs. 75, 76). The major goal in studies of both types is to define a molecular pathway that regulates expression of a given gene.

Gene expression can be studied by quantitation of the cell content or steady-state level of a specific mRNA or protein. Assay of specific proteins is most important, because proteins are the major effectors of cell structure and function. On the other hand, documentation that a specific mRNA is increased provides the information that regulation is pretranslational, at least in part (Fig. 4-4). Furthermore, molecular biology techniques have made it simpler to quantify a specific mRNA than the protein it encodes, in many cases. In the absence of protein quantitation, however, one must keep in mind the formal possibilities that an mRNA may not be translated into protein or that the rate of protein degradation may exceed the rate of protein synthesis (e.g., see ref. 77).

In our work, we have quantified the mRNA encoding a nuclear proto-oncogene (c-*myc*), the actin iso-mRNAs, and the cardiac MHC iso-mRNAs. The MHC iso-proteins have been measured also. The following sections will overview the methods used for the mRNA and protein measurements. A useful source for detailed protocols, the multiple possible variations of the techniques, and background information is the manual edited by Ausubel et al.[78]

DATA ANALYSIS IN STUDIES OF GENE EXPRESSION

The goal is to determine the per-cell content of particular gene products, whether protein or RNA (see preceding discussion). Therefore, equal numbers of myocytes in different treatment groups are taken for assay, and data are expressed on a per-cell basis. Data also could be expressed relative to total protein or total RNA (which is largely rRNA). It seems preferable, however, to use the cell as the basic unit of study. In addition, determination of cell number is more precise than assay of protein and RNA, in our experience. Furthermore, because there may be important effects of cell density in culture, and because the proportion of contaminating nonmuscle cells in a primary culture must be known, it is essential to determine cell number in any case.

The final product in most mRNA assays is a ^{32}P-labeled band or spot on a gel. The bands can be quantified by densitometry after autoradiography. Band density, however, should be shown to be linearly related to counts in the band, and this is often not the case for heavily exposed or dense bands. Therefore, a more accurate method may be to excise the bands from the gel and to count them in a scintillation counter. Data in counts per minute (cpm) for treated groups, normalized on a per-cell basis, are expressed typically relative to those for vehicle-treated controls. Treated/control ratios from multiple experiments with independent culture preparations are averaged for statistical analysis.

EXTRACTION OF RNA

Total RNA is extracted using a technique to avoid mRNA degradation. Cells are disrupted and ribonucleases are inactivated using urea, and the RNA is precipitated with lithium chloride. After further purification, the RNA is quantified by ultraviolet (UV) spectrophotometry. Under our conditions, each control myocyte contains ~10 pg of total RNA. Experiments with labeled RNA indicate that RNA recovery is over 85% (author's unpublished data).

NORTHERN ANALYSIS FOR ACTIN mRNAs

The mRNAs encoding the major isoform groups of actin can be quantified by Northern analysis (Fig. 4-5).[62] Individual RNA species in total RNA samples from treated and control cells are separated using a denaturing agarose gel. Then 10 to 20 μg of total RNA from equal numbers of cells are applied to each lane of the gel. After separation, the denatured RNAs are transferred to a nitrocellulose filter by capillary action ("blotted"). A single DNA plasmid probe that hybridizes with both the sarcomeric (α) and cytoskeletal (β/γ) isoforms of actin is labeled with ^{32}P by nick-translation, which replaces nonlabeled nucleotides with labeled ones. The labeled probe is hybridized with the RNA bound on the filter. The resulting bands are quantified after autoradiography.

Although sensitivity depends on the amount and specific activity of the probe, autoradiography for 1 to 2 days should be able to detect an mRNA present at 0.0001% of total RNA (10 pg of the mRNA in 10 μg of total RNA).[78] Sensitivity can be increased if only mRNA (poly-A$^+$ RNA) is analyzed rather than total RNA. Because mRNA is only a small fraction of total cell RNA (~3%[79]), however, many more cells are required to obtain sufficient quantities of poly-A$^+$ RNA, and recoveries may be more variable among groups than is the case for extraction of total RNA.

A unique feature of the Northern assay is that it allows sizing of the intact mRNA. RNA size is determined by comparing the relative mobility of labeled DNA fragments of known size or of RNAs of known size, e.g., 18S or 28S rRNA, 1869 and 4712 bases, respectively (visualized by UV after ethidium bromide staining of the gel). An additional advantage of the Northern technique is that the filters can be stripped of radioactivity and rehybridized with other probes.

Total RNA samples can also be applied directly to filters and hybridized with labeled probes, without prior electrophoretic separation ("dot" or "slot" blots). Because mRNA size is not revealed by the dot-blot technique, however, problems of nonspecific probe binding can be missed.

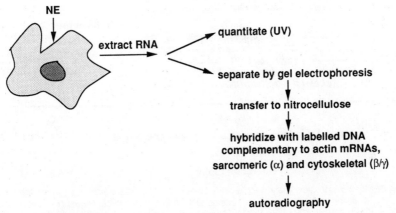

Fig. 4–5.   Northern analysis for actin mRNAs. Northern analysis can be used to separate and quantify the mRNAs encoding sarcomeric (α) and cytoskeletal (β/γ) actins.[62] A fundamentally similar scheme can be used for DNA (Southern) or for protein (Western). (mRNA, messenger RNA; NE, norepinephrine; UV, ultraviolet.)

A disadvantage of the Northern technique is that iso-mRNAs of similar size are not resolved on agarose gels and cannot be quantified if the probe used is not specific. For example, this gel technique does not separate the cytoskeletal actin (β- and γ-actin) mRNAs or the skeletal and cardiac iso-mRNAs of α-actin.

PRIMER EXTENSION ANALYSIS OF α-ACTIN ISO-mRNAS

The cardiac and skeletal iso-mRNAs of α-actin can be quantified by Northern analysis, using DNA probes that are complementary to divergent sequences in the 3′ ends of the two mRNAs.[31,80] This analysis, however, requires two separate gels for each sample or stripping and reprobing a single nitrocellulose filter, because the two iso-mRNAs are not separated on the gels.

A primer extension analysis can be used to quantify the skeletal and cardiac iso-mRNAs simultaneously in the same RNA sample[62] (Fig. 4-6). A single-stranded 18-base DNA oligonucleotide primer is synthesized, which is complementary to an identical sequence found near the 5′ end of the mRNAs of both skeletal α-actin and cardiac α-actin. This primer is ^{32}P labeled at the 5′ end using polynucleotide kinase. The labeled primer is then hybridized in solution to total RNA (up to 20 μg) from equal numbers

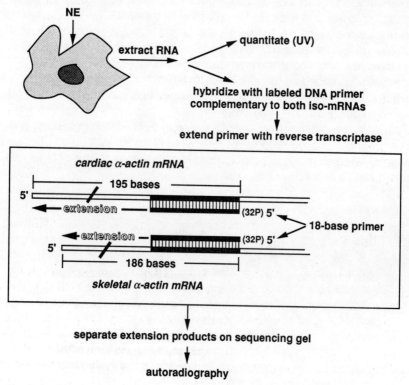

Fig. 4–6. Primer extension analysis for α-actin iso-mRNAs. The cross-hatching between the shaded areas indicates hybridization of the labeled DNA primer to identical sequences in exon 2 of cardiac and skeletal α-actin iso-mRNAs. The primer is extended to the 5′ end of the mRNAs (arrows) with reverse transcriptase in the presence of nonlabeled deoxynucleotides. The resulting extension products are of different lengths (195 and 186 bases for cardiac α-actin iso-mRNA and skeletal α-actin iso-mRNA, respectively) and are separated on a sequencing gel.[62] (mRNA, messenger RNA; NE, norepinephrine; UV, ultraviolet.)

of treated and control cells; total RNA in each sample can be equalized with transfer RNA (tRNA). After hybridization of the primer to skeletal and cardiac α-actin mRNAs in the total RNA sample, DNA copies of the two mRNAs are obtained by extending the primer in the 3' direction (to the 5' end of the mRNAs), using reverse transcriptase and nonlabeled deoxynucleotides. The resulting end-labeled extension products are separated on a denaturing polyacrylamide/urea sequencing gel.

Separate extension product bands are seen after autoradiography, because of a small difference in length (9 bases) of the 5' untranslated regions of the skeletal and cardiac α-actin iso-mRNAs (Fig. 4-6). The number of counts in each band reflects the number of molecules of each mRNA in the total RNA sample. The primer extension technique can detect an mRNA that is 0.001% of total RNA.[78] The specificity of the assay is demonstrated by analyzing RNA from tissues with known patterns of expression of the α-actin isoforms (skeletal α-actin in adult muscle, cardiac α-actin in adult heart, neither in brain). Linearity is confirmed by quantifying mRNA levels in different amounts of total RNA.

Thus, the primer extension technique can be used to quantify mRNAs that are not measured easily by Northern analysis. An additional use of this technique is to map the 5' end of an mRNA. The main problem encountered is pausing or termination of the reverse transcriptase before extending to the 5' end of the mRNA, producing bands of low molecular weight on the gel. The chances of this occurrence are reduced if the primer is selected such that the predicted extension product is no longer than 100 to 200 bases.

NUCLEASE S_1 PROTECTION ASSAY FOR MYOSIN HEAVY CHAIN ISO-MRNAS

The α- and β-iso-mRNAs of MHC can be quantified by Northern analysis, using DNA probes complementary to divergent sequences in the 3' ends of the two iso-mRNAs.[81] As in the case of the α-actin iso-mRNAs, however, separate gels or reprobing of the same filter is required.

A nuclease S_1 assay can be used to quantify both iso-mRNAs in the same sample[77,82, 82a] (Fig. 4-7). See also Izumo et al.,[28] for an S_1 nuclease assay using a different DNA probe. A single-stranded DNA oligonucleotide is synthesized and labeled with ^{32}P-adenosine triphosphate (ATP) at its 3' end using terminal transferase. This 62-base probe is complementary to 41 bases of α-MHC mRNA and 56 bases of β-MHC mRNA (Fig. 4-7). Six bases at the 5' end of the probe are not complementary to either mRNA. The labeled probe is hybridized in solution to α- and β-MHC mRNAs in total RNA samples extracted from equal numbers of treated and control cells (once again, total RNA in each sample is equalized using tRNA). The mixture is then digested with S_1 nuclease. Single-stranded (nonhybridized) DNA and RNA molecules are digested by S_1 nuclease, whereas DNA–RNA hybrids are protected from digestion. The protected probe fragments are separated on a denaturing polyacrylamide/urea sequencing gel.

Separate bands are seen after autoradiography of the gel, reflecting the different fragments of the probe protected from S_1 digestion by formation of DNA–RNA hybrids, 41 bases for α-MHC mRNA and 56 bases for β-MHC mRNA. Any undigested probe appears as a 62-base band. The number of counts in each band reflects the number of molecules of each mRNA in the total RNA sample. The sensitivity of this technique should be similar to that of primer extension technique. The specificity of the assay is demonstrated by analyzing RNA from tissues with known patterns of expression in the α-MHC isoforms (α-MHC in hearts from hyperthyroid rats, β-MHC in hearts from hy-

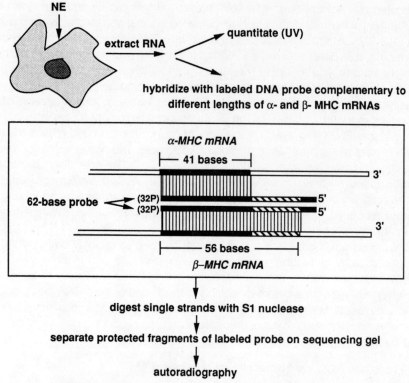

NE

extract RNA → quantitate (UV)

hybridize with labeled DNA probe complementary to different lengths of α- and β- MHC mRNAs

α-MHC mRNA

|— 41 bases —|

62-base probe → (32P) (32P)

56 bases

β-MHC mRNA

digest single strands with S1 nuclease

separate protected fragments of labeled probe on sequencing gel

autoradiography

Fig. 4–7. Nuclease S_1 protection assay for myosin heavy chain iso-mRNAs. Cross-hatching indicates that the synthetic DNA probe is complementary to different lengths of α-myosin heavy chain (MHC) mRNA (41 bases) and β-MHC mRNA (56 bases).[77,82,82a] DNA–RNA duplex formation protects different lengths of labeled probe from S_1 nuclease digestion of single-stranded nucleic acids.

pothyroid rats). Linearity is confirmed by quantifying mRNA levels in different amounts of total RNA. For this method to be considered quantitative, it is essential that the probe be present in excess relative to the amount of mRNA to be measured.

MYOSIN HEAVY CHAIN ISO-PROTEIN CONTENTS

The MHC proteins encoded by the α- and β-MHC iso-mRNAs are paired to form three different myosin molecules in cardiac myocytes. V_3 myosin, which has lower ATPase activity, is a ββ-MHC homodimer, whereas V_1 is an αα-MHC homodimer and V_2 is an αβ-MHC heterodimer.[83] The method for quantitation of MHC iso-protein levels uses both denaturing and nondenaturing gel electrophoresis. Samples from equal numbers of control and treated cells are divided into aliquots. Myosin is extracted in one aliquot, and myosin iso-proteins are separated on nondenaturing gels.[84] The fraction of each iso-protein in the total myosin sample is determined by densitometry of Coomassie blue-stained bands. Total MHC protein in the other aliquot is resolved by sodium dodecylsulfate polyacrylamide gel electrophoresis (SDS-PAGE) and quantified by densitometry after Coomassie blue staining. The density of sample bands is related to that of myosin standards run on the same gel. The amount of each MHC iso-protein is calculated in picograms from the amount of total MHC and the fraction of each myosin iso-protein in that sample. These iso-protein data are normalized on a per-cell basis.

RIBONUCLEASE PROTECTION ASSAY FOR C-*myc* MRNA

A ribonuclease (RNase) protection assay can be employed to quantify the mRNA encoding c-*myc*, a nuclear proto-oncogene that may be a transcription factor.[21,85] The RNase protection technique is fundamentally similar to the S_1 nuclease protection assay described previously. Several differences, however, make the RNase protection assay for mRNA more sensitive. A DNA template of the c-*myc* mRNA to be assayed is inserted into a plasmid next to a bacteriophage promoter. An RNA complementary to the c-*myc* mRNA is then transcribed using bacteriophage RNA polymerase. Synthesis is in the presence of a ^{32}P-labeled nucleotide trisphosphate, so that the complementary RNA (cRNA) probe is labeled throughout its length. Probe-specific activity is higher than with nick translation and is much higher than with end labeling.

The cRNA probe is hybridized in solution to total RNA extracted from treated and control cells. After hybridization of the c-*myc* mRNA to the probe, single-stranded RNA is digested with ribonucleases. The probe fragments protected from digestion by hybridization to c-*myc* mRNA are separated on a denaturing polyacrylamide/urea sequencing gel. The counts in the resultant bands reflect the number of c-*myc* mRNAs. Two bands are seen, since *myc* mRNAs can be transcribed from two different start sites.

Because of the higher probe-specific activity, the sensitivity of the RNase protection assay may be ten- to hundredfold greater than for the other techniques. Complementary RNA probes can be used in Northern analyses and in in-situ hybridizations, in addition to their use in the RNase protection assay.

GENE EXPRESSION IN α_1-STIMULATED CARDIAC MYOCYTE HYPERTROPHY

Application of the previously outlined methods to the cardiac myocytes in the culture model indicates that the α_1 receptor is coupled to specific alterations of gene expression.

α_1-Receptor stimulation induces selective changes in the per-cell contents of MHC iso-proteins and their cognate iso-mRNAs.[77,82a,86] β-MHC iso-protein is increased approximately 4-fold versus control, and a comparable increase occurs in the level of β-MHC iso-mRNA. Thus, the induction of the β-MHC iso-protein can be ascribed to regulation at the pretranslational level. In contrast, α_1 stimulation does not change the per-cell content of α-MHC iso-protein or α-MHC iso-mRNA. β-MHC iso-mRNA accounts for 40% of total MHC mRNA in control cultured myocytes, versus 69% of total after α_1 stimulation. Therefore, β-MHC is induced in α_1-stimulated hypertrophy, and α-MHC is not altered.

α_1 Stimulation also produces selective changes in actin iso-mRNAs[62] (Fig. 4-8). Sarcomeric (α) actin mRNA content per cell is increased greater than fourfold vs. control, whereas cytoskeletal (β/γ) actin mRNA is elevated less than twofold. Furthermore, disparate changes occur in the two α-actin iso-mRNAs. Skeletal α-actin iso-mRNA content per cell is elevated almost 11-fold vs. control, whereas cardiac α-actin is increased just over twofold. Skeletal α-actin iso-mRNA makes up 18% of total α-actin mRNA in control cells, vs. 50% of total after α_1 stimulation. Therefore, skeletal α-actin iso-mRNA is up-regulated to a much greater extent than is cardiac α-actin iso-mRNA in α_1-stimulated hypertrophy.

The increases in β-MHC and skeletal α-actin indicate that α_1-stimulated cardiac myocyte hypertrophy in culture is characterized by selective up-regulation of isogenes that are expressed predominantly during early developmental stages.

α_1 Stimulation of the cultured myocytes also produces a prompt and dramatic in-

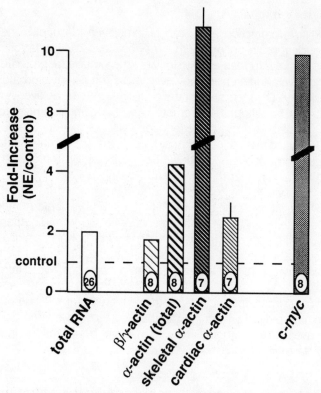

Fig. 4–8. Gene expression in α_1-stimulated cardiac myocyte hypertrophy. The figure summarizes the results of several experiments (numbers within the bars) on per-cell contents of total RNA and various mRNAs after α_1-adrenergic stimulation of the cultured cardiac myocytes. Note that all mRNAs studied are not increased to the same extent, indicating that the α_1-receptor mediates selective changes in gene expression. Primary data are in references 62 and 85. (NE, norepinephrine.)

crease in c-*myc* mRNA[85] (Fig. 4-8). Whether the c-*myc* protein is increased is not known at present (for discussion, see ref. 20).

In summary, skeletal α-actin mRNA, β-MHC mRNA and protein, and c-*myc* mRNA are induced selectively in α_1-stimulated hypertrophy in culture. These same genes are up-regulated in pressure-load myocardial hypertrophy in vivo.[27–29,31,87] One possible implication of these findings is that similar molecular mechanisms are operative in one form of hypertrophy in vivo and in the culture model. Much more must be learned, however, about the mechanisms of hypertrophy and gene expression in culture before this possibility can be tested rigorously in vivo.

Transcriptional Regulation of Gene Expression

The increases in specific mRNA contents measured in the culture model and in vivo indicate that gene expression is regulated at the pretranslational level, at least in part. The mRNA changes, however, may not reflect increases in mRNA synthesis but could be accounted for entirely by changes in mRNA processing or stability (Fig. 4-4). To provide a focus for studies on mechanism, it was critical to determine whether selective modulation of actin isogene transcription occurred in α_1-stimulated hypertrophy in culture.

THE NUCLEAR RUN-ON TRANSCRIPTION ASSAY

A technique known as the nuclear run-on assay can be used to determine whether transcription of the actin isogenes is increased by α_1-adrenergic stimulation[64] (Fig. 4-9). The nuclear run-on technique quantifies the level of transcription in the intact cell at the time nuclei are isolated. The basic procedure is as follows. Nuclei are isolated from control and treated cells at various times after stimulation with NE. During nuclear isolation, RNA polymerase complexes remain attached to genes that are being actively transcribed. Transcription by these RNA polymerase complexes can resume when the nuclei are reconstituted in an appropriate buffer, and new initiation does not occur. Elongation (run-on) of the previously initiated transcripts in vitro in the presence of ^{32}P-nucleotide trisphosphates (RNA precursors) produces labeled RNAs. The labeled RNA produced is purified and counted, providing an index of total transcriptional activity.

The labeled run-on RNA is then hybridized with gene-specific DNA probes dotted onto a nitrocellulose filter. The probes are complementary to the mRNAs for skeletal α-actin, cardiac α-actin, and β-actin. After autoradiography, the labeled dots are cut out and counted. The specific counts in run-on RNA that hybridize to a given actin probe provide an index of the number of RNA polymerase II molecules on that gene at the time of nuclear isolation, and therefore quantify the rate of initiation of transcription.[88] The data for gene-specific hybridization are generally presented in parts

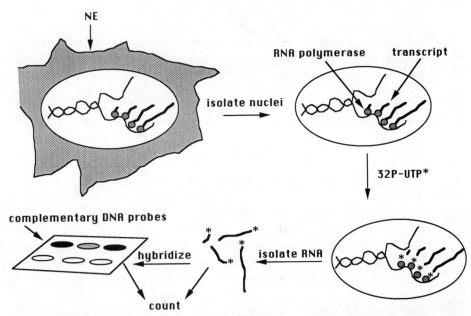

Fig. 4–9. Nuclear run-on assay for actin isogene transcription. This technique measures the level of transcription in the intact cell at the time nuclei are isolated. Transcripts initiated by RNA polymerase in the intact cell are elongated (run on) with radiolabeled RNA precursors when the isolated nuclei are incubated in vitro. Each nascent transcript incorporates the same number of labeled precursors. Thus, incorporation of counts is proportional to the number of RNA polymerase complexes that have initiated transcription. The RNA produced is purified and counted as an index of total transcriptional activity (the activities of RNA polymerases I, II, and III). The labeled RNA is hybridized with gene-specific probes to quantify actin isogene transcription[64] (Fig. 4-10). (NE, norepinephrine; UTP, uridine triphosphate.)

per million (ppm), calculated by dividing the counts per minute hybridized to each probe by the total counts per minute included in the hybridization reaction.

Isolation of nuclei that will resume transcription in vitro requires care. Nuclei can be released by gentle detergent lysis of the cell membrane and homogenization and purified through sucrose gradient centrifugation. Intact nuclei are counted to determine recovery and to use equal numbers of nuclei in each in-vitro reaction.

The in-vitro transcription reaction must be shown to be linear with time and with numbers of input nuclei. Inclusion of α-amanitin (1 μg/mL) in the transcription reaction is an important control. At this dose, α-amanitin specifically inhibits RNA polymerase II, the polymerase responsible for mRNA synthesis.[89] Elimination of labeled run-on RNA hybridization to the DNA probes by α-amanitin indicates that the hybridized counts are specific for mRNA transcription by RNA polymerase II. α-Amanitin at 1 μg/mL also can be used to estimate the contribution to total transcriptional activity of mRNA synthesis (α-amanitin-sensitive activity of RNA polymerase II), in comparison with rRNA and tRNA synthesis (α-amanitin-insensitive activity of RNA polymerases I and III, respectively).

The hybridized counts are shown to increase linearly as the input counts of run-on RNA are increased, indicating that the DNA probe is present in excess. Efficiency of hybridization to the probes under the conditions used (~30% in our experiments) is determined using a cRNA synthesized with a ^{3}H-labeled nucleotide trisphosphate (so that the cRNA counts hybridized can be distinguished from those of ^{32}P-run-on RNA). Ideally, the DNA probe is single stranded in the anti-sense (3'-to-5') direction, rather than double stranded, so that possible anti-sense transcription does not produce a hybridization signal. It is also desirable to use hybridization probes for both the 5' and 3' ends of the mRNA, to detect possible transcriptional pausing or premature termination.

SELECTIVE TRANSCRIPTION OF α-ACTIN ISOGENES IN α_1-STIMULATED HYPERTROPHY

Use of the in-vitro nuclear run-on assay to quantify the rates of initiation of transcription of different RNA species at various times after treatment of the cultured myocytes with NE reveals a complex pattern of transcriptional induction following stimulation of the α_1-receptor.[64]

α_1 Stimulation increases selectively the transcription of the skeletal α-actin isogene, as compared with transcription of cardiac α-actin, β-actin (a nonsignificant 1.5-fold increase, not shown), or total RNA (Fig. 4-10). An increase in transcription is detected within 1 hour after α_1 stimulation, preceding the increases in RNA content. Inhibition of protein synthesis with cycloheximide does not block the increase in transcription.

Interestingly, cardiac α-actin transcription returns to the control level when NE-stimulated skeletal α-actin transcription is still increasing (Fig. 4-10). As a result, the duration of the increase in skeletal α-actin isogene transcription is much greater than that of cardiac α-actin. The asynchronous transcription of the skeletal and cardiac α-actin isogenes is surprising, but asynchronous transcription of closely related genes also is seen in response to cyclic adenosine monophosphate (cAMP).[90]

Therefore, the selective effect of α_1-adrenergic stimulation seen at the level of skeletal α-actin iso-mRNA content is reflected in a selective effect on skeletal α-actin isogene transcription. Induction of an early developmental isogene in this model of hypertrophy reflects a fundamental change in the transcriptional program of the cardiac myocyte nucleus.

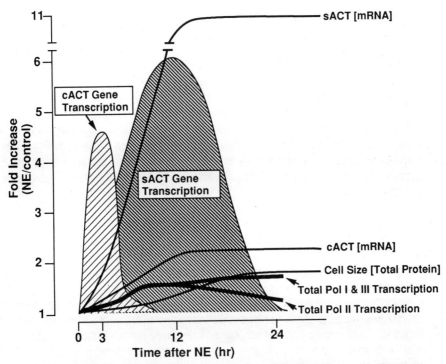

Fig. 4–10. Regulation of transcription in α_1-stimulated hypertrophy. The diagram summarizes the time course of activation of transcription and accumulation of specific mRNAs [mRNA] and total protein following stimulation of the cultured neonatal rat heart myocytes with norepinephrine (NE). Note the selective effect of α_1 stimulation on transcription of skeletal α-actin (sACT) as compared to cardiac α-actin (cACT). Primary data are in references 55, 60, 62, and 64. (Pol, RNA polymerase.) (Modified with permission from Long, C.S., Ordahl, and Simpson, P.C.: Alpha-1 adrenergic receptor stimulation of sarcomeric actin isogene transcription in hypertrophy of cultured rat heart muscle cells. J. Clin. Invest., *83*:1078, 1989.)

   The varied kinetics of transcription of the different RNA species after stimulation of the single receptor deserves emphasis. First, the time course of skeletal α-actin isogene transcription differs from that of cardiac α-actin, as just noted. Second, activation of mRNA transcription by α_1 stimulation is transient, whereas transcription of total RNA is sustained. In particular, studies with α-amanitin indicate that rRNA and tRNA transcription are still increasing when both total mRNA and specific mRNA transcription have returned to baseline values (Fig. 4-10). Transient activation of mRNA transcription is seen in other systems, and attenuation of transcription has been attributed to synthesis of a transcriptional repressor.[91] Disparate intracellular mechanisms appear to regulate synthesis of mRNA and rRNA after activation of the α_1 receptor.

   These results are the first to demonstrate that the α_1-adrenergic receptor is a molecular mediator of transcriptional changes underlying an isogene switch associated with cardiac myocyte hypertrophy. Furthermore, rather than effecting a general, simultaneous increase in overall gene transcription, stimulation of the α_1-adrenergic receptor leads to a distinctive temporal sequence of transcriptional activation. It will be interesting to see if conditions that induce myocardial hypertrophy in vivo also stimulate the same complex temporal sequence of gene transcription as does stimulation of the α_1-adrenergic receptor.

Our observations, and those of Chien and colleagues,[92] indicate that α_1-stimulated hypertrophy is produced, at least in part, by regulation at the level of transcription. Evidence also shows that thyroid hormone modulates transcription of cardiac myocyte genes,[33,93] although there is controversy regarding the direct effects of thyroid hormone on myocardial hypertrophy.[30,94] Documentation of transcriptional regulation in α_1-stimulated hypertrophy certainly does not exclude the possibility of regulation at other levels, such as translation. It does, however, provide the foundation to define an intracellular pathway connecting the α_1-adrenergic receptor in the plasma membrane to activation of RNA polymerase II on the skeletal α-actin gene in the cardiac myocyte nucleus. Furthermore, the fact that transcription is activated in α_1-stimulated hypertrophy means that work in this system can be viewed in terms of the proto-oncogene paradigm (Fig. 4-2).

An Intracellular Pathway for α_1-Stimulated Transcription

Current work is directed toward defining an intracellular pathway connecting the α_1-adrenergic receptor at the cell surface to activation of transcription in the myocyte nucleus. On the basis of studies in other systems, the working model shown in Figure 4-11 can be proposed.

Critical to this model is the hypothesis that the skeletal α-actin gene contains one or more α_1-response elements, a DNA sequence element of ~10 to 20 bp that is required for activation of transcription in response to α_1 stimulation. DNA sequence elements involved in the regulation of transcription are generally located in the region 5' to the start site of transcription and are respectively designated promoters or enhancers if they are contiguous with the start site or are at a variable distance from it. These *cis* elements serve as binding sites for transcription factors. Some *cis* elements and their cognate transcription factors are found in all genes, such as the TATA box and its binding proteins, which appear to be required for accurate initiation by RNA

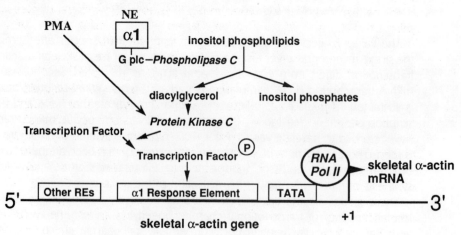

Fig. 4-11. This simplified working model for α_1-stimulated transcription suggests that transcription of skeletal α-actin is increased through the action of a transcription factor that binds to an α_1-response element. The transcription factor is activated by protein kinase C-mediated phosphorylation. (α_1, α_1-adrenergic receptor; G plc, guanine nucleotide regulatory protein believed to couple the α_1-receptor to phospholipase C; NE, norepinephrine; PMA, phorbol myristate acetate, an activator of protein kinase C; RE, response element; TATA, TATA box, a *cis* element involved in positioning of RNA polymerase II; +1, start site of transcription.)

polymerase II. Other *cis* elements and transcription factors may be limited to particular genes or families of genes, such as those expressed only in myocytes.[95-97] A third group of *cis* elements and *trans* factors may be involved in transcription in response to hormones or growth factors,[98-101] although the extent to which the second and third groups are distinct or related is unclear at present. Transcription factors are believed to form a complex with RNA polymerase II and to somehow initiate its transcription of the gene. (Current concepts of transcription factors are discussed further in refs. 21, 102, and 103).

In many cases, the binding of a transcription factor to its *cis* element, and/or its function in activating transcription, appear to depend on post-translational modifications.[98,102-105] That is, a change in conformation of a pre-existing transcription factor may be important, rather than new synthesis. Of particular relevance to α_1-stimulated transcription in the cardiac myocytes is the suggestion that protein kinase C (PKC) can mediate activation of a transcription factor.[101,104,106-108]

Therefore, in analogy with other hormonal response elements, the α_1-response element is proposed to be a specific DNA sequence, a promoter or an enhancer, that is required for activation of skeletal α-actin transcription in response to α_1 stimulation. The α_1-response element is visualized as binding a transcription factor. It is postulated further that this transcription factor is activated by PKC and in turn activates RNA polymerase II.

Circumstantial evidence supports this hypothesis, in addition to analogy with other systems. First, new protein synthesis is not required for activation of transcription in response to α_1 stimulation.[64] Second, α_1 stimulation activates PKC in the cultured myocytes,[109] perhaps via generation of diacylglycerol from inositol phospholipids, as shown in Figure 4-11.[110,111] Other mechanisms of PKC activation are possible also.[112] Third, activation of PKC by phorbol myristate acetate reproduces certain of the effects of α_1 stimulation in the cultured cardiac myocytes, including hypertrophy and gene expression.[71,77,85,109] (also our unpublished results). The behavior of PKC is very complex in the cardiac myocytes (discussed in ref. 113), so that additional or alternate mechanisms for activation of transcription should be kept in mind. The model shown in Figure 4-11, however, is a reasonable basis for further study.

TRANSFECTION ANALYSIS FOR DNA SEQUENCES INVOLVED IN REGULATION OF TRANSCRIPTION

To test the model shown in Figure 4-11, it is necessary to ask whether the skeletal α-actin gene contains an α_1-response element. One approach to this question is transfection analysis[95,114] (Fig. 4-12). The basic principle of this technique is as follows: A hybrid gene is constructed and inserted into a plasmid vector. The hybrid gene contains the sequences of the skeletal α-actin gene 5' to the start site of transcription, the region most likely to bear the putative α_1-response element, fused to the coding sequence for chloramphenicol acetyltransferase (CAT). CAT is a bacterial enzyme not found in eukaryotic cells and can be measured easily and with high sensitivity by immunofluorescence or enzyme activity. Other "reporter" genes can be used also, such as those encoding β-galactosidase or luciferase. The hybrid gene contains in addition a sequence necessary for correct processing of transcribed mRNA (SV40 polyA signal).

Plasmids bearing the hybrid gene are introduced into the cultured cells ("transfected") using one of a variety of techniques, such as calcium phosphate microprecipitates to stimulate cellular uptake. The plasmids localize in the nucleus, by unknown mechanisms; and the hybrid genes respond to endogenous transcription factors. Thus,

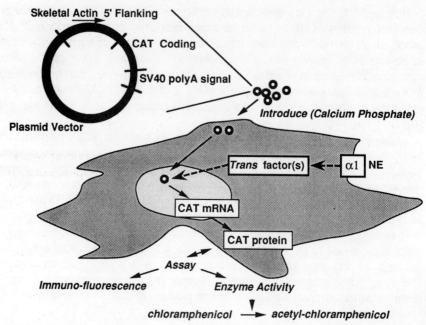

Fig. 4–12. Transfection analysis for DNA sequences involved in regulation of transcription. The transfection approach can be used to identify a putative α_1-response element, a DNA sequence that binds a transcription factor (*trans* factor) activated by α_1-adrenergic stimulation. A hybrid gene is constructed containing the coding sequence for chloramphenicol acetyltransferase (CAT), preceded by a portion of the skeletal α-actin gene 5' to the start site of skeletal α-actin transcription, the region that may contain the α_1-response element (Fig. 4-11). This hybrid gene, which contains in addition a sequence required for proper messenger RNA (mRNA) processing (SV40 polyA signal), is inserted into a plasmid vector and introduced into the cultured myocytes ("transfected") using calcium phosphate microprecipitates to stimulate uptake. The transfected cells are treated with an α_1-agonist. If a transcription factor is activated, it can bind to the α_1-response element on the hybrid gene and induce synthesis of CAT mRNA, which is translated into CAT protein. See text for further discussion. (NE, norepinephrine.)

if α_1 stimulation of the transfected cell activates a transcription factor, it can bind to an α_1-response element contained in the hybrid gene, initiating transcription of CAT mRNA. CAT mRNA is translated into CAT protein, which is assayed. Several controls must be done to ensure that the effect of the α_1-agonist is on CAT transcription and not on other processes that might alter CAT protein, such as plasmid uptake or CAT translation.

Once induction of CAT transcription by α_1 stimulation has been demonstrated, new hybrid genes are constructed bearing systematic deletions of portions of the skeletal α-actin gene 5' flanking region. A specific deletion that eliminates CAT induction by α_1 stimulation identifies the deleted segment as potentially containing the α_1-response element. Other methods can then be used to provide additional confirmation and to define the α_1-response element more precisely, such as the gel retardation or footprinting techniques—methods that identify a DNA-protein interaction.[93,115] It is then feasible to isolate the transcription factor that interacts with the α_1-response element, using the DNA sequence as an affinity ligand for purification.[78] The function of an isolated protein can be confirmed by reconstituting transcription in vitro, with added RNA polymerase II and a DNA template.

Although the approaches outlined in the preceding paragraphs are difficult and time-consuming, their application has identified several transcription factors.[116] How they work and how they are activated by extracellular signals are questions at the forefront of research on regulation of transcription.[118]

COMMENTS

Myocardial hypertrophy is an important problem in clinical cardiology. A major component of myocardial hypertrophy in vivo is hypertrophy of cardiac myocytes, a growth process that varies both quantitatively and qualitatively. A cell culture model has made it possible to identify catecholamines as growth factors producing cardiac myocyte hypertrophy, and the α_1-adrenergic receptor as a growth-factor receptor. The hypertrophy-promoting effect of α_1-receptor stimulation is independent of contractile activity.

Stimulation of the α_1-adrenergic receptor by the catecholamine NE is coupled to selective increases in the myocyte contents of specific mRNAs and proteins and to the transcription of specific genes, those encoding early developmental isoforms of contractile proteins. Thus, transcriptional regulation may be a key event in hypertrophy in this model. A potential intracellular pathway connecting the α_1-receptor at the cell surface to transcription in the nucleus can be proposed and can be investigated. PKC may be an important transducer in this pathway.

The approach outlined in this chapter can be used to identify and study other soluble and mechanical growth factors for cardiac myocytes. Once other growth factors, their effects on gene expression, and their transduction mechanisms have been identified, it will be feasible to investigate how they are integrated, first in simple systems, and then in vivo. Elucidation of the signals for cardiac myocyte hypertrophy should make it possible to manipulate this important growth process.

Acknowledgments

This work was supported by grants from the National Institutes of Health, the Veterans Administration Research Service, and the California Affiliate of the American Heart Association. PCS is a Clinical Investigator of the Veterans Administration. We thank Charles P. Ordahl for his major contribution to the work on gene expression.

REFERENCES

1. Grossman, W.: Cardiac hypertrophy: useful adaptation or pathologic process? Am. J. Med., *69*:576, 1980.
2. Perloff, J.K.: Pathogenesis of hypertrophic cardiomyopathy: hypotheses and speculations. Am. Heart J., *101*:219, 1981.
3. Sugrue, D.D. et al.: Cardiac histologic findings in patients with life-threatening ventricular arrhythmias of unknown origin. J. Am. Coll. Cardiol., *4*:952, 1984.
4. McLenachan, J.M., Henderson, E., Morris, K.I., and Dargie, H.J.: Ventricular arrhythmias in patients with hypertensive left ventricular hypertrophy. N. Engl. J. Med., *317*:787, 1987.
5. Harizi, R.C., Bianco, J.A., and Alpert, J.S.: Diastolic function of the heart in clinical cardiology. Arch. Intern. Med., *148*:99, 1988.
6. Schaible, T.F., Malhotra, A., Ciambrone, G.J., and Scheuer, J.: Chronic swimming reverses cardiac dysfunction and myosin abnormalities in hypertensive rats. J. Appl. Physiol., *60*:1435, 1986.
7. Schaible, T.F., Ciambrone, G.J., Capasso, J.M., and Scheuer, J.: Cardiac conditioning ameliorates cardiac dysfunction associated with renal hypertension in rats. J. Clin. Invest., *73*:1086, 1984.
8. Schaible, T.F., and Scheuer, J.: Response of the heart to exercise training. In Growth of the Heart in Health and Disease Edited by R. Zak. New York, Raven Press, 1984.
9. Shapiro, L.M., Moore, R.B., Logan-Sinclair, R.B., and Gibson, D.G.: Relation of regional echo amplitude to left ventricular function and the electrocardiogram in left ventricular hypertrophy. Br. Heart J., *52*:99, 1984.

10. Fagard, R. et al.: Assessment of stiffness of the hypertrophied left ventricle of bicyclists using left ventricular inflow doppler velocimetry. J. Am. Coll. Cardiol., *9*:1250, 1987.
11. Kelly, P., Thigpen, T., and Ginzton, L.: Ventricular hypertrophy in endurance athletes is not associated with complex ventricular ectopy (abstract). Circulation, *72*:III-46, 1985.
12. Zak, R.: Development and proliferative capacity of cardiac muscle cells. Circ. Res., *34/35*(suppl. 2):17, 1974.
13. Fanburg, B.L.: Experimental cardiac hypertrophy. N. Engl. J. Med., *282*:723, 1970.
14. Baserga, R.: Growth in size and cell DNA replication. Exp. Cell Res., *151*:1, 1984.
15. Fine, L.G., Holley, R.W., Nasri, H., and Badie-Dezfooly, B.: BSC-1 growth inhibitor transforms a mitogenic stimulus into a hypertrophic stimulus for renal proximal tubular cells: relationship to Na + /H + antiport activity. Proc. Natl. Acad. Sci. USA, *82*:6163, 1985.
16. Nag, A.C.: Study of non-muscle cells of the adult mammalian heart: a fine structural analysis and distribution. Cytobios., *28*:41, 1980.
17. Olivetti, G., Anversa, P., and Loud, A.V.: Morphometric study of early postnatal development in the left and right ventricular myocardium of the rat. 2. Tissue composition, capillary growth, and sarcoplasmic alterations. Circ. Res., *46*:503, 1980.
18. Clubb, F.J., Jr., and Bishop, S.P.: Formation of binucleated myocardial cells in the neonatal rat: an index for growth hypertrophy. Lab. Invest., *50*:571, 1984.
19. Clubb, F.J., Jr., Bell, P.D., Kriseman, J.D., and Bishop, S.P.: Myocardial cell growth and blood pressure development in neonatal spontaneously hypertensive rats. Lab. Invest. *56*:189, 1987.
20. Simpson, P.C.: Role of proto-oncogenes in myocardial hypertrophy. Am. J. Cardiol. *62*:13, 1988.
21. Simpson, P.C.: Proto-oncogenes and cardiac hypertrophy. Annu. Rev. Physiol., *51*:189, 1989.
22. Smith, S.H., McCaslin, M., Sreenan, C., and Bishop, S.P.: Regional myocyte size in two-kidney, one clip renal hypertension. J. Mol. Cell. Cardiol., *20*:1035, 1988.
23. Mahdavi, V., Periasamy, M., and Nadal-Ginard, B.: Molecular characterization of two myosin heavy chain genes expressed in the adult heart. Nature, *297*:659, 1982.
24. Zakut, R. et al.: Nucleotide sequence of the rat skeletal muscle actin gene. Nature, *238*:857, 1982.
25. Scheuer, J., and Bahn, A.K.: Cardiac contractile proteins: adenosine triphosphatase activity and physiological function. Circ. Res., *45*:1, 1979.
26. Vandekerckhove, J., Bugaisky, G., and Buckingham, M.: Simultaneous expression of skeletal muscle and heart actin proteins in various striated muscle tissues and cells: a quantitative determination of the two actin isoforms. J. Biol. Chem., *261*:1838, 1986.
27. Nagai, R. et al.: Myosin isozyme synthesis and mRNA levels in pressure-overloaded rabbit hearts. Circ. Res., *60*:692, 1987.
28. Izumo, S. et al.: Myosin heavy chain messenger RNA and protein isoform transitions during cardiac hypertrophy: interaction between hemodynamic and thyroid hormone-induced signals. J. Clin. Invest., *79*:970, 1987.
29. Izumo, S., Nadal-Ginard, B., and Mahdavi, V.: Protooncogene induction and reprogramming of cardiac gene expression produced by pressure overload. Proc. Natl. Acad. Sci. USA, *85*:339, 1988.
30. Korecky, B, Zak, R., Schwartz, K., and Aschenbrenner, V.: Role of thyroid hormone in regulation of isomyosin composition, contractility, and size of heterotopically isotransplanted rat heart. Circ. Res., *60*:824, 1987.
31. Schwartz, K. et al.: α-Skeletal muscle actin mRNAs accumulate in hypertrophied adult rat hearts. Circ. Res., *59*:551, 1986.
32. Schwartz, K., Mercadier, J.J., Swynghedauw, B., and Lompre, A.M.: Modifications of gene expression in cardiac myocyte hypertrophy. Heart Failure, *4*:154, 1988.
33. Umeda, P.K. et al.: Control of myosin heavy chain expression in cardiac hypertrophy. Am. J. Cardiol., *59*:49, 1987.
34. Scheuer, J. et al.: Physiologic cardiac hypertrophy corrects contractile protein abnormalities associated with pathologic hypertrophy in rats. J. Clin. Invest., *70*:1300, 1982.
35. Baserga, R.: The Biology of Cell Reproduction. Cambridge, MA: Harvard University Press, 1985.
36. Heldin, C.H., and Westermark, B.: Growth factors: mechanism of action and relation to oncogenes. Cell, *37*:9, 1984.
37. Marshall, C.J.: Oncogenes and growth control 1987. Cell, *49*:723, 1987.
38. Terracio, L., Miller, B., and Borg, T.K.: Effects of cyclic mechanical stimulation of the cellular components of the heart: in vitro. In Vitro Cell Dev. Biol., *24*:53, 1988.
39. Kent, R.L., Hoober, J.K., and Cooper, G., IV: Load responsiveness of protein synthesis in adult myocardium: role of cardiac deformation linked to sodium influx. Circ. Res., *64*:74, 1989.
40. Kira, Y., Kochel, P.J., Gordon, E.E., and Morgan, H.E.: Aortic perfusion pressure as a determinant of cardiac protein synthesis. Am. J. Physiol., *246*:C247, 1984.
41. Vandenburgh, H.H., and Kaufman, S.: Stretch-induced growth of skeletal myotubes correlates with activation of the sodium pump. J. Cell. Physiol., *109*:205, 1981.
42. Vandenburgh, H.H.: Cell shape and growth regulation in skeletal muscle: exogenous versus endogenous factors. J. Cell. Physiol., *116*:363, 1983.

43. Sachs, F.: Biophysics of mechanoreception. Membr. Biochem., *6*:173, 1986.
44. Bishop, J.M.: Viral oncogenes. Cell, *42*:23, 1985.
45. Bishop, J.M.: The molecular genetics of cancer. Science, *235*:305, 1987.
46. Ostman-Smith, I.: Cardiac sympathetic nerves as the final common pathway in the induction of adaptive cardiac hypertrophy. Clin. Sci., *61*:265, 1981.
47. Laks, M.M., and Morady, F.: Norepinephrine—the myocardial hypertrophy hormone? Am. Heart J., *91*:674, 1976.
48. Rossi, M.A., and Carillo, S.V.: Does norepinephrine play a central causative role in the process of cardiac hypertrophy? Am. Heart J., *109*:622, 1985.
49. Tarazi, R.C., Sen, S., Saragoca, M., and Khairallah, P.: The multifactorial role of catecholamines in hypertensive cardiac hypertrophy. Eur. Heart J., *3*(suppl. A):103, 1982.
50. Simpson, P.C.: α_1-Adrenergic stimulated hypertrophy in neonatal rat heart muscle cells. *In* Hypertrophic Cardiomyopathy, Cardiomyopathy Update: Series 2. Edited by H. Toshima and B.J. Maron. Tokyo, University of Tokyo Press, 1989.
51. Simpson, P.C.: Molecular mechanisms in myocardial hypertrophy. Heart Failure, *5*:113, 1989.
52. Sporn, M.B., and Roberts, A.B.: Peptide growth factors are multifunctional. Nature, *332*:217, 1988.
53. Simpson, P., McGrath, A., and Savion, S.: Myocyte hypertrophy in neonatal rat heart cultures and its regulation by serum and by catecholamines. Circ. Res., *51*:787, 1982.
54. Simpson, P., and Savion, S.: Differentiation of rat myocytes in single cell cultures with and without proliferating nonmyocardial cells: cross-striations, ultrastructure, and chronotropic response to isoproterenol. Circ. Res., *50*:101, 1982.
55. Simpson, P.: Stimulation of hypertrophy of cultured neonatal rat heart cells through an α_1-adrenergic receptor and induction of beating through an α_1- and β_1-adrenergic receptor interaction: evidence for independent regulation of growth and beating. Circ. Res., *56*:884, 1985.
56. Simpson, P.C.: Comments on "Load regulation of the properties of adult feline cardiocytes: The role of substrate adhesion" which appeared in Circ. Res., *58*:692–705, 1986 (letter). Circ. Res., *62*:864, 1988.
57. Simpson, P.C.: Measurement of pharmacological effects in isolated myocytes. *In* Biology of Isolated Adult Cardiac Myocytes. Edited by W.A. Clark, R.S. Decker, and T.K. Borg. New York, Elsevier Science, 1988.
58. Hyde, A. et al.: Homo- and heterocellular junctions in cell cultures: an electrophysiological and morphological study. Prog. Brain Res., *31*:283, 1969.
59. Henrich, C.J., and Simpson, P.C.: Neonatal rat heart nonmuscle cells produce factor(s) causing muscle cell hypertrophy (abstract). J. Cell. Biochem. [Suppl.], *12A*:132, 1988.
60. Simpson, P.: Norepinephrine-stimulated hypertrophy of cultured rat myocardial cells is an alpha$_1$-adrenergic response. J. Clin. Invest., *72*:732, 1983.
61. Doorey, A.J., and Barry, W.H.: The effects of inhibition of oxidative phosphorylation and glycolysis on contractility and high-energy phosphate content in cultured chick heart cells. Circ. Res., *53*:192, 1983.
62. Bishopric, N.H., Simpson, P.C., and Ordahl, C.P.: Induction of the skeletal α-actin gene in α_1-adrenoceptor-mediated hypertrophy of rat cardiac myocytes. J. Clin. Invest., *80*:1194, 1987.
63. Bevan, J.A.: Norepinephrine and the presynaptic control of adrenergic transmitter release. Fed. Proc., *39*:187, 1978.
64. Long, C.S., Ordahl, C.P., and Simpson, P.C.: Alpha-1 adrenergic receptor stimulation of sarcomeric actin isogene transcription in hypertrophy of cultured rat heart muscle cells. J. Clin. Invest., *83*:1078, 1989.
65. Meidell, R.S. et al.: α_1-Adrenergic stimulation of rat myocardial cells increases protein synthesis. Am. J. Physiol., *251*:H1076, 1986.
66. Sherline, P., and Mascardo, R.: Catecholamines are mitogenic in 3T3 and bovine aortic endothelial cells. J. Clin. Invest., *74*:483, 1984.
67. Cruise, J.L., Houck, K.A., and Michalopoulous, G.K.: Induction of DNA synthesis in cultured rat hepatocytes through stimulation of alpha-1 adrenoceptors by norepinephrine. Science, *227*:749, 1985.
68. Burch, R.M. et al.: Alpha-1-adrenergic stimulation of arachidonic acid release and metabolism in a rat thyroid cell line: mediation of cell replication by prostaglandin E2. J. Biol. Chem., *261*:11236, 1986.
69. Blaes, N., and Boissel, J.P.: Growth-stimulating effect of catecholamines on rat aortic smooth muscle cells in culture. J. Cell. Physiol., *116*:167, 1983.
70. Simpson, P.C.: Calcium entry blockers inhibit catecholamine-induced beating but not catecholamine-stimulated hypertrophy of cultured rat heart cells (abstract). Clin. Res., *33*:90, 1984.
71. Simpson, P.C., and Karliner, J.S.: Regulation of cardiac myocyte hypertrophy by a tumor-promoting phorbol ester (abstract). Clin. Res., *33*:229, 1985.
72. White, N., Tsao, T., and Simpson, P.: Contractile protein content is increased in hypertrophy stimulated by norepinephrine but not by a phorbol ester: an example of pathological hypertrophy in cultured neonatal rat heart muscle cells (abstract). J. Am. Coll. Cardiol., *7*:122, 1986.
73. Lindemann, J.P.: α-Adrenergic stimulation of sarcolemmal protein phosphorylation and slow responses in intact myocardium. J. Biol. Chem., *261*:4860, 1986.
74. Tohse, N., Hattori, Y., Nakaya, H., and Kanno, M.: Effects of α-adrenoceptor stimulation on electrophysiological properties and mechanics in rat papillary muscles. Gen. Pharmacol., *18*:539, 1987.

75. Seidman, C.E. et al.: Cis-acting sequences that modulate atrial natriuretic factor gene expression. Proc. Natl. Acad. Sci. USA, *85*:4104, 1988.

76. La Pointe, M.C., Wu, J.P., Greenberg, B., and Gardner, D.G.: Upstream sequences confer atrial-specific expression on the human atrial natriuretic factor gene. J. Biol. Chem., *263*:9075, 1988.

77. Waspe, L.E., Ordahl, C.P., and Simpson, P.C.: Altered myosin gene expression in phorbol ester-induced hypertrophy of cultured heart cells (abstract). Circulation, *78*(suppl. II):II-562, 1988.

78. Ausubel, F.M. et al.: Current Protocols in Molecular Biology. New York, Greene Publishing Assoc. and Wiley-Interscience, 1987.

79. Alberts, B. et al.: Molecular Biology of the Cell. New York, Garland Publishing, 1983, p. 1146.

80. Hayward, L.J., and Schwartz, R.J.: Sequential expression of chicken actin genes during myogenesis. J. Cell Biol., *102*:1485, 1986.

81. Gustafson, T.A., Markham, B.E., and Morkin, E.: Effects of thyroid hormone on α-actin and myosin heavy chain gene expression in cardiac and skeletal muscles of the rat: measurement of mRNA content using synthetic oligonucleotide probes. Circ. Res., *59*:194, 1986.

82. Bishopric, N.H. et al.: The molecular biology of cardiac myocyte hypertrophy: studies using a cell culture model. *In* Cellular and Molecular Biology of Muscle Development. UCLA Symposium on Molecular and Cellular Biology, New Series, vol. 93. Edited by L.H. Kedes and F.E. Stockdale. New York, Alan R. Liss, 1989.

82a. Waspe, L.E., Ordahl, C.P., and Simpson, P.C.: The cardiac β-myosin heavy chain isogene in induced selectivity in α$_1$-adrenergic receptor-stimulated hypertrophy of cultured rat heart myocytes. J. Clin. Invest., in press, 1990.

83. Swynghedauw, B.: Developmental and functional adaptation of contractile proteins in cardiac and skeletal muscle. Physiol. Rev., *66*:710, 1986.

84. Hoh, J.F.Y., McGrath, P.A., and Hale, P.T.: Electrophoretic analysis of multiple forms of rat cardiac myosin: effects of hypophysectomy and thyroxine replacement. J. Mol. Cell. Cardiol., *10*:1053, 1978.

85. Starksen, N.F. et al.: Cardiac myocyte hypertrophy is associated with c-myc proto-oncogene expression. Proc. Natl. Acad. Sci. USA, *83*:8348, 1986.

86. White, N., Tsao, T., and Simpson, P.: Differential regulation of myosin isoenzymes in alpha-1 and thyroid hormone stimulated hypertrophy in cultured neonatal rat heart muscle cells (abstract). Clin. Res., *34*:16, 1986.

87. Mulvagh, S.L. et al.: A hemodynamic load in vivo induces cardiac expression of the cellular oncogene, c-myc. Biochem. Biophys. Res. Commun., *147*:627, 1987.

88. McKnight, G.S., and Palmiter, R.D.: Transcriptional regulation of the ovalbumin and conalbumin genes by steroid hormones in chick oviduct. J. Biol. Chem., *254*:9050, 1979.

89. Kedinger, C. et al.: Alpha-amanitin: a specific inhibitor of one of two DNA-dependent RNA polymerase activities from calf thymus. Biochem. Biophys. Res. Commun., *38*:165, 1970.

90. Milsted, A., Cox, R.P., and Nilson, J.H.: Cyclic AMP regulates transcription of the genes encoding human chorionic gonadotropin with different kinetics. DNA, *6*:213, 1987.

91. Larner, A.C., Chaudhuri, A., and Darnell, J.E., Jr.: Transcriptional induction by interferon: new protein(s) determine the extent and length of the induction. J. Biol. Chem., *261*:453, 1986.

92. Lee, J.R. et al.: α$_1$-Adrenergic stimulation of cardiac gene transcription in neonatal rat myocardial cells: Effects on myosin light chain-2 gene expression. J. Biol. Chem., *263*:7352, 1988.

93. Gustafson, T.A., Markham, B.E., Bahl, J.J., and Morkin, E.: Thyroid hormone regulates expression of a transfected α-myosin heavy chain fusion gene in fetal heart cells. Proc. Natl. Acad. Sci. USA, *84*:3122, 1987.

94. Klein, I., and Hong, C.: Effects of thyroid hormone on cardiac size and myosin content of the heterotopically transplanted rat heart. J. Clin. Invest., *77*:1694, 1986.

95. Mar, J.H., and Ordahl, C.P.: A conserved CATTCCT motif is required for skeletal muscle-specific activity of the cardiac troponin T gene promoter. Proc. Natl. Acad. Sci. USA, *85*:6404, 1988.

96. Konieczny, S.F., and Emerson, C.P., Jr.: Complex regulation of the muscle-specific contractile protein (troponin I) gene. Mol. Cell. Biol., *7*:3065, 1987.

97. Miwa, T., Boxer, L.M., and Kedes, L.: CArG boxes in the human cardiac α-actin gene are core binding sites for positive trans-acting regulatory factors. Proc. Natl. Acad. Sci. USA, *84*:6702, 1987.

98. Yamamoto, K.R.: Steroid receptor regulated transcription of specific genes and gene networks. Annu. Rev. Genet., *19*:209, 1985.

99. Izumo, S., and Mahdavi, V.: The thyroid hormone receptor alpha isoforms generated by alternative splicing differentially activate myosin heavy chain gene transcription. Nature, *334*:539, 1988.

100. Montminy, M.R., and Bilezikjian, L.M.: Binding of a nuclear protein to the cyclic AMP response element of the somatostatin gene. Nature, *328*:175, 1987.

101. Angel, P. et al.: Phorbol ester-inducible genes contain a common cis element recognized by a TPA-modulated trans-acting factor. Cell, *49*:729, 1987.

102. Maniatis, T., Goodbourn, S., and Fischer, J.A.: Regulation of inducible and tissue-specific gene expression. Science, *236*:1237, 1987.

103. Ptashne, M.: How eukaryotic transcriptional activators work. Nature, *335*:683, 1988.

104. Imagawa, M., Chiu, R., and Karin, M.: Transcriptional factor AP-2 mediates induction by two different signal-transduction pathways: protein kinase C and cAMP. Cell, *51:*251, 1987.
105. Jackson, S.P., and Tijan, R.: O-Glycosylation of eukaryotic transcription factors: implications for mechanisms of transcriptional regulation. Cell, *55:*125, 1988.
106. Bohmann, D. et al.: Human proto-oncogene c-jun encodes a DNA binding protein with structural and functional properties of transcription factor AP-1. Science, *238:*1386, 1987.
107. Elsholtz, H.P. et al.: Two different cis-active elements transfer the transcriptional effects of both EGF and phorbol esters. Science, *234:*1552, 1986.
108. Sen, R., and Baltimore, D.: Inducibility of κ immunoglobin enhancer-binding protein NF-κB by a post-translational mechanism. Cell, *47:*921, 1986.
109. Henrich, C.J., and Simpson, P.C.: Differential acute and chronic response of protein kinase C in cultured neonatal rat heart myocytes to α_1-adrenergic and phorbol ester stimulation. J. Mol. Cell. Cardiol., *20:*1081, 1988.
110. Karliner, J.S. et al.: α_1-Adrenoceptor regulation of phosphoinositide turnover in hypertrophied myocardial cells (abstract). Circulation, *27*(suppl. III):III-182, 1985.
111. Brown, J.H., Buxton, I.L., and Brunton, L.L.: α_1-Adrenergic and muscarinic cholinergic stimulation of phosphoinositide hydrolysis in adult rat cardiomyocytes. Circ. Res., *57:*532, 1985.
112. Besterman, J.M., Duronio, V., and Cuatrecasas, P.: Rapid formation of diacylglycerol from phosphatidylcholine: a pathway for generation of a second messenger. Proc. Natl. Acad. Sci. USA, *83:*6785, 1986.
113. Simpson, P.C. et al.: Transcription of early developmental isogenes in cardiac myocyte hypertrophy. J. Mol. Cell. Cardiol., *21,* Suppl. V:79, 1989.
114. Antin, P.B., Mar, J.H., and Ordahl, C.P.: Single cell analysis of transfected gene expression in primary heart cultures containing multiple cell types. Biotechniques, *6:*640, 1988.
115. Markham, B.E., Bahl, J.J., Gustafson, T.A., and Morkin, E.: Interaction of a protein factor within a thyroid hormone-sensitive region of rat α-myosin heavy chain gene. J. Biol. Chem., *26:*12856, 1987.
116. Wingender, E.: Compilation of transcription regulating proteins. Nucleic Acids Res., *16:*1879, 1988.
117. Jones, K.A. et al.: A cellular DNA-binding protein that activates eukaryotic transcription and DNA replication. Cell, *48:*79, 1987.
118. Mitchell, P.J., and Tjian, R.: Transcriptional regulation in mammalian cells by sequence-specific DNA binding proteins. Science, *245:*371, 1989.

Chapter 5

REGULATED EXPRESSION OF THE ATRIAL NATRIURETIC FACTOR GENE

Christine E. Seidman

Over the past several years, our knowledge of the mechanism by which the heart participates in regulation of cardiovascular homeostasis has increased significantly. Research in the fields of biochemistry and molecular biology has shown that this organ is not only a sophisticated pump but also an endocrine gland that produces a small peptide hormone called atrial natriuretic peptide or factor (ANF). The physiologic properties of ANF are that of a diuretic and vasodilatory substance. As such, this hormone can play a significant role in modulating volume and pressure. Further, a myocardial site of synthesis optimizes regulation of hormone production in concordance with cardiac status. Understanding the regulated expression of the ANF gene in euvolumic and pathophysiologic states has been aided by molecular biologic technologies. This chapter reviews the use of such methods in delineating physiologic ANF gene expression and how this is affected by cardiac pathologies.

The gene encoding ANF has been cloned and characterized from several mammalian species,[1-3] including humans.[1,4,5] This was facilitated by isolation of small amounts of purified rat ANF, from which a partial amino acid sequence was determined.[6] Using reverse genetics, the oligonucleotides that correspond to these amino acids were used as probes from which the ANF gene and a complementary DNA corresponding to ANF messenger RNA (mRNA) were isolated. Characterization of the ANF gene[1-5] and complementary DNA (cDNA)[7-9] from several species demonstrated that the structure and nucleic acid sequence from all mammals has been highly conserved. The human ANF gene contains three exons, separated by two intervening sequences. Transcription of the ANF gene yields an RNA species that encodes a precursor hormone, preproANF. Isolation and characterization of the ANF cDNA and gene has greatly facilitated research into the molecular basis of regulated ANF gene expression.

REGULATION OF ANF GENE EXPRESSION

The ability of a cell to transcribe the ANF gene is stringently regulated. In the adult mammal, ANF mRNA is quite abundant in both atria where it accounts for 3% of all atrial mRNA species.[10,11] ANF mRNAs in nonatrial tissues, however, are rare. Low-level gene transcription is seen in ventricular tissues, and very small amounts of ANF mRNAs

Parts of this chapter are based on an article published in *Heart Failure*: Seidman CE: Expression of Atrial Natriuretic Factor in the Normal and Hypertrophied Heart. 1989;5(3):130–134. With permission of Haymarket Doyma Inc.

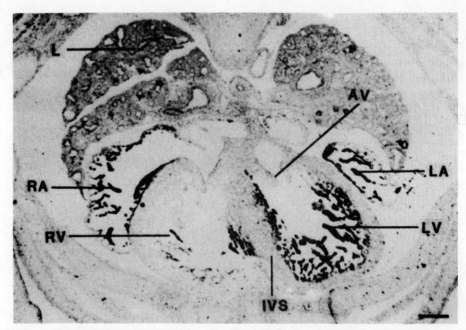

Fig. 5–1. Transillumination photograph of a transverse section through a day-14 mouse embryo hybridized to a [³⁵S]UTP-labeled atrial natriuretic factor (ANF) anti-sense RNA probe, showing localized ANF gene expression in the fetal heart. Comparable levels of ANF messenger RNA transcripts are present in the right and left atria (RA, LA) and left ventricle (LV). Lower levels are seen in the right ventricle (RV) and none is detected in the lung (L), aortic valve (AV), or interventricular septum (IVS). (Reprinted with permission from Zeller, R. et al.: Localized expression of the atrial natriuretic factor gene during cardiac embryogenesis. Genes Dev., *1*:693, 1987.)

have been detected in the lung, aorta,[12] and central nervous system.[13] Hence, in the adult, ANF is expressed in a cardiac- and chamber-specific fashion. Myocardial expression is further regulated during development. In embryonic rodents, ANF mRNA transcripts are present by gestational day 8 in both primitive atrial and ventricular cells.[14,15] Prenatal ventricular expression of the ANF gene continues throughout cardiac embryogenesis, and before birth, the levels of ventricular ANF transcripts are comparable to the atrial level (Fig. 5-1). This pattern changes abruptly with birth, and during the first week of life, ventricular ANF gene expression falls rapidly while atrial expression increases. This developmental pattern has been confirmed in humans.[16] Ventricular tissues from human fetuses contain significant levels of ANF mRNAs, whereas very low levels exist in the normal adult ventricle.

An increase in adult ventricular levels of ANF can be induced by conditions producing cardiac hypertrophy. Animal models in which states of chronic volume and fluid overload are produced demonstrate a dramatic increase in left ventricular ANF gene transcription.[17–19] This phenomenon can occur in both ventricles. In a genetic line of rats with spontaneous biventricular hypertrophy, both the right and left ventricular chambers contain high levels of ANF transcripts, as compared to nonhypertrophied controls.[20] Furthermore, conditions known to produce predominantly right ventricular hypertrophy (chronic hypoxia) increase ANF gene transcription in only the right ventricular chamber.[21] Altered gene transcription in response to cardiac hypertrophy is not limited to the ANF gene. In rodents, hypertrophy has been shown to alter the myosin isoforms expressed in ventricular tissues so that the fetal V_3 isoform predominates[22] (see also Chap. 3 of this book). Similarly, re-expression of a fetal α-skeletal actin occurs in adult rodent ventricles in which hypertrophy has been induced[23] (see also Chap. 4). The re-expression of fetal genes in response to cardiac hypertrophy thus appears to be a more general phenomenon.

Cis-ACTING SEQUENCES

The mechanism by which myocardial hypertrophy occurs and alters gene transcription is poorly understood. Catecholamines are elevated, and cardiac sympathetic activity is enhanced in conditions that cause hypertrophy, suggesting a role for humoral factors in this process (see Chap. 4). Altered mechanics of stretch and relaxation also have been postulated to alter cardiac protein synthesis to yield a hypertrophied phenotype. Expression of proto-oncogenes by the heart has been shown to be an early genetic response to imposed hemodynamic stress, and these may be responsible for modifying expression of fetal isogenes[24] (see also Chap. 4).

Transcription of cardiac genes is therefore not a static process but one that is regulated in response to developmental and physiologic signals that today are poorly understood. The molecular basis for ANF gene expression lies in nucleic acid sequences that do not encode the hormone or its precursor but that modulate gene transcription. Such regulatory regions are present in the nucleic acids that flank all genes and are termed *cis*-acting sequences.[25,26] Our laboratory has cloned and characterized the *cis*-acting sequences, which are 5′ to the ANF gene (Fig. 5-2).[27] Because the nucleic acid sequence of this region revealed minimal homology to previously characterized regulatory sequences, we used a functional assay to analyze these sequences. Putative regulatory sequences were fused with a marker gene, encoding chloramphenicol acetyltransferase (CAT). These were introduced into primary cultures of rat atrial or ventricular cells derived from neonatal animals. If ANF regulatory signals were contained within the sequences introduced, the marker gene CAT should be expressed in a pattern comparable to ANF. Extracts of transfected cells were prepared and assayed for CAT gene expression, which was detected by acetylation of [^{14}C]chloramphenicol. Figure 5-3 shows an autoradiogram of CAT activity detected in neonatal atrial cells, but not ventricular cells, transfected with ANF–CAT hybrid genes. CAT expression is absent in all cells transfected with hybrid genes lacking regulatory sequences (pO-CAT) and is present in both atrial and ventricular cells transfected with constructs bearing the simian virus 40 early promoter and enhancer driving the CAT gene (pSV$_2$-CAT). This approach delineated a region of 2.5 kilobases (kb) flanking the ANF gene that promotes atrial, but not ventricular, expression of a marker gene in neonatal cultures. This same region promotes ventricular expression of the gene when cells are derived from embryonic animals. Collectively, these data suggest that a regulatory region has been defined that directs atria-specific expression of the ANF gene and developmental control of ventricular ANF transcription.

Trans-ACTING FACTORS

Cis-acting sequences alone are insufficient to promote gene transcription.[25,26] Introduction of ANF-regulatory sequences fused to a marker gene, or ANF-coding sequences, into fibroblasts will not direct gene transcription. Extensive research in other systems suggests that humoral and nuclear factors (also called *trans*-acting factors)

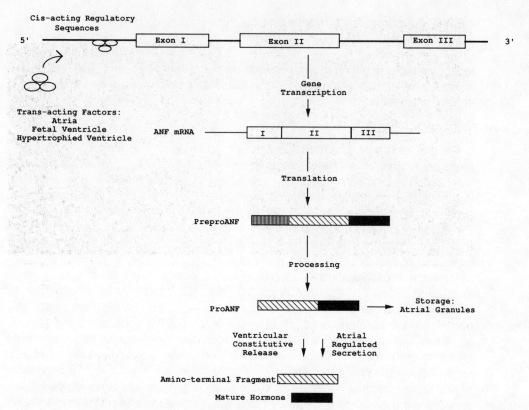

Fig. 5–2. Schematic of atrial natriuretic factor (ANF) gene transcription and translation: Expression of the ANF gene is controlled by regulatory sequences that interact with *trans*-acting factors present in atrial and fetal or hypertrophied ventricular cells. The ANF gene contains 3 exons (boxed), which are transcribed and spliced together to yield a mature mRNA. Translation of ANF mRNA yields a preprohormone that is processed intracellularly to proANF. This is stored within granules in the atria. At the time of hormone release (from either atrial or ventricular cells), maturation of the hormone is completed, and both the amino-terminal fragment and mature ANF are secreted into the circulation. (Reprinted with permission from Seidman, C.E.: Expression of atrial natriuretic factor in the normal and hypertrophied heart. Heart Failure, *5*:130, 1989.)

bind *cis*-acting sequences to permit and modulate the level of gene transcription. Cells that lack appropriate *trans*-acting factors cannot transcribe a given gene. This suggests a mechanism by which ventricular expression of the ANF gene changes throughout development (Fig. 5-2). During cardiac embryogenesis, appropriate nuclear factors are available in both the atria and ventricles to promote ANF gene transcription. At birth, ventricular synthesis of these factors declines, and consequently ANF transcription also falls. Alternatively, at birth a new factor is produced within ventricular cells that suppresses ANF gene transcription. By extrapolation, we can predict that with cardiac hypertrophy, a new complement of nuclear factors appears and activates (or ceases the repression of) ANF gene transcription in ventricular cells. The *trans*-acting factors induced with cardiac hypertrophy may be either identical to or different from those present in embryonic ventricular cells. Further, these factors may activate uniquely only one gene or may activate several genes to yield the multiple genotypic changes that concurrently appear in cardiac hypertrophy.[24]

Circulating humoral substances may similarly interact with *cis*-acting sequences to

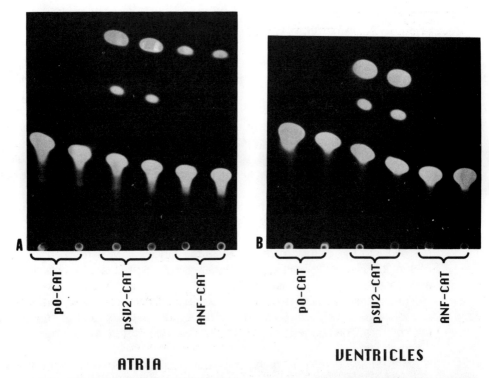

ATRIA

VENTRICLES

Fig. 5–3. Transient expression of transfected chloramphenicol acetyltransferase (CAT) genes by primary cultured cardiocytes derived from atria and ventricles from 1-day-old rats. DNA was transfected by electroporation and CAT gene expression was assayed 60 hours later. Transfected hybrid genes include pO-CAT, which contains CAT-coding sequences without eukaryotic promoter sequences; pSV₂-CAT, which contains the simian virus 40 early promoter and enhancer driving the CAT gene; and ANF–CAT, which contains 3.4 kb of sequences 5′ to the ANF gene driving the CAT gene. (Reprinted from Seidman, C.E. et al.: Cis-acting sequences that modulate atrial natriuretic factor gene expression. Proc. Natl. Acad. Sci. USA, *85*:4104, 1988.)

promote gene transcription. Thyroid hormone increases the levels of the ANF mRNA[28,29] and the circulating ANF. These changes have been documented in both intact animals and in vitro, using primary cultures of rodent atrial and ventricular cells. The latter suggests that thyroid hormone can act directly to affect gene transcription, rather than via secondary mechanisms that are operative in the whole animal.

Glucocorticoids have also been shown to increase ANF levels in cultured atrial and ventricular cells by affecting gene transcription[27,30,31] and also to increase plasma ANF levels in vivo. Dexamethasone can induce ventricular expression of the ANF gene via a mechanism that is additive to that produced by cardiac hypertrophy. These data suggest that circulating hormones may interact through their receptors with unique cis-acting sequences to augment ANF gene expression. Characterization of the hormone-responsive and hypertrophy-responsive regulatory regions of the ANF gene is an area of active research.

POST-TRANSCRIPTIONAL REGULATION OF ANF PRODUCTION

ANF gene transcription in atrial and ventricular cells can be differentially affected by development, hormones, and pathology. Further differences between these chambers lie in the way they store and secrete the peptide hormone, thereby affecting circulating levels of ANF in health and disease. In both atrial and ventricular cells, ANF mRNA is translated into a precursor hormone, preproANF (Fig. 5-1). This contains a hydrophobic leader segment commonly found in secreted peptides, which is rapidly processed intracellularly to yield proANF.[7-9] ProANF contains the mature ANF hormone at its carboxyl end, and an amino-terminal precursor peptide of approximately 110 amino acids.

In atrial myocytes, the precursor proANF is stored in abundant perinuclear electron-dense granules. This provides a ready store for bolus release into the circulation in response to appropriate stimuli. Ventricular cells contain few if any granules. Even neonatal ventricular cells appear to lack the apparatus necessary to store the hormone precursor. This observation led to an analysis of the secretory pathway of atrial and ventricular cells in culture.[31] Tissue culture media supplemented with radiolabeled amino acids permitted newly synthesized ANF to be tagged. A time course analysis of hormone synthesis in atrial cells showed that half of the radiolabeled ANF was released into the culture medium, while the other half was retained intracellularly, presumably in storage granules. These data predict that atrial cells actively synthesize and store the hormone until its release is triggered by a secretagogue. A strikingly different pattern of secretion was seen when ventricular cells were analyzed. Almost all radiolabeled ANF was secreted into the culture medium, and none was retained intracellularly. This suggests that ventricular cells do not store the hormone but release it constitutively.

Atrial storage of ANF in granules permits regulated secretion of the hormone in response to appropriate stimuli. The most potent stimuli for atrial secretion of ANF is stretch, which produces a significant increase in circulating levels of the hormone and a prompt diuresis and natriuresis in the intact animal. Recent studies have documented that ouabain can also cause ANF secretion from cultured atrial cardiocytes.[32] Ventricular myocytes neither store ANF nor release the hormone in a regulated fashion. In vitro, these cells are not responsive to secretagogues such as cardiac glycosides. Regulation of ventricular production of the hormone appears to depend on the level of gene transcription, translation into protein, and precursor maturation and processing. Conditions in which the ANF gene is actively transcribed by ventricular cells may then directly increase circulating levels of hormone. These observations appear to be substantiated by clinical studies. Ventricular ANF transcripts are present in both human fetuses and neonatal infants, and elevated circulating levels of ANF have been documented in the latter.[16,33] In patients with cardiac hypertrophy,[34] congestive heart failure,[35] and cardiomyopathies,[36] the levels of both ventricular ANF mRNA and serum hormone are also increased, suggesting a ventricular contribution to the circulating hormone level. It is unclear if ventricular expression of a hormone with diuretic and vasodilatory properties represents a compensatory process to augment cardiac function or whether ventricular production simply signifies that the myocardium is resuming a fetal phenotype.

Several mechanisms may therefore account for the observed increases in circulating levels of ANF with cardiac pathologies. An increase in intra-atrial pressures that is due to altered hemodynamics will augment atrial secretion of ANF. The frequent use of cardiac glycosides in patients with heart disease may similarly increase atrial secretion of ANF. This may be a novel explanation for the diuresis that frequently accompanies cardiac glycoside administration in patients with heart failure. Finally, the induction of ventricular transcription of the ANF gene coupled with this tissue's constitutive release of the hormone will also augment circulating hormone levels.

Clinical studies in patients with cardiac disease suggest that post-translational processing of this hormone may be altered by cardiac pathology.[37] In healthy subjects, circulating ANF exists as a 28 amino acid peptide, whereas intracellular (atrial or ventricular) hormone is almost entirely proANF. Cleavage of the precursor to the active hormone appears to occur at the time of secretion, although the precise cellular location for this process has not been identified. Recent work has identified a seryl protease that converts proANF into the circulating hormone.[38] This enzyme has been partially purified and appears to be localized in the microsomal fraction of bovine atria. This subcellular localization would permit intracellular storage of the prohormone in granules and would provide two potential mechanisms in the atria for control of circulating hormone levels: processing and secretion.

A ventricular processing enzyme has not yet been identified. In individuals with severe cardiac disease, there are circulating levels of the prohormone in addition to the active 28 amino acid peptide, suggesting that processing of ANF in failing hearts differs from that in normal hearts.[37] This may be due to (1) augmented hormone production with an insufficient reserve in the processing machinery, (2) direct effects on the processing enzyme produced by disease, and/or (3) an absence of this enzyme in ventricular cells in which ANF gene expression has been induced. Further characterization of the proANF-processing enzyme in atrial and ventricular cells should improve our knowledge of how ANF secretion can be altered in the failing heart.

COMMENTS

Research aimed at delineating the regulation of ANF gene transcription and translation in atrial and ventricular cells has changed our understanding of cardiac participation in pressure and volume homeostasis. Characterization of changes that occur in these processes during cardiac pathologies has provided insights into the molecular and biochemical bases for altered cardiac function. Future studies should improve our knowledge of how physiologic events are altered by disease processes and expand our therapeutic options for improving or restoring cardiac function.

ACKNOWLEDGMENTS

This work was supported in part by NIH grants HL35624, HL19259, and HL41474. CES is the recipient of a Clinician Scientist Award from the American Heart Association.

REFERENCES

1. Seidman, C.E. et al.: Nucleotide sequences of the human and mouse atrial natriuretic factor genes. Science, *226*:1206, 1984.
2. Argentin S. et al.: The gene for rat atrial natriuretic factor. J. Biol. Chem., *260*:4568, 1985.
3. Vlasuk, G.P. et al.: Structure and analysis of the bovine atrial natriuretic peptide precursor gene. Biochem. Biophys. Res. Commun., *136*:396, 1986.
4. Greenburg, B.D. et al.: Nucleotide sequence of the gene encoding human atrial natriuretic factor precursor. Nature, *312*:656, 1984.
5. Nemer, M. et al.: Gene structure of human cardiac hormone precursor, pronatriodilatin. Nature, *312*:654, 1984.
6. Currie, M.G. et al.: Purification and sequence analysis of bioactive atrial peptides (atriopeptins). Science, *223*:67, 1984.
7. Seidman, C.E. et al.: The structure of rat preproatrial natriuretic factor as defined by a complementary DNA clone. Science, *226*:324, 1984.
8. Yamanaka, M. et al.: Cloning and sequence analysis of the cDNA for the rat atrial factor precursor. Nature, *309*:719, 1984.

9. Maki, M. et al.: Structure of rat atrial natriuretic factor precursor deduced from cDNA sequence. Nature, *309:*722, 1984.

10. Lewicki, J.A. et al.: Cloning sequence analysis and processing of the rat and human atrial natriuretic peptide precursors. Fed. Proc., *45:*2086, 1986.

11. Seidman, C.E. et al.: Molecular studies of the atrial natriuretic factor gene. Hypertension, *7:*1–31, 1985.

12. Gardner, D. G. et al.: Extra-atrial expression of the gene for atrial natriuretic factor. Proc. Natl. Acad. Sci. USA, *83:*6697, 1986.

13. Saper, C.B. et al.: Atriopeptin-immunoreactive neurons in the brain: presence in cardiovascular regulatory areas. Science, *227:*1047, 1985.

14. Thompson, R.P., Simson, J.A.V., and Currie, M.G.: Atriopeptin distribution in the developing rat heart. Anat. Embryol., *175:*227, 1986.

15. Zeller, R. et al.: Localized expression of the atrial natriuretic factor gene during cardiac embryogenesis. Genes Dev., *1:*693, 1987.

16. Kikuchi, K. et al.: Ontogeny of atrial natriuretic polypeptide in the human heart. Acta Endocrinol. (Copenh), *115:*211, 1987.

17. Nemer, M. et al.: Expression of atrial natriuretic factor gene in heart ventricular tissue. Peptides, *7:*1147, 1986.

18. Kato, J. et al.: Atrial natriuretic polypeptide (ANP) in the development of spontaneously hypertensive rats (SHR) and stroke-prone SHR (SHRSP). Biochem. Biophys. Res. Commun., *143:*316, 1987.

19. Lattion, A.L. et al.: Myocardial recruitment during ANF mRNA increase with volume overload in the rat. Am. J. Physiol., *251:*H890, 1986.

20. Lee, R.T. et al.: Atrial natriuretic factor gene expression in ventricles of rats with spontaneous biventricular hypertrophy. J. Clin. Invest., *1:*431, 1988.

21. Stockmann, P.T. et al.: Reversible induction of right ventricular atriopeptin synthesis in hypertrophy due to hypoxia. Circ. Res., *63:*207, 1988.

22. Lompre, A.M. et al.: Myosin isoenzyme redistribution in chronic heart overload. Nature, *282:*105, 1979.

23. Schwartz, K. et al.: α-Skeletal muscle actin mRNAs accumulate in hypertrophied adult rat hearts. Circ. Res., *59:*551, 1986.

24. Izumo, S., Nadal-Ginard, B., and Mahdavi, V.: Protooncogene induction and reprogramming of cardiac gene expression produced by pressure overload. Proc. Natl. Acad. Sci. USA, *85:*339, 1988.

25. McKnight, S., and Tjian, R.: Transcriptional selectivity of viral genes in mammalian cells. Cell, *46:*795, 1986.

26. Maniatis, T., Goodbourn, S., and Fischer, J.A.: Regulation of inducible and tissue-specific gene expression. Science, *236:*1237, 1987.

27. Seidman, C.E. et al.: Cis-acting sequences that modulate atrial natriuretic factor gene expression. Proc. Natl. Acad. Sci. USA, *85:*4104, 1988.

28. Gardner, D.G., Gertz, B.J., and Hane, S.: Thyroid hormone increases rat atrial natriuretic peptide messenger ribonucleic acid accumulation in vivo and in vitro. Mol. Endocrinol., *1:*260, 1987.

29. Kohno, M. et al.: Atrial natriuretic polypeptide in atria and plasma in experimental hyperthyroidism and hypothyroidism. Biochem. Biophys. Res. Commun., *134:*178, 1986.

30. Day, M.L. et al.: Ventricular atriopeptin: unmasking of mRNA and peptide synthesis by hypertrophy of dexamethasone. Hypertension, *9:*485, 1987.

31. Bloch, K.D. et al.: Neonatal atria and ventricles secrete atrail natriuretic factor via tissue-specific pathways. Cell, *47:*695, 1986.

32. Bloch, K.D. et al.: Oubain induces secretion of proatrial natriuretic factor by rat atrial cardiocytes. Am. J. Physiol., *255:*E383, 1988.

33. Weil, J. et al.: Comparison of plasma atrial natriuretic peptide levels in healthy children from birth to adolescence and in children with cardiac diseases. Pediatr. Res., *20:*1328, 1986.

34. Hara, H. et al.: Plasma atrial natriuretic peptide level as an index for the severity of congestive heart failure. Clin. Cardiol., *10:*437, 1987.

35. Sugawara, A. et al.: Synthesis of atrial natriuretic polypeptide in human failing hearts. Evidence for altered processing of atrial natriuretic polypeptide precursor and augmented synthesis of β-human ANP. J. Clin. Invest., *81:*1962, 1988.

36. Haass, M. et al.: Role of right and left atrial dimensions for release of atrial natriuretic peptide in left-sided valvular heart disease and idiopathic dilated cardiomyopathy. Am. J. Cardiol., *62:*764, 1988.

37. Marumo, F. et al.: Changes of molecular forms of atrial natriuretic peptide after treatment for congestive heart failure. Klin. Wochenschr., *66:*675, 1988.

38. Imada, T., Takayanagi, R., and Inagami, T.: Atrioactivase, a specific peptidase in bovine atria for the processing of proatrial natriuretic factor. J. Biol. Chem., *263:*9515, 1988.

Chapter 6

MOLECULAR BIOLOGY OF THE VASCULAR SMOOTH MUSCLE CONTRACTILE APPARATUS

Mark B. Taubman

The smooth muscle cell (SMC) is the major contractile element of the blood vessel wall. Abnormalities of vascular SMC growth (hypertrophy and proliferation) and contractility have been implicated in such pathologic states as atherosclerosis and hypertension.[1,2] The availability of cultured vascular smooth muscle cells[3] has greatly enhanced the studies of SMC growth and the SMC contractile system. These cultured vascular SMCs possess many of the characteristics of SMCs in vivo.[1,3] In particular, cultured SMCs can display both "synthetic" and "contractile" phenotypes. The change from a contractile to a synthetic phenotype is analogous to that seen in the intact blood vessel at the site of injury or atheromatous plaque.[1,2] The molecular bases for the changes in SMC phenotype, or "phenotypic modulation," seen both in vivo and in cell culture are largely unknown. Work has focused both on the regulation of SMC growth (in particular, the response of SMC to growth factors) and the regulation of contractile protein expression. This chapter focuses on the recent advances in the molecular biology of the SMC contractile apparatus. It begins with brief reviews of the SMC contractile apparatus and of the phenotypic changes that SMCs undergo both in vivo and in tissue culture. This is followed by an overview of the role of recombinant DNA technology in studying both the structure and regulation of the genes and proteins involved in smooth muscle contraction. The remainder of the chapter focuses in detail on the molecular biology of each specific protein component of the smooth muscle contractile system.

THE CONTRACTILE APPARATUS OF SMOOTH MUSCLE

The contractile apparatus of smooth muscle includes two major proteins, actin and myosin, which form thin and thick filaments, respectively. Contraction is associated with the cyclic attachment (crossbridging) and detachment of the globular portion, or "head," of the myosin to the actin filament. This results in the sliding of the filaments past each other. The energy for the actin–myosin interaction is derived from the hydrolysis of adenosine triphosphate (ATP) by an Mg^{2+} ATPase residing on the myosin molecule. Unlike striated muscle, where the actin and myosin filaments form part of a highly organized structure (i.e., the sarcomere), smooth muscle is characterized by thin and thick filaments scattered throughout the cytoplasm (Fig. 6-1) of the SMC.[3] Smooth muscle also differs from sarcomeric muscle in that the regulation of contraction appears to be based on the thick filament rather than the thin filament.[4,5]

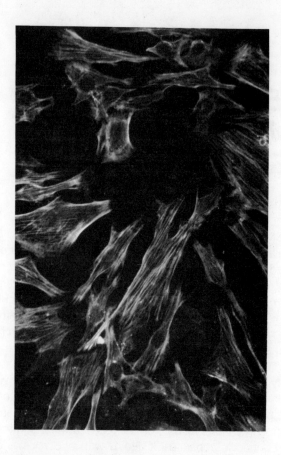

Fig. 6–1. Fluorescent staining of rat pulmonary artery smooth muscle cells with antibody directed against smooth muscle myosin. Myosin-containing filaments are scattered throughout the cytoplasm of the smooth muscle cell. (Courtesy Dr. Abraham Rothman, University of California, San Diego.)

Smooth Muscle Contractile Proteins

Actin, the major component of the thin filament, is a 43,000-dalton globular protein that polymerizes to form double helical filaments under physiologic conditions. Actin plays several roles in smooth muscle contraction: It serves as an important structural element and as an activator of the myosin-associated Mg^{2+} ATPase. Myosin, the major component of the thick filament, consists of a pair of heavy chains (molecular weight $\approx$ 200,000 daltons) and several pairs of light chains (reviewed in ref. 6). Mammalian muscle contains two types of myosin light chains. One type appears to be required for myosin ATPase activity and is referred to as the essential, or alkali, light chain (molecular weight $\approx$ 17,000 daltons). A second type is often referred to as the regulatory light chain (molecular weight $\approx$ 20,000 daltons) because of its role in regulating SMC contraction. Smooth muscle myosin consists of one pair of essential and one pair of regulatory light chains. Skeletal muscle myosin consists of one or two pairs of alkali light chains (myosin light chain 1 MLC_1 and MLC_3) and one pair of regulatory light chains (MLC_2). The myosin molecule has an unusual structure (Fig. 6-2A). The carboxylterminal portion forms a long α-helical rod, while the aminoterminal portion forms a globular head. The myosin light chains are located in the globular portion. As noted previously, the globular portion interacts with the actin filament and contains the Mg^{2+} ATPase activity. In addition to actin and myosin, striated and smooth muscles contain tropomyosin (molecular weight $\approx$ 66,000 daltons). This molecule is composed of two

A

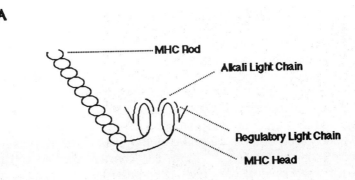

B

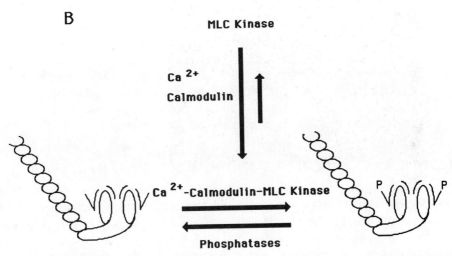

Fig. 6–2.   A. Schematic drawing of smooth muscle myosin structure showing the two myosin heavy chains (MHCs) forming an α-helical rod and globular heads. The myosin alkali and regulatory light chains are situated on the MHC heads. B. Phosphorylation of the myosin regulatory light chains by the myosin light chain (MLC) kinase. The MLC kinase is activated in the presence of Ca^{2+} and calmodulin. Phosphorylation of the MLCs allows for interaction with actin and activation of the Mg^{2+} adenosine triphosphatase, resulting in contraction. Dephosphorylation of the light chains by phosphatases results in relaxation.

α-helical subunits that form a coiled coil dimer. The tropomyosin dimer lies in a groove in the thin filament created by the two strands of the actin double helix. In striated muscle, tropomyosin interacts with actin and a group of proteins known as troponins to regulate the actin–myosin interaction (reviewed in ref. 7). The role of tropomyosin in smooth muscle is unclear.

The Regulation of Smooth Muscle Contraction

In contrast to sarcomeric muscle contraction, which is regulated by the thin filament, smooth muscle contraction is regulated by the thick filament (reviewed in refs. 4 and 5). The central feature of this thick filament regulation is the phosphorylation of a specific serine residue on the myosin regulatory light chain. This phosphorylation appears to be required for the activation of the myosin Mg^{2+} ATPase by actin and thus is essential for providing the energy for contraction. Phosphorylation occurs via a specific MLC kinase.[8] This kinase, and thus smooth muscle contraction, is regulated by calcium. Calcium binds to the Ca^{2+}-binding protein, calmodulin. The resulting Ca^{2+}–calmodulin complex binds to the inactive MLC kinase, thereby activating the kinase and resulting in regulatory light chain phosphorylation (Fig. 6-2B). The MLC kinase is also regulated by a cyclic adenosine monophosphate (cAMP)-dependent protein kinase.[9] Phosphorylation of the MLC kinase by this enzyme results in its having a decreased binding constant for the Ca^{2+}-calmodulin complex, thus favoring the inactive form of the light chain kinase. Finally, several phosphatases can dephosphorylate the regulatory light chain[10,11] and thus may play a role in smooth muscle relaxation.

The following summarizes the most prevalent theory concerning SMC contraction.[4,5,10,12,13] Contraction of the SMC is initiated by an increase in Ca^{2+} levels, activating the MLC kinase. This results in phosphorylation of the regulatory light chain and the formation of crossbridges between actin and myosin. These phosphorylated crossbridges cycle rapidly and are associated with "fast" responses, such as those seen during phasic contraction in response to vasoconstrictor hormones. Subsequently, a reduction in Ca^{2+} levels leads to inactivation of the MLC kinase and to dephosphorylation of the regulatory light chain by myosin phosphatases. This results in crossbridges that have a markedly reduced rate of cycling. This reduced cycling has been referred to as the "latch" state and appears to be characteristic of tonic contractions such as those necessary to maintain tone in vascular smooth muscle.[10,12] Further reductions in Ca^{2+} levels result in relaxation.[13]

PHENOTYPIC MODULATION OF SMOOTH MUSCLE CELLS IN CULTURE

The SMC can express a variety of phenotypes (reviewed in ref. 3). Cells whose function is predominantly that of contraction, such as the small muscular arteries or teniae coli, possess the "contractile phenotype." This is characterized by a cytoplasm comprised predominantly of thin and thick filaments, the hallmarks of the smooth muscle contractile apparatus, and relatively small amounts of synthetic organelles. In contrast, SMCs whose function is predominantly that of synthesis (the "synthetic phenotype") contain few filaments and large amounts of rough endoplasmic reticulum and other synthetic organelles. Such cells are seen in a variety of pathologic states, such as those involving vessel wall injury or atherosclerosis, and during development.

SMCs in culture also demonstrate a range of phenotypes that mimic those seen in vivo.[3,14,15] During the first few days in primary culture, SMCs from a variety of sources express the contractile phenotype and resemble SMC in vivo. These cells contract spontaneously and/or in response to vasoactive agents.[3,16–19] Morphologically, they contain large numbers of myofilaments and few synthetic organelles. Few SMCs synthesize significant amounts of DNA or divide.[20] After 5 to 7 days in culture, SMCs lose their ability to contract either spontaneously or in response to vasoconstrictors.[3] Concomitant with this is the loss of cytoplasmic filaments, a decrease or loss in staining with antibodies to smooth muscle myosin,[16,21,22] an increase in the number of synthetic organelles, and an increase in DNA synthesis.[3,14] A major feature of the change from the contractile to the synthetic phenotype is the loss of the cytoplasmic filaments that make up the smooth muscle contractile apparatus.[16,21,22] Thus, the regulation of SMC growth appears to be intimately connected with the regulation of the contractile apparatus and most likely of the individual contractile proteins.

RECOMBINANT DNA TECHNOLOGY IN THE STUDY OF SMC CONTRACTION

The advances in recombinant DNA technology have provided powerful tools for studying the molecular biology of the contractile proteins. Much of this work has focused on the sarcomeric (skeletal and cardiac) muscle systems. More recently, the smooth muscle contractile proteins have also received considerable attention. As will be detailed, the contractile proteins exist as groups of "isoforms," which are expressed in tissue-specific and developmentally specific fashion. In some cases (e.g., actins), these isoforms have remarkably similar structures; while in others (e.g., myosin heavy chains), the sarcomeric, smooth, and nonmuscle isoforms appear to have larger differences. Several important issues need to be addressed regarding smooth muscle:

1. Do smooth muscle-specific isoforms exist for each type of contractile protein? If so, how many?
2. Are the same contractile protein isoforms always expressed in smooth muscle, or are different isoforms expressed at different times during the life of the SMC? Are smooth muscle isoforms developmentally regulated? Are different isoforms expressed when SMCs modulate from the "contractile" to the "synthetic" phenotype?
3. By what mechanism(s) are the smooth muscle contractile protein isoforms generated? Do separate genes encode each isoform, or are multiple isoforms derived from the same gene?
4. How does the SMC know which isoform to synthesize? Which regions of the contractile protein genes are important in determining tissue specificity and developmental specificity? What are the factors produced by SMCs that regulate the expression of the smooth muscle contractile protein genes?
5. What are the structural differences between the smooth and sarcomeric contractile protein isoforms? What are the functional and biochemical consequences of these structural differences vis-à-vis the smooth muscle contractile apparatus? Could sarcomeric or nonmuscle contractile protein isoforms substitute functionally for smooth muscle isoforms?
6. Does the change from "synthetic" to "contractile" phenotype involve the coordinate regulation of the contractile protein genes? If so, does a specific factor(s) trigger this process, and do common regions in the contractile protein genes recognize this factor(s)?

During the last 5 years, messenger RNAs (mRNAs) encoding many of the contractile protein isoforms in a number of mammalian and nonmammalian systems have been identified, and clones containing complementary (cDNAs) corresponding to these mRNAs have been isolated and characterized. In a number of instances, the genes encoding the contractile protein mRNAs have also been isolated and characterized. This has provided considerable information about the nucleotide sequences of the mRNAs encoding the contractile proteins and has allowed investigators to conclude that each type of contractile protein is encoded by a multigene family whose members are closely related and likely evolved from a common ancestral gene. By comparing the nucleotide sequences of mRNAs encoding different isoforms and by analysis of the corresponding gene(s), the mechanisms for generating isoform diversity for a number of the contractile proteins have been established. What has emerged is a fascinating story in which some isoforms (e.g., actins, MHCs) are generated by transcription of distinct isoform-specific genes, while others (e.g., α-tropomyosins, myosin alkali light chains) are generated by producing multiple or alternatively spliced transcripts from a single gene (see following discussion).

In addition to providing information about isoform diversity, the cDNA clones have been used to deduce the amino acid sequences of the corresponding contractile proteins. This has greatly enhanced our knowledge of the structure of the contractile proteins and the possible biologic and functional significance of the different isoforms. The cDNA clones have been used as probes to study the expression of the corresponding mRNAs in cell culture using the techniques of RNA blot hybridization and S_1 nuclease mapping. This has allowed for detailed analyses of contractile protein isoform expression during development, in pathologic states, and in cell culture during phenotypic modulation and in response to hormones and other agents. Most recently, work has begun on dissecting the regulatory regions of the contractile protein genes. These regions, particularly those located near the promoters, contain the sequences ("*cis*-acting" sequences) that determine tissue and developmental specificity.

ACTIN

Actins are a major component of the cytoskeleton of all eukaryotic cells. In addition, they are an important part of the contractile apparatus of all types of muscle.[23] Mammalian cells contain at least six actin isoforms.[24–26] These isoforms have been grouped into three classes, designated as α-, β-, and γ-actins, on the basis of their migration on isoelectric focusing gels.[24,27] Muscle tissues and cells contain α-actins, designated as α-skeletal, α-cardiac, and α-smooth muscle, based on the cell type in which they predominate.[24,25,28] In the case of smooth muscle, the α-smooth muscle isoform is seen principally in vascular tissues, while a γ-smooth muscle actin predominates in tissues of the genital and gastrointestinal tracts.[28–31] Two nonmuscle β- and γ-actins ("cytoplasmic" actins) predominate in nonmuscle tissues.[25,28]

The Actin Multigene Family

Protein sequence analysis has demonstrated that all of the actin isoforms have similar primary structures and share >90% amino acid homology throughout the entire length of the molecule, with differences occurring predominantly at the amino terminus. In particular, the four muscle actins (α-cardiac, α-skeletal, α-smooth, and γ-smooth) have remarkable sequence conservation, even between different species.[28] For example, bovine aortic α-actin and chicken gizzard γ-actin differ by only three amino acid substitutions, all located in the amino terminus. These two smooth muscle actins differ from bovine α-skeletal actin by no more than eight residues. In contrast, the two cytoplasmic actins differ from the muscle actins by between 22 and 25 residues.

A number of actin genes have been isolated and structurally characterized. These include genes encoding striated muscle α-actins,[32–37] cytoplasmic actins,[38–41] and α-smooth muscle actins.[42–44] These studies have confirmed that the actins are encoded by a multigene family whose members show a high degree of sequence homology, and it is likely that they arose by duplication and subsequent divergence from a common ancestral gene. Each actin mRNA isoform has a unique 3'-untranslated region, and therefore isoform-specific cDNA probes can be constructed. For each specific isoform, however, domains within the 3'-untranslated region show striking evolutionary conservation.[34,35,37,45,48] The exception to this is the α-smooth muscle actin gene, which does not show any sequence conservation in the 3' untranslated region between human and chicken.[43] The significance of these conserved regions or the lack of these conserved regions in α-smooth muscle actin is unknown.

Actin Gene Expression in Sarcomeric Tissues

The expression of striated muscle actin isoforms is regulated in both a tissue-specific and developmentally specific fashion. Proliferating skeletal myoblasts contain predominantly β- and γ-cytoplasmic actins and small amounts of α-actin. On cessation of growth and cell fusion, a marked decline in the amount of β- and γ-actin and a rapid switch to the α-skeletal and α-cardiac muscle isoforms occur.[49–52] A variety of experiments[53–60] have demonstrated that elements in the 5' flanking region of the actin gene (*cis*-acting elements) as well as soluble factors ("*trans*-acting factors") regulate the developmental expression of the striated muscle actin genes.

Actin Gene Expression in Smooth Muscle

The expression of actin isoforms in smooth muscle appears to be regulated in a similar fashion as in striated muscle. Saborio et al.[61] have found that gizzards from 8-day-old chick embryos contain almost exclusively β-actin, whereas after 8 days, a progressive increase in the amount of γ-smooth muscle actin and a decrease in β-actin occur. Similarly, aortas from newborn rats contain predominantly β-actin, whereas those from adult rats contain mostly α-smooth muscle actin.[62,63] These changes appear to be regulated at the transcriptional level.[61,64] Cultured SMCs undergo similar changes in actin isoform expression.[64–69] While proliferating, cultured rat aortic SMCs express predominantly β-actin. On growth arrest, however, either by reaching confluency or by incubation in serum-free medium, these cells switch to the α-smooth muscle isoform. The marked increase in the amount of α-actin protein in density-arrested postconfluent cultures is not accompanied by a significant increase in the level of α-actin mRNA. Thus, the changes in actin isoform expression in cell culture appear to be regulated at the level of translation, rather than transcription.[64,69] The expression of α-smooth muscle actin by confluent SMCs can be repressed by the addition of serum or growth factors, such as platelet-derived growth factor.[68,69]

Changes in actin isoform expression also appear to be features of several pathologic states. Gabbiani and co-workers[64,65,70] have demonstrated that aortic SMCs from human atheromatous plaques or from experimentally induced rat aortic intimal thickening contain almost exclusively β- and γ-cytoplasmic actins. This is in contrast to cells from normal aortas, which contain mainly α-smooth muscle actin and small amounts of β-actin. Thus, the change from a "contractile" to a "synthetic" SMC, such as is seen at the site of vessel wall injury or atherosclerotic plaque,[1] is associated with a switch in actin isoform expression, with those isoforms normally seen in proliferating cells predominating. The switch from α-smooth muscle to β- and γ-cytoplasmic actins is regulated largely at the transcriptional level.[64] A recent report states that actin gene expression in smooth muscle is also affected by steroid hormones.[71] After treatment with estradiol, the levels of β- and γ-cytoplasmic as well as α-smooth muscle actin mRNAs were increased in rat uteri. The induction kinetics of the cytoplasmic and smooth muscle isoforms were significantly different, with the α-smooth muscle actin peaking considerably later (8 to 12 hours vs. 4 hours for the cytoplasmic isoforms).

Regulation of Smooth Muscle Actin Gene Expression

The isolation of the complete chicken smooth muscle α-actin gene[43] has provided the material for studying the mechanisms by which tissue specificity is regulated. While predominantly expressed in smooth muscle, small amounts of smooth muscle α-actin mRNA have also been detected in fibroblasts and skeletal myoblasts.[72] Carroll and co-workers[72] have made a series of deletion constructs containing parts of the promoter region (5′ flanking sequences) of the smooth muscle α-actin gene inserted into the chloramphenicol acetyltransferase (CAT) expression vector, pSV0CAT. By transfecting this vector into fibroblasts and skeletal myoblasts and assaying for CAT activity, these investigators determined which regions near the promoter were important in conveying activity and tissue specificity to the α-actin promoter. These studies determined that a "core" promoter is able to direct high levels of transcription in both fibroblasts and skeletal myoblasts. This core promoter is regulated in fibroblasts and myoblasts by "governor" element(s) also located in the 5′ flanking sequences. One such evolutionarily conserved 29-nucleotide element (negative regulatory element) was identified that was able to suppress the expression of the core promoter in skeletal myoblasts. Presumably, *trans*-acting soluble factors are present in these myoblasts; these factors bind to the negative element, thereby suppressing the expression of the smooth muscle α-actin gene. Isolation of such *trans*-acting factors would help further elucidate how the smooth muscle α-actin gene is regulated.

Biologic Significance of Actin Isoform Switching

The functional and biologic significance of changes in actin isoform expression duriing development and cell differentiation is unknown. Presumably, however, the actin isoforms have specific properties that make them suited for specific physiologic functions. In the case of skeletal muscle cells, the shift from nonmuscle to α-skeletal muscle actin is associated with the development of the mature contractile unit (i.e., sarcomere) and thus with a major change in the structure and function of the actin.[27,50] Owens and Thompson[67] have noted that increases in α-smooth muscle actin in aortic SMCs correlate with increases in force-generating capacity. These authors have also demonstrated differential turnover of actin isoforms in SMCs and have suggested that this implies differences in structure and/or intracellular processing of these actins.[67] With the availability of a variety of actin genes and with rapidly improving techniques for introducing these genes into cells as well as inactivating native genes,[53–60] it should be possible to study the functional roles of the various actin isoforms in much greater detail.

α-TROPOMYOSIN

Tropomyosin (TM) is found in both nonmuscle and muscle cells and exists in three different forms (α, β, and nonmuscle-cytoplasmic).[73,74] Skeletal muscle contains both α- and β-TM, the proportions varying with the fiber type. In adult cardiac muscle, α-TM predominates. In striated muscle, α-TM forms a coiled coil dimer that lies in the major groove of the thin filament.[73] α-TM plays a fundamental role in Ca^{2+}-activated contraction in striated muscle through its interaction with actin and the Ca^{2+}-regulatory troponin proteins.[73,75] The smooth muscle α- and β-TMs differ in their physical and chemical properties from the striated forms.[76,77] The role of TMs in Ca^{2+}-activated smooth muscle contraction is not as well understood. The presence of α-TM does not appear to be necessary for Ca^{2+}-mediated phosphorylation of the regulatory light chain or the resultant activation of the Mg^{2+} ATPase. The function of TMs in smooth muscle and nonmuscle cells is unknown. It has been postulated, however, that TM might play a role in determining cell architecture and modulating contraction[78] or motility.[79]

TMs are members of a family of related isoforms that have a tissue-specific and developmentally specific pattern of expression. Protein and cDNA sequencing of smooth muscle TMs[80–84] have demonstrated that the smooth muscle isoforms are distinct from those of striated muscle. TM isoform diversity is generated by two different mechanisms. Recent data suggest that in mammals there are several distinct TM genes, which include those encoding the fast striated muscle α- and β-TMs.[84–88] Different isoforms, however, including those encoding the smooth and nonmuscle TMs, are generated from these genes by a tissue-specific alternative splicing mechanism.[81,83,85,89–92]

Alternative Splicing and α-Tropomyosin Diversity

Alternative splicing has been shown to be an important method of generating isoform diversity for a variety of proteins, including several components of the contractile apparatus, troponin-T and α-TM (reviewed in ref. 93). As outlined in Chapter 1, the general scheme for producing mammalian proteins is first to transcribe a large heterogeneous nuclear RNA molecule from the gene. This RNA is essentially a copy of one of the two strands of the gene's DNA and contains both exons and introns. The introns are then removed, and the exons are joined together to form the mRNA, a process known as "splicing." This mRNA undergoes further processing, including the addition of polyadenosine to its 3' end, and is then secreted into the cytoplasm, where it ultimately attaches to polyribosomes and is used as a template to synthesize the corresponding peptide. In most cases, splicing involves the removal of all of the introns and joining of all of the exons in one configuration, producing a single mRNA species. In some cases, however, a variety of different mRNAs can be generated by the removal of different introns and the joining of different exons. In the case of the rat α-TM gene, at least six different mRNA isoforms (two striated, one smooth, and three nonmuscle) are generated by this process of alternative splicing.[92] These mRNA isoforms encode proteins that have slightly different primary structures and may have different structural or functional properties. For example (Fig. 6-3), the smooth muscle isoform of rat α-TM contains exons 1, 2, 4 to 6, 8 to 10, and 13, the nonmuscle isoform exons 1, 3, 4 to 6, 8 to 10, and 13, and one of the striated isoforms, exons 1, 3, 4 to 6, and 8 to 12. In most cases, exon 3 is preferentially selected over exon 2. In no instance, however, are both exons 2 and 3 spliced together. This splicing behavior is referred to as *mutually exclusive.*

The distinct rat α-TM isoforms are expressed specifically in nonmuscle, smooth, and striated muscle cells, and thus their splicing appears to be directed by tissue-specific *trans*-acting factors.[92] In addition, different isoforms are expressed in the same tissue during different stages of development. For example, during differentiation of skeletal myoblasts[92] or BC3H1 cells,[94] a switch from the nonmuscle to the striated muscle α-TM isoform occurs. These changes in α-TM isoform expression during myoblast differentiation suggest that, in addition to tissue-specific factors, there are also developmentally specific *trans*-acting factors that direct α-TM splicing. The molecular mechanisms responsible for the alternative splicing of α-TM and the factors that direct the tissue-specific and developmentally specific expression of the different TM isoforms are

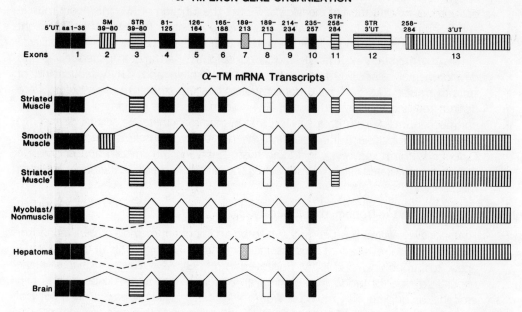

Fig. 6–3. Diagram of the alternative splicing of the rat α-tropomyosin (TM) gene. Boxes represent exons as follows: common exons (black), striated muscle (STR) or striated muscle–nonmuscle exons (horizontal stripes), smooth muscle (SM)–nonmuscle exons (vertical stripes), alternatively spliced exons for amino acids 189 to 213 (white and stippled). UT denotes the untranslated regions. The numbers above the exons indicate the codons in each exon. Solid lines between boxes represent confirmed splicing patterns, whereas broken lines indicate probable, but not directly proven, splice patterns. (Reprinted with permission from Wieczorek, D.F., Smith, C.W.J., and Nadal-Ginard, B.: The rat α-tropomyosin-gene generates a minimum of six different mRNAs coding for striated, smooth, and nonmuscle isoforms by alternative splicing. Mol. Cell. Biol., *8:*679, 1988.)

the focus of current research. As an example, recent work by Smith and Nadal-Ginard[95] suggests a mechanism for the mutually exclusive splicing of exons 2 and 3. These investigators have shown an absolute incompatibility between the exon 2 donor and the exon 3 acceptor. The basis for this is that the intron separating exons 2 and 3 has an unusual structure in which the "branch point" necessary for splicing the two exons is too close to the exon 2 donor. When the distance between the putative branch point and exon 2 is increased, the two exons can be spliced together.

Biologic Significance of α-Tropomyosin Isoforms

The functional significance and physiologic roles of the alternatively spliced TM isoforms remain to be determined. It is intriguing to speculate that the different isoforms may be responsible for the different properties seen in smooth muscle, striated muscle, and nonmuscle cells vis-à-vis filament assembly and binding to actin and Ca^{2+}-regulatory proteins. All six rat α-TM isoforms encode proteins that contain the same number (284) of amino acids. These isoforms also contain the sequences involved in head-to-tail polymerization and those involved in the interaction with troponin-T.[92] Thus, the structural differences between these isoforms are not obvious and will require further study. The presence of sequences involved in the interaction of α-TM with troponin-T is particularly striking, given the fact that troponin-T is not found in smooth or nonmuscle cells.[75] The availability of cDNAs encoding a number of the tropomyosin isoforms should provide ample material for elucidating the functional and structural differences among the TM isoforms.

MYOSIN HEAVY CHAINS

As noted previously, smooth muscle myosin, the major component of the thick filament, consists of two identical heavy chains (molecular weight ≈200,000 daltons) and two sets of light chains, the 20,000-dalton regulatory light chain and the 17,000-dalton alkali light chain (review ref. 6).

Striated muscle MHCs belong to very large multigene families, whose members (as many as 31 in chicken, ref. 96) are expressed in a tissue-specific and developmentally specific fashion. Several excellent reviews[97,98] have detailed the current information concerning these large sarcomeric MHC gene families. In contrast, mammalian smooth muscle and nonmuscle MHCs appear to have only one or two isoforms.

Smooth muscle MHCs are immunologically and electrophoretically distinct from MHCs found in striated muscle and nonmuscle cells.[19,21,99-101] In addition, striated muscle MHC cDNA probes do not hybridize to RNA from SMCs, suggesting striking differences on the nucleotide level as well. Despite these differences, the basic structures of smooth muscle and striated muscle MHCs appear to be quite similar. Both comprise an amino-terminal globular head, which binds two (smooth muscle) or three (striated muscle) sets of light chains, and a long α-helical tail extending to the carboxy terminal. Both contain an Mg^{2+} ATPase located in the head region, which is activated by interaction with actin.

Expression of MHC Isoforms in Smooth Muscle

The expression of MHC isoforms in cultured SMCs has been a topic of considerable interest. Chamley-Campbell and others [3,20] have demonstrated, by both electron microscopy and staining with smooth muscle myosin-specific antibodies, that as part of the phenomenon of "phenotypic modulation," SMCs lose their myosin-containing thick filaments concomitant with entering the proliferative state. Despite this, SMCs continue to produce myosin at fairly constant levels[21,22,102] as determined by sodium dodecyl sulfate polyacrylamide gel electrophoresis. Larson et al.[21] noted that the immunologic properties of the MHCs expressed at subconfluence and postconfluence differed. While intact uterus stained only with antibody directed against human uterine myosin (anti-smooth muscle myosin, [ASMM]), subconfluent rat mesenteric artery cells stained with antibodies directed against platelet myosin (APM) in addition to ASMM. At postconfluency, staining with ASMM diminished, whereas staining with APM remained constant. These authors suggested that the differences in staining might be due to either conformational changes in the myosins or the presence of different isoforms.

The number of MHC isoforms expressed in smooth muscle cells and tissues has not been firmly established. Several laboratories have reported that smooth muscle tissues contain two MHCs of different molecular weights.[101-104] In addition, cultured SMCs recently have been shown to contain different MHC isoforms and to vary their expression with the growth state of the cells.[103,105] Subconfluent, proliferating cultures contain a form of MHC (nonmuscle MHC) that is not found in rat aorta cells in vivo and is identical antigenically, electrophoretically, and by peptide mapping with the MHC seen in nonmuscle cells. Small amounts of two MHC forms found in intact smooth muscle tissue (smooth muscle MHCs) are also seen in these subconfluent cells. In contrast, postconfluent, growth-arrested cells express increased levels of the smooth muscle

MHCs and continue to express large amounts of the nonmuscle MHC. Furthermore, differences in the expression of the nonmuscle and smooth muscle MHCs are paralleled by changes in the rates of synthesis of these proteins.[105] These changes are therefore similar to changes in the expression of the actin isoforms, as described previously.[61-67] Although both a "nonmuscle" and at least one "smooth muscle" MHC isoform clearly are expressed in cultured SMCs, it has not been firmly established that the two "smooth muscle" MHCs found in intact aorta or in cell culture are true isoforms rather than one being the proteolytic breakdown product of the other.[102,104]

Characterization of Smooth Muscle MHC Genes

Several laboratories have recently isolated cDNA clones encoding smooth muscle MHCs.[106,107] Yanagisawa et al.[106] have reported the complete sequence of the embryonic chicken gizzard smooth muscle MHC. Genomic Southern blot analysis suggests that the gizzard MHC is encoded by a single-copy gene with no closely related members, rather than a multigene family similar to that encoding the sarcomeric MHCs. Their work has not ruled out the existence of smooth muscle MHC gene(s) with homologies too low to be detected by Southern blot analysis. Northern blot analysis demonstrates the presence of mRNA hybridizing to the gizzard MHC probe in both gizzard and aorta but not in striated muscle. The sequence homology between the chicken gizzard smooth muscle MHC and the MHCs from chicken sarcomeric muscle are considerably lower than the homologies among sarcomeric MHCs from different species, including nematodes. This suggests[106] that the smooth muscle MHC gene diverged from the ancestral sarcomeric MHC gene before the emergence of vertebrates and before the gene duplications that generated the sarcomeric multigene family. Despite these differences, the smooth muscle MHC maintains some of the properties of its sarcomeric relatives. For example, a seven-residue hydrophobic periodicity as well as 28- and 196-residue repeats previously described in nematode myosin is found in the gizzard myosin rod. The regions around the ATP-binding site and reactive thiol groups in the head are highly conserved as well, presumably reflecting important functional constraints on MHC evolution.[106]

Nagai et al.[107] have reported isolation of a cDNA clone (SMHC-29) encoding a 1571-nucleotide portion of the rabbit uterus MHC, including the 3' untranslated region. This cDNA has <50% homology with the MHCs from rat skeletal muscle or from nematode. Comparisons of the amino acid sequences deduced from SMHC-29 with those of rat and rabbit skeletal muscle MHCs, nematode unc-54 MHCs and slime mold nonmuscle MHC show between 25 and 31% homology.[107] In sharp contrast, the homology between the amino acid sequence of SMHC-29 and the chicken gizzard MHC is >95%, suggesting that the smooth muscle MHCs are highly conserved between species. S_1 nuclease mapping using probes derived from SMH-29 demonstrated the presence of homologous, or "fully protected," mRNA only in smooth muscle tissues. In addition, a "partially protected" species was noted in smooth muscle tissues, suggesting the presence of a second smooth muscle MHC isoform. Genomic Southern blot analysis was inconclusive as to whether the rat uterine MHC was derived from a single-copy gene. It remains to be determined whether the second mRNA species noted by Nagai et al.[107] encodes the second smooth muscle MHC previously detected by others[102-105] and, if so, whether this MHC is derived from a second smooth muscle MHC gene or from alternative splicing of a single gene.

As noted previously SMCs in culture express a nonmuscle-type MHC as well as one or two smooth muscle MHCs. Several laboratories[108,109] have recently reported the isolation of clones encoding chicken and rat nonmuscle MHCs. The amino acid sequences derived from the rod portion of these MHCs are highly homologous with each other and with the corresponding regions of the rat and chicken smooth muscle MHCs. Both nonmuscle MHCs appear to be encoded by single-copy genes. The smooth muscle and nonmuscle MHC genes thus appear to be far more closely related than they are to the sarcomeric MHC genes. As would be expected of a nonmuscle MHC, S_1 nuclease mapping has demonstrated homologous mRNA in nonmuscle, striated muscle, and smooth muscle cells.

MYOSIN REGULATORY LIGHT CHAINS

As noted previously, the 20,000-dalton regulatory light chain (RLC) plays a central role in the regulation of smooth muscle contraction through phosphorylation of a specific serine residue (serine 19) by the enzyme MLC kinase. Both skeletal and cardiac muscle contain phosphorylatable RLCs, which possess similar amino acid sequences flanking the phosphorylated serine.[110] Phosphorylation of the serine, however, does not result in the dramatic changes in ATPase activity and does not appear to be necessary for skeletal or cardiac muscle contraction (see ref. 4). Thus, the RLC appears to play a unique role in smooth muscle.

RLC Genes

RLC genes have been isolated from skeletal,[111] cardiac,[112,113] and smooth muscle.[114–116] Unlike the actins, these RLCs differ significantly in both their nucleotide and amino acid sequences. Thus, the amino acid sequence of the rat smooth muscle RLCs has 70 and 52% homology with those of the skeletal and cardiac muscle RLCs, respectively. The cardiac and skeletal muscle RLCs share only 67% of their amino acids. The differences in sequence are most striking at both the amino and carboxy termini of the RLCs. It has been reported[117] that the sequence at the amino terminus of the smooth muscle RLC is critical to the activity of the MLC kinase on the RLC. Differences in this region might provide a means by which the light chains can be distinguished by their respective kinases. Although the exact role of the carboxy terminus in RLC function is not known, its structure may also be important in providing the RLCs with their differentiated functions. Despite the differences between the sarcomeric and smooth muscle RLCs, in several areas the sequences are highly conserved (>75% homology). This argues that, like the actins, the RLCs all originated by duplication of the same ancestral gene. Apart from variations in the amino acid sequences of RLCs derived from different types of muscles, significant species differences also exist between RLCs of the same muscle type. This is least true of the smooth muscle RLCs, where the chicken gizzard RLC has 94 and 98% homology with the rat aortic and human umbilical artery RLCs, respectively.[114,116] This suggests that the specialized function of the smooth muscle RLC vis-à-vis contraction may place greater constraints on the amino acid sequence of the RLC than are placed on the sarcomeric RLCs.

Rat aortic SMCs apparently contain two closely related RLCs (RLC-A and RLC-B). These RLCs are encoded by distinct genes,[118] which are located on the same chromosome and separated by only 12 kilobases (kb) (Fig. 6-4). The coding regions of these two genes are highly homologous, and the derived amino acid sequences differ in only six positions, one of which is located near the amino terminus (Grant and Taubman, unpublished results). In contrast to the conservation of the coding region, the 3' un-

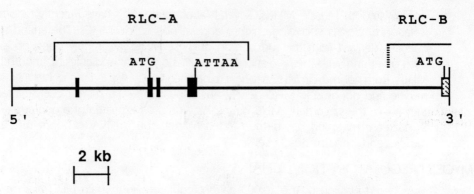

Fig. 6–4. Partial map of the rat aortic regulatory light chain (RLC) genes, representing a composite of several overlapping genomic clones (Grant and Taubman, unpublished results). RLC-A comprises four exons (black boxes) and is expressed in sarcomeric muscle, smooth muscle, and nonmuscle tissues. RLC-B is expressed in smooth muscle and nonmuscle tissues only. The location of a 307-base pair exon (checked box) containing part of the 5′ untranslated region and 5′ coding region is shown. The location of the remaining exons encoding RLC-B remains to be determined.

translated regions are strikingly different. This has allowed for the generation of cDNA probes that are specific for the two RLCs and that enable studies of the expression of these RLC genes. One RLC is expressed in sarcomeric, smooth muscle, and nonmuscle tissues (RLC-A), while the other (RLC-B) is expressed only in smooth muscle and nonmuscle tissues. Extensive screening of rat smooth muscle cDNA and genomic libraries has failed to identify other smooth muscle RLC cDNAs or additional nonmuscle or smooth muscle RLC genes (Grant and Taubman, unpublished observations). Thus, in contrast to the cardiac and skeletal muscle RLCs, there may not be a truly smooth muscle-specific RLC. The biologic significance of two almost identical RLC isoforms in smooth muscle is unknown. However, the fact that an amino acid substitution is in the amino terminus raises the possibility that the isoforms may have different binding and/or activation kinetics vis-à-vis the MLC kinase and thus may have different contractile properties.

MYOSIN ALKALI LIGHT CHAINS

In addition to the 20,000-dalton RLC, smooth muscle myosin also contains a pair of 17,000-dalton alkali light chains (ALC). The ALCs are located in the globular head portion of the myosin molecule. The function of the ALCs in smooth as well as sarcomeric muscle is not well understood. It has been suggested, however, that they play a role in the interaction of the myosin head with actin and the activation of the myosin Mg^{2+}-ATPase.[119] Like the other elements of the contractile system, distinct ALC isoforms are also present in different muscle and nonmuscle cells (for review, see ref. 119). Fast skeletal muscle contains two ALCs (an MLC_{1F} of $\approx23,000$ daltons and an MLC_{3F} of $\approx17,000$ daltons), while slow skeletal muscle contains one or two MLCs (MLC_{1S}) of a size comparable to MLC_{1F}. Cardiac muscle contains one major ventricular isoform (MLC_{1V}) and, in mammals, one atrial isoform (MLC_{1A}); both MLC_{1V} and MLC_{1A} are of a size comparable to MLC_{1F}. Smooth muscle and nonmuscle cells contain one isoform of identical size, comparable to MLC_{3F}.

Sarcomeric ALC Genes

Considerable progress has been made in identifying the genes encoding the ALCs and in determining the mechanisms by which many of the ALC isoforms are generated.

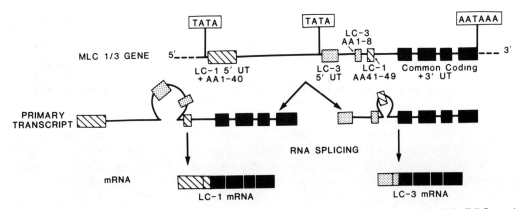

Fig. 6–5. Representation of the alternative splicing of the rat skeletal muscle alkali myosin light chain (MLC$_{1F}$ and MLC$_{3F}$) gene. Transcription originates from two different promoters (represented by the TATA sequences). The MLC$_{1F}$ exons are shown as striped boxes, and the MLC$_{3F}$ specific exons are shown as dotted boxes. The common coding and 3' untranslated regions (UT) are shown as black boxes. The mature mRNAs (LC-1 and LC-3) differ in their 5' ends: LC-1 contains sequences (striped boxes) encoding 49 amino acids (LC-1 AA 1 to 40, 41 to 49), and LC-3 contains sequences (second dotted box) encoding 8 amino acids (LC-3 AA 1 to 8). (Reprinted with permission from Periasamy, M., et al.: Fast skeletal muscle myosin light chains 1 and 3 are produced from a single gene by a combined process of differential RNA transcription and splicing. *J. Biol. Chem., 259:*13595, 1984.)

The genes encoding the chicken,[120] rat,[121] and mouse[122] fast skeletal muscle light chains, MLC$_{1F}$ and MLC$_{3F}$, have been isolated and characterized. It has been determined that both the MLC$_{1F}$ and MLC$_{3F}$ isoforms are produced from a single gene by an unusual process involving transcription from two different start sites (promoters) as well as alternative splicing of exons, similar to that described for α-TM (Fig. 6-5). The results, in the case of the rat, are two proteins that contain identical carboxy-terminal sequences but that differ at their amino-terminal ends, where MLC$_{1F}$ contains 49 amino acids (derived from exons 1 and 4), whereas MLC$_{3F}$ contains 8 amino acids (derived from exons 2 and 3).

Smooth Muscle/Nonmuscle ALC Genes

The amino acid sequence of the smooth muscle ALC from chicken gizzard has been determined, and the structure of this light chain appears similar to that of skeletal MLC$_{3F}$.[123] In addition, the electrophoretic mobilities of the ALCs derived from chick fibroblasts and gizzard appear to be identical,[98] suggesting that the nonmuscle and smooth muscle ALCs are closely related. Recently, Nabeshima et al.[124] have isolated cDNA clones encoding ALCs from chicken gizzard and fibroblast cells. Sequence analysis has shown that the two cDNAs are identical in the 3' and 5' untranslated regions as well as in most of the coding region. The only exception is a 39-nucleotide insertion in the fibroblast cDNA, 26 nucelotides of which encode nine amino acids of the carboxyterminal. The position of this nucleotide insertion corresponds to an exon–intron junction and, together with DNA blot analysis, suggests that the smooth muscle and nonmuscle ALCs are generated from a single gene by alternative splicing.[124] Furthermore, splicing appears to be tissue specific, in that the nonmuscle mRNA was found only in fibroblasts and the smooth muscle mRNA predominantly in gizzard (a small amount of smooth muscle mRNA was found in fibroblasts as well). Evidence that two ALC proteins are present in aortic smooth muscle has recently been presented.[125]

Although these likely represent the nonmuscle and smooth muscle isoforms corresponding to the alternatively splied mRNAs, this has yet to be proven.

Embryonic ALC Genes

In addition to the ALCs found in adult tissues, an embryonic light chain has been isolated from developing chickens,[126–128] rats,[129] and humans.[130,131] This light chain, initially identified in chicken embryos[126–128] and referred to as either MLC_{emb} or L_{23} (molecular weight of 23,000 daltons), has been shown to be structurally related to the fast skeletal muscle MLC_F. In addition, electrophoretic and immunologic studies in chicken embryos have demonstrated that the MLC_{emb} is expressed in developing smooth muscle as well as in sarcomeric muscles. Recently, a cDNA clone encoding the chick MLC_{emb} has been isolated and characterized.[132] The amino acid sequence for the MLC_{emb} deduced from this cDNA clone has demonstrated that this light chain is closely related to other ALCs. RNA blot analysis has revealed that the chicken MLC_{emb} mRNA is transiently expressed in embryonic smooth, cardiac, and skeletal muscles but not in adult sarcomeric or smooth muscles. In contrast, the MLC_{emb} is expressed in the brain of chickens at all stages of development. Furthermore, Southern blot analysis of chicken genomic DNA has suggested that the MLC_{emb} is encoded by a single gene. The pattern of expression of the mammalian MLC_{emb} appears to differ from that of chicken, in that the mammalian MLC_{emb} appears to be indistinguishable from the adult cardiac atrial muscle MLC_A.[133,134] Moreover, recent evidence has suggested that the mammalian MLC_{emb} and MLC_A are encoded by a single gene.[135]

Recent advances in recombinant DNA technology have provided a wealth of information on the organization of the ALC gene family and the mechanisms for generating ALC isoform diversity. The availability of both genes and cDNA clones encoding the ALCs should provide important tools for studying the functional significance of the ALC family as a whole as well as for elucidating the particular biologic roles of the tissue-specific and developmentally specific ALC isoforms.

MYOSIN LIGHT CHAIN KINASE

As previously discussed, the regulation of SMC contraction is based on phosphorylation of the myosin regulatory light chain by a specific MLC kinase. This phosphorylation is necessary for stimulation of the myosin Mg^{2+} ATPase and, as a consequence, smooth muscle contraction.[8] MLC kinases have been isolated from a variety of sources, including smooth muscle,[136,137] skeletal muscle,[138,139] cardiac muscle,[140,141] and nonmuscle cells.[142–145] The enzymes appear to be specific for myosin regulatory light chains and, although they can act on all types of regulatory light chains, they preferentially phosphorylate light chains from homologous tissues.[146] All of the MLC kinases require calmodulin and Ca^{2+} for activity. The smooth[136] and nonmuscle MLC kinases[145] are inhibited by phosphorylation, but the cardiac[147] and skeletal forms[148] are not. More importantly, MLC kinase-directed phosphorylation of the sarcomeric regulatory light chains (MLC_{2s}) does not appear to be necessary for sarcomeric muscle contraction. In fact, the role of MLC_2 phosphorylation in the sarcomeric systems remains to be established.

Considerable progress has recently been made in elucidating the primary structure of the MLC kinase from a number of tissues. Such information should ultimately lead to a better understanding of the differences between the sarcomeric, smooth muscle, and nonmuscle MLC kinase. In addition, isolating and characterizing the genes encoding the smooth muscle MLC kinase should provide important insights into how the contractile apparatus is regulated on a molecular genetic level in smooth muscle. Recently, a cDNA clone has been isolated encoding the carboxy terminal 60% of the chicken gizzard MLCK.[149,150] This clone contains the nucleotides encoding the catalytic domain of the MLC kinase, as well as the calmodulin-binding domain, and the cAMP-dependent protein kinase phosphorylation sites. Although the derived amino acid sequence contains a region of 54% homology (in the catalytic domain) with the rabbit skeletal muscle MLC kinase, both the primary structure and the location of particular domains differ significantly (e.g., the calmodulin-binding region). These results demonstrate that while the smooth and skeletal muscle enzymes have similarities, their sequences show a divergence that is at least as great as, if not greater than, that found for their corresponding RLCs. This may reflect the different roles that MLC phosphorylation plays in the smooth and sarcomeric muscle systems. It would be interesting to determine whether the smooth muscle MLC kinases show a high level of sequence conservation among various species similar to that noted for the RLCs.

COMMENTS

The smooth muscle contractile system differs significantly from that of sarcomeric muscle, not only in its lack of an organized contractile unit (i.e., the sarcomere) but also in the actual mechanism and kinetics of contraction. Although all vertebrate muscles contain the same types of contractile proteins (actin, myosin, tropomyosin, and MLC kinase), these proteins exist in multiple isoforms, some of which are specific to smooth muscle and presumably have features that are important for smooth muscle contraction. Using recombinant DNA technology, it has been shown that these contractile protein isoforms are derived from multigene families whose members range from 2 to as many as 31 genes. The mechanisms employed in generating isoform

Table 6–1. Summary of the Predominant Contractile Protein Isoforms and the Mechanisms of Their Generation in Smooth Muscle

Protein	Predominant Isoforms in Smooth Muscle	Mechanism(s) of Isoform Generation
Actin	α-Smooth muscle β- and γ-cytoplasmic	Each isoform transcribed from a unique gene
Myosin heavy chain	One (?) NM One or two (?) SM	Each isoform (NM and SM) transcribed from a unique gene (?) Alternative splicing of a single SM gene to produce two SM isoforms
Myosin regulatory light chain	RLC-A: Expressed in all cells RLC-B: Expressed in SM/NM cells only	Each isoform transcribed from a unique gene
Myosin alkali light chain	NM and SM isoforms	Alternative splicing of a single NM/SM gene
α-Tropomyosin	NM and SM isoforms	Alternative splicing of a single gene to produce sarcomeric, NM, SM isoforms
Myosin light chain kinase	(?) Smooth muscle-specific isoform	(?) Transcribed from a unique gene

NM, nonmuscle; RLC, regulatory light chain; SM, smooth muscle.
(?) denotes where the information has not yet been firmly established.

diversity are varied. In some cases (e.g., actin, MHC), each isoform is transcribed from a single gene. In others (e.g., α-TM, myosin ALCs), several different isoforms are transcribed from a single gene by alternative splicing or the use of multiple start sites.

Table 6-1 summarizes the predominant isoforms of each type of contractile protein found in vascular smooth muscle cells and the mechanisms employed in generating these isoforms. Most of the work described in this chapter has been of a descriptive nature and has focused on identifying the smooth muscle-specific contractile protein isoforms and the genes encoding them. With the identification and isolation of these genes, it should now be possible to study smooth muscle gene regulation in the same detail as has been done in the sarcomeric systems. Of particular interest will be the identification of *cis*-acting genomic elements and soluble *trans*-acting factors that govern smooth muscle-specific gene expression and that are responsible for directing the changes in contractile protein isoforms seen in cell culture during "phenotypic modulation" or in vivo in response to blood vessel injury.

ACKNOWLEDGMENT

I thank Dr. Bernardo Nadal-Ginard of Children's Hospital, Boston, for reading and criticizing the manuscript.

REFERENCES

1. Ross, R.: Atherosclerosis: a problem of the biology of arterial wall cells and their interactions with blood components. Arteriosclerosis, *1*:293, 1981.
2. Benditt, E.P., and Gown, A.M.: Atheroma: the artery wall and the environment. Int. Rev. Exp. Pathol., *21*:55, 1980.
3. Chamley-Campbell, J.H., Campbell, G.R., and Ross, R.: The smooth muscle cell in culture. Physiol. Rev., *59*:1, 1979.
4. Adelstein, R.S., and Eisenberg, E.: Regulation and kinetics of the actin-myosin-ATP interaction. Annu. Rev. Biochem., *49*:921, 1980.
5. Somlyo, A.V., et al.: The contractile apparatus of smooth muscle: an update. *In* Vascular Smooth Muscle Contraction. Edited by N.L. Stephens. New York, Marcel Dekker, 1984. p. 1.
6. Harrington, W.F., and Rodgers, M.E.: Myosin. Annu. Rev. Biochem., *53*:35, 1984.
7. Weber, A., and Murray, J.M.: Molecualr control in contraction. Physiol. Rev., *53*:612, 1973.
8. Adelstein, R.S., and Klee, C.B.: Smooth muscle myosin light chain kinase. *In* Calcium and Cell Function. Edited by W.Y. Cheung. New York, Academic Press, 1980. Vol. 1, pp. 167–182.
9. Conti, M.A., and Adelstein, R.S.: Phosphorylation by cyclic adenosine 3':5'-monophosphate-dependent protein kinase regulates myosin light chain kinase. Fed. Proc., *39*:1569, 1980.
10. Aksoy, M.O., Mras, S., Kamm, K.E., and Murphy, R.A.: Ca^{2+}, cAMP, and changes in myosin phosphorylation during contraction of smooth muscle. Am. J. Physiol., *245*:C255, 1983.
11. Pato, M.D., and Adelstein, R.S.: Dephosphorylation of the 20,000 dalton light chain of myosin by two different phosphatases from smooth muscle. J. Biol. Chem., *255*:6535, 1980.
12. Chatterjee, M., and Murphy, R.A.: Calcium-dependent stress maintenance without myosin phosphorylation in skinned smooth muscle. Science, *221*:464, 1983.
13. Gerthoffer, W.T., and Murphy, R.A.: Ca^{2+}, myosin phosphorylation, and relaxation of arterial smooth muscle. Am. J. Physiol., *245*:C271, 1983.
14. Thyberg, J. et al.: Phenotype modulation in primary cultures of arterial smooth muscle cells. On the role of platelet-derived growth factor. Differentiation, *25*:156, 1983.
15. Campbell, G.R., and Campbell, J.H.: Phenotypic modulation of smooth muscle cells in primary culture. *In* Vascular Smooth Muscle in Culture. Edited by J.H. Campbell and G.R. Campbell. Boca Raton, FL, CRC Press, 1987. Vol. 1.
16. Chamley, J.H., Campbell, G.R., McConnell, J.D., and Groschel-Stewart, U.: Comparison of vascular smooth muscle cells from adult human, monkey and rabbit in primary culture and in subculture. Cell Tissue Res., *177*:503, 1977.
17. Hermsmeyer, K., De Cino, P., and White, R.: Spontaneous contractions of dispersed vascular muscle in cell culture. In Vitro, *12*:628, 1976.
18. Mauger, J.P., Worcel, M., Tassin, J., and Courtois, Y.: Contractility of smooth muscle cells of rabbit aorta in tissue culture. Nature, *255*:337, 1975.
19. Groschel-Stewart, U., Chamley, J.H., Campbell, G.R., and Burnstock, G.: Changes in myosin distribution

in dedifferentiating and redifferentiating smooth muscle cells in tissue culture. Cell Tissue Res., *165*:13, 1975.

20. Chamley-Campbell, J.H., Campbell, G.R., and Ross, R.: Phenotype-dependent response of cultured aortic smooth muscle to serum mitogens. J. Cell Biol., *89*:379, 1981.
21. Larson, D.M., Fujiwara, K., Alexander, R.W., and Gimbrone, M.A., Jr.: Heterogeneity of myosin antigenic expression in vascular smooth muscle in vivo. Lab. Invest., *50*:401, 1984.
22. Larson, D.M., Fujiwara, K., Alexander, R.W., and Gimbrone, M.A., Jr.: Myosin in cultured vascular smooth muscle cells: immunofluorescence and immunochemical studies of alterations in antigenic expression. J. Cell Biol., *99*:1582, 1984.
23. Pollard, T.D., and Weihing, R.R.: Actin and myosin and cell movement. Crit Rev. Biochem., *2*:1, 1974.
24. Garrels, J.L., and Gibson, W.: identification and characterization of multiple forms of actin. Cell, *9*:793, 1976.
25. Vanderckhove, J., and Weber, K.: At least six different actins are expressed in a higher mammal: an analysis based on the amino acid sequence of the amino-terminal tryptic peptide. J. Mol. Biol, *126*:783, 1978.
26. Chang, K.S. et al.: Isolation and characterization of six different chicken actin genes. Mol. Cell. Biol., *4*:2498, 1984.
27. Wallevik, K., and Jansenius, J.C.: A simple and reliable method for drying polyacrylamide slab gels. J. Biochem. Biophys. Methods, *6*:17, 1982.
28. Vanderckhove, J., and Weber, K.: The complete amino acid sequence of actin from bovine aorta, bovine heart, bovine fast skeletal muscle, and rabbit slow skeletal muscle. A protein-chemical analysis of muscle actin differentiation. Differentiation, *14*:123, 1979.
29. Vanderckhove, J., and Weber, K.: Actin typing on total cellular extracts. A highly sensitive protein-chemical procedure able to distinguish different actins. Eur. J. Biochem., *113*:595, 1981.
30. Vanderckhove, J., and Weber, K.: Chordate muscle actins differ distinctly from invertebrate muscle actins: the evolution of the different vertebrate muscle actins. J. Mol. Biol., *179*:291, 1984.
31. Gabbiani, G. et al.: Vascular smooth muscle cells differ from other smooth muscle cells: predominance of vimentin fibers and specific α-type actin. Proc. Natl. Acad. Sci. USA, *78*:298, 1981.
32. Fornwald, J.A., Kuncio, G., Peng, I., and Ordahl, C.P.: The complete nucleotide sequence of the chick α-actin gene and its evolutionary relationship to the actin gene family. Nucleic Acids Res., *10*:3861, 1982.
33. Zakut, R. et al.: Nucleotide sequence of the rat skeletal muscle actin gene. Nature. *298*:857, 1982.
34. Hu, M.C.-T., Sharp, S.B., and Davidson, N.: The complete sequence of the mouse skeletal α-actin gene reveals several conserved and inverted repeat sequences outside of the protein coding region. Mol. Cell. Biol., *6*:15, 1986.
35. Chang, K.S., Rothblum, K.N., and Schwartz, R.J.: The complete sequence of the chicken α-cardiac actin gene: a highly conserved vertebrate gene. Nucleic Acids Res., *13*:1223, 1985.
36. Hamada, H., Petrino, M.G., and Kakunaga, T.: Molecular structure and evolutionary origin of human cardiac muscle actin gene. Proc. Natl. Acad. Sci. USA, *79*:5901, 1982.
37. Eldridge, J., Zehner, Z., and Paterson, B.M.: Nucleotide sequence of the chick cardiac α-actin gene. Gene, *36*:55, 1985.
38. Kost, T.A., Theodorakis, N., and Hughes, S.H.: The nucleotide sequence of the chick cytoplasmic β-actin gene. Nucleic Acids Res., *11*:8287, 1983.
39. Nudel, U. et al.: The nucleotide sequence of the rat cytoplasmic β-actin gene. Nucleic Acids Res., *11*:1759, 1983.
40. Ng, S.-Y. et al.: Evolution of the human β-actin gene and its multipseudogene family: conservation of noncoding regions and chromosomal dispersion of pseudogenes. Mol. Cell. Biol., *5*:2720, 1985.
41. Bergsma, D.J., Chang, K.S., and Schwartz, R.J.: Novel chicken actin gene: third cytoplasmic isoform. Mol. Cell. Biol., *5*:1151, 1985.
42. Ueyama, H., Hamada, H., Battula, N., and Kakunaga, T.: Structure of a human smooth muscle actin gene (aortic type) with a unique intron site. Mol. Cell. Biol, *4*:1073, 1984.
43. Carroll, S.L., Bergsma, D.J., and Schwartz, R.S.: Structure and complete nucleotide sequence of the chicken α-actin smooth muscle (aortic) actin gene: an actin gene which produces multiple mRNAs. J. Biol. Chem., *261*:8965, 1986.
44. McHugh, K.M., and Lessard, J.L.: The nucleotide sequence of a rat vascular smooth muscle α-actin cDNA. Nucleic Acids Res., *16*:4167, 1988.
45. Gunning, P. et al.: Evolution of the human sarcomeric-actin genes: evidence for units of selection within the 3′ untranslated regions of the mRNAs. J. Mol. Evol., *20*:202, 1984.
46. Yaffe, D., Nudel, U., Mayer, Y., and Neuman, S.: Highly conserved sequences in the 3′ untranslated region of mRNAs coding for homologous proteins in distantly related species. Nucleic Acids Res., *13*:3723, 1985.
47. Ponte, P. et al.: Evolutionary conservation in the untranslated regions of actin mRNAs: DNA sequence of a human beta-actin cDNA. Nucleic Acids Res., *12*:1687, 1984.

48. Ordahl, C.P., and Cooper, T.A.: Strong homology in the promoter and 3' untranslated regions of chick and rat α-actin genes. Nature, *263:*211, 1983.
49. Whalen, R.G., Butler-Browne, G.S., and Gros, F.: Protein synthesis and actin heterogeneity in calf muscle cells in culture. Proc. Natl. Acad. Sci. USA, *73:*2018, 1976.
50. Devlin, R.B., and Emerson, C.P., Jr.: Coordinate regulation of contractile protein synthesis during myoblast differentiation. Cell, *13:*599, 1978.
51. Minty, A.J., Alonso, S., Caravatti, M., and Buckingham, M.E.: A fetal skeletal muscle actin mRNA in the mouse and its identity with cardiac actin mRNA. Cell, *30:*185, 1982.
52. Mohun, T.J. et al.: Cell type specific activation of actin genes in the early amphibian embryo. Nature, *311:*716, 1984.
53. Grichnik, J.M., Bergsma, D.J., and Schwartz, R.J.: Tissue restricted and stage specific transcription is maintained within 411 nucleotides flanking the 5' end of the chicken α-skeletal actin gene. Nucleic Acids Res., *14:*1683, 1986.
54. Bergsma, D.J., Grichnik, J.M., Gossett, L.M.A., and Schwartz, R.J.: Delimitation and characterization of *cis*-acting DNA sequences required for the regulated expression and transcriptional control of the chicken skeletal α-actin gene. Mol. Cell. Biol., *6:*2462, 1986.
55. Melloul, D. et al.: Developmental expression of chimeric genes containing muscle actin DNA sequences in transfected myogenic cells. EMBO J., *3:*983, 1984.
56. Seiler-Tuyns, A., Eldridge, J.D., and Paterson, B.: Expression and regulation of chicken actin genes introduced into mouse myogenic and nonmyogenic cells. Proc. Natl. Acad. Sci. USA, *81:*2980, 1984.
57. Minty, A., and Kedes, L.: Upstream regions of the human cardiac actin gene that modulate its transcription in muscle cells: presence of an evolutionarily conserved repeated motif. Mol. Cell. Biol., *6:*2125, 1986.
58. Shani, M.: Tissue-specific and developmentally-regulated expression of a chimeric actin-globin gene in transgenic mice. Mol. Cell. Biol., *6:*2624, 1986.
59. Hardeman, E.C., Chiu, C.-P., Minty, A., and Blau, H.M.: The pattern of actin expression in human fibroblast x mouse muscle heterokaryons suggests that human muscle regulatory factors are produced. Cell, *47:*123, 1986.
60. Minty, A., Blau, H., and Kedes, L.: Two-level regulation of cardiac actin gene transcription: muscle-specific modulating factors can accumulate before gene activation. Mol. Cell. Biol., *6:*2137, 1986.
61. Saborio, J.S. et al.: Differential expression of gizzard actin genes during chick embryogenesis. J. Biol. Chem., *254:*11119, 1979.
62. Owens, G.K., Loeb, A., Gordon, D., and Thompson, M.M.: Expression of smooth muscle-specific α-isoactin in cultured vascular smooth muscle cells: relationship between growth and cytodifferentiation. J. Cell Biol., *102:*343, 1986.
63. Kocher, O. et al.: Cytoskeletal features of rat aortic cells during development. An electron microscopic, immunohistochemical, and biochemical study. Circ. Res., *56:*829, 1985.
64. Kocher, O., and Gabbiani, G.: Analysis of α-smooth-muscle actin mRNA expression in rat aortic smooth-muscle cells using a specific cDNA probe. Differentiation, *34:*201, 1987.
65. Gabbiani, G. et al.: Actin expression in smooth muscle cells of rat aortic intimal thickening, human atheromatous plaque, and cultured rat aortic media. J. Clin. Invest., *73:*148, 1984.
66. Kocher, O., and Gabbiani, G.: Expression of actin mRNAs in rat aortic smooth muscle cells during development, experimental intimal thickening and culture. Differentiation, *32:*245, 1986.
67. Owens, G.K., and Thompson, M.M.: Developmental changes in isoactin expression in rat aortic smooth muscle cells in vivo. J. Biol. Chem., *261:*13373, 1986.
68. Blank, R.S., Thompson, M.M., and Owens, G.K.: Cell cycle versus density dependence of smooth muscle alpha actin expression in cultured rat aortic smooth muscle cells. J. Cell Biol., *107:*299, 1988.
69. Corjay, M.H., Thompson, M.M., Lynch, K.R., and Owens, G.K.: Differential effect of platelet-derived growth factor-versus serum-induced growth on smooth muscle α-actin and nonmuscle β-actin mRNA in cultured rat aortic smooth muscle cells. J. Biol. Chem., *264:*10501, 1989.
70. Kocher, O., and Gabbiani, G.: Cytoskeletal features of human normal and atheromatous arterial smooth muscle cells. Hum. Pathol., *17:*875, 1986.
71. Hsu, C.-Y.J., and Frankel, F.R.: Effect of estrogen on the expression of mRNAs of different actin isoforms in immature rat uterus. J. Biol. Chem., *262:*9594, 1987.
72. Carroll, S.L., Bergsma, D.J., and Schwartz, R.J.: A 29-nucleotide DNA segment containing an evolutionarily conserved motif is required in *cis* for cell-type-restricted repression of the chicken α-smooth muscle actin gene core promoter. Mol. Cell. Biol., *8:*241, 1988.
73. Smillie, L.: Structure and functions of tropomyosins from muscle and non-muscle sources. Trends Biochem. Sci., *4:*151, 1979.
74. Fine, R.E., and Blitz, A.L.: A chemical comparison of tropomyosins from muscle and non-muscle tissues. J. Mol. Biol., *95:*447, 1975.
75. Marston, S.B., and Smith, C.W.J.: The thin filaments of smooth muscles. J. Muscle Res. Cell Motil., *6:*669, 1985.
76. Cummins, P., and Perry, S.V.: The subunits and biological activity of polymorphic forms of tropomyosin. Biochem. J., *133:*765, 1973.

77. Cummins, P., and Perry, S.V.: Chemical and immunochemical characteristics of tropomyosins from striated and smooth muscle. Biochem. J., *141*:43, 1974.
78. Cote, G.P.: Structural and functional properties of the non-muscle tropomyosins. Mol. Cell. Biochem., *57*:127, 1983.
79. Leavitt, J. et al.: Tropomyosin isoform switching in tumorigenic human fibroblasts. Mol. Cell. Biol., *7*:2721, 1986.
80. Helfman, D.M., Feramisco, J., Ricci, W., and Hughes, S.: Isolation and sequence of a cDNA clone that contains the entire coding region for chicken smooth-muscle α-tropomyosin. J. Biol. Chem., *259*:14136, 1984.
81. Ruiz-Opazo, N., Weinberger, J., and Nadal-Ginard, B.: Comparison of α-tropomyosin sequences from smooth and striated muscle. Nature, *315*:67, 1985.
82. Sanders, C., and Smillie, L.: Amino acid sequence of chicken gizzard β-tropomyosin. J. Biol. Chem., *260*:7264, 1985.
83. Lau, S., Sanders, C., and Smillie, L.: Amino acid sequence of chicken gizzard γ-tropomyosin. J. Biol. Chem., *260*:7257, 1985.
84. Ruiz-Opazo, N., and Nadal-Ginard, B.: α-Tropomyosin gene organization. J. Biol. Chem., *262*:4755, 1987.
85. MacLeod, A. et al.: A muscle-type tropomyosin in human fibroblasts: evidence for expression by an alternative RNA splicing mechanism. Proc. Natl. Acad. Sci. USA, *82*:7835, 1985.
86. Reinach, F., and MacLeod, A.: Tissue-specific expression of the human tropomyosin gene involved in the generation of the *trk* oncogene. Nature, *322*:648, 1986.
87. Helfman, D.M. et al.: Nonmuscle and muscle tropomyosin isoforms are expressed from a single gene by alternative RNA splicing and polyadenylation. Mol. Cell. Biol., *6*:3582, 1986.
88. Yamawaki-Kataoka, Y., and Helfman, D.M.: Isolation and characterization of cDNA clones encoding a low molecular weight nonmuscle tropomyosin isoform. J. Biol. Chem., *262*:10791, 1987.
89. Fiszman, M., Kardami, E., and Leomonnier, M.: A single gene codes for gizzard and skeletal muscle α-tropomyosin. *In* Molecular Biology of Muscle Development. Edited by C. Emerson, D.A. Fishman, B. Nadal-Ginard, and M.A.Q. Siddiqui. New York, Alan R. Liss, p. 457.
90. Flach, J. et al.: Analysis of tropomyosin cDNAs isolated from skeletal and non-muscle mRNA. Nucleic Acids Res., *14*:9193, 1986.
91. Hallauer, P. et al.: Closely related α-tropomyosin mRNAs in quail fibroblasts and skeletal muscle cells. J. Biol. Chem., *262*:3590, 1987.
92. Wieczorek, D.F., Smith, C.W.J., and Nadal-Ginard, B.: The rat α-tropomyosin gene generates a minimum of six different mRNAs coding for striated, smooth, and nonmuscle isoforms by alternative splicing. Mol. Cell. Biol., *8*:679, 1988.
93. Breitbart, R., Andreadis, A., and Nadal-Ginard, B.: Alternative splicing: a ubiquitous mechanism for the generation of multiple protein isoforms from a single gene. Annu. Rev. Biochem., *56*:467, 1987.
94. Taubman, M.B. et al.: The expression of sarcomeric muscle-specific contractile protein genes in BC3H1 cells: BC3H1 cells resemble skeletal myoblasts that are defective for commitment to terminal differentiation. J. Cell Biol., *108*:1799, 1989.
95. Smith, C.W.J., and Nadal-Ginard, B.: Mutually exclusive splicing of α-tropomyosin exons enforced by an unusual lariat branch point location; implications for constitutive splicing. Cell, *56*:749, 1989.
96. Robbins, J., Horan, T., Gulick, J., and Kropp, K.: The chicken myosin heavy chain family. J. Biol. Chem., *261*:6606, 1986.
97. Mahdavi, V. et al.: Sarcomeric myosin heavy chain gene family: organization and pattern of expression. *In* Molecular Biology of Muscle Development. Edited by C. Emerson, D.A. Fishman, B. Nadal-Ginard, and M.A.Q. Siddiqui. New York, Alan R. Liss, 1986, p. 345.
98. Emerson, C.P., and Bernstein, S.I.: Molecular genetics of myosin. Annu. Rev. Biochem., *56*:695, 1987.
99. Burridge, K.: A comparison of fibroblast and smooth muscle myosins. FEBS Lett., *45*:14, 1974.
100. Groschel-Stewart, U.: Comparative studies of human smooth and striated muscle myosins. Biochim. Biophys. Acta, *229*:322, 1971.
101. Burridge, K., and Bray, D.: Purification and structural analysis of myosin from brain and other non-muscle tissues. J. Mol. Biol., *99*:1, 1975.
102. Rovner, A.S., Thompson, M.M., and Murphy, R.A.: Detection of myosin heavy chain variants in smooth muscle by SDS polyacrylamide gel electrophoresis. Fed. Proc. *43*:427, 1984.
103. Rovner, A.S., Thompson, M.M., and Murphy, R.A.: Two different heavy chains are found in smooth muscle myosin. Am. J. Physiol., *250*:C861, 1986.
104. Kawamoto, S., and Adelstein, R.S.: Characterization of myosin heavy chains in cultured aorta smooth muscle cells. J. Biol. Chem., *262*:7282, 1987.
105. Rovner, A.S., Murphy, R.A., and Owens, G.K. Expression of smooth muscle and nonmuscle myosin heavy chains in cultured vascular smooth muscle cells. J. Biol. Chem., *261*:14740, 1986.
106. Yanagisawa, M. et al.: Complete primary structure of vertebrate smooth muscle myosin heavy chain deduced from its complementary cDNA sequence. J. Mol. Biol., *198*:143, 1987.
107. Nagai, R., Larsen, D.M., and Periasamy, M.: Characterization of a mammalian smooth muscle myosin

heavy chain cDNA clone and its expression in various smooth muscle types. Proc. Natl. Acad. Sci. USA, *85:*1047, 1988.

108. Arnold, L.W., Taubman, M.B., and Nadal-Ginard, B.: Characterization of a non-muscle myosin heavy chain cDNA from rat aortic smooth muscle. Circulation, *78:*11–65, 1988.

109. Shohet, R.V., Brill, D.A., Preston, Y.A., and Adelstein, R.S.: Cloning of a vertebrate nonmuscle myosin heavy chain. J. Biochem., *12:*332, 1988.

110. Pearson, R.B. et al.: Comparison of substrate specificity of myosin kinase and cyclic AMP dependent protein kinase. Biochim. Biophys. Acta, *786:*261, 1984.

111. Nudel, U., Calvo, J.M., Shani, M., and Levy, Z.: The nucleotide sequence of a rat myosin light chain 2 gene. Nucleic Acids Res., *12:*7175, 1984.

112. Kumar, C.C. et al.: Heart myosin light chain 2 gene. Nucleotide sequence of full length complementary DNA and expression in normal and hypertensive rat. J. Biol. Chem., *261:*2866, 1986.

113. Winter, B. et al., Isolation and characterization of the chicken cardiac myosin chain (L-2A) gene. Evidence for two additional N-terminal amino acids. J. Biol. Chem., *260:*4478, 1985.

114. Taubman, M.B., Grant, J.W., and Nadal-Ginard, B.: Cloning and characterization of mammalian myosin regulatory light chain (RLC) cDNA: the RLC gene is expressed in smooth, sarcomeric, and non-muscle tissues. J. Cell Biol., *104:*1505, 1987.

115. Zavodny, P.J. et al.: The nucleotide sequence of chicken smooth muscle myosin light chain 2. Nucleic Acids Res., *16:*1214, 1988.

116. Kumar, C.C. et al.: Characterization and differential expression of human vascular smooth muscle myosin light chain 2 isoform in nonmuscle cells. Biochemistry *28:*4027, 1989.

117. Kemp, B.E., Pearson, R.B., and House, C.: Role of basic residues in the phosphorylation of synthetic peptides by myosin light chain kinase. Proc. Natl. Acad. Sci. USA, *80:*7471, 1983.

118. Grant, J.W., Benutto, B., and Taubman, M.B.: The aortic myosin regulatory light chain is a member of a multigene family. Circulation, *78:*11–65, 1988.

119. Barton, P.J.R., and Buckiingham, M.E.: The myosin alkali light chain proteins and their genes. Biochem. J., *231:*249, 1985.

120. Nabeshima, Y.-I., Fujii-Kuriyama, Y., Muramatsu, M., and Ogata, K.: Alternative transcription and two modes of splicing result in two myosin light chains from one gene. Nature, *308:*333, 1984.

121. Periasamy, M. et al.: Fast skeletal muscle myosin light chains 1 and 3 are produced from a single gene by a combined process of differential RNA transcription and splicing. J. Biol. Chem., *259:*13595, 1984.

122. Robert, B. et al.: A single locus in the mouse encodes both myosin light chains 1 and 3: a second locus corresponds to a related pseudogene. Cell, *39:*129, 1984.

123. Matsuda, G. et al.: Amino-acid sequences of the cardiac L-2A, L-2B, and gizzard 17,000 molecular weight light chains of chicken muscle myosin. FEBS Lett., *135:*232, 1981.

124. Nabeshima, Y., Nabeshima, Y.-I., Nonomura, Y., and Fujii-Kuriyama, Y.: Nonmuscle and smooth muscle myosin light chain mRNAs are generated from a single gene by the tissue-specific alternative RNA splicing. J. Biol. Chem., *262:*10608, 1987.

125. Hasegawa, Y., Ueno, H., Horie, K., and Morita, F.: Two isoforms of 17-kdA essential light chain of aorta media smooth muscle myosin. J. Biochem. (Tokyo), *103:*15, 1988.

126. Katoh, N., and Kubo, S.: Light chains of chicken embryonic gizzard myosin. Biochim. Biophys. Acta., *535:*401, 1978.

127. Takano-Ohmuro, H., Obinata, T., Mikawa, T., and Masaki, T.: Changes in myosin isozymes during development of chicken gizzard. J. Biochem. (Tokyo), *93:*903, 1983.

128. Takano-Ohmuro, H. et al.: Embryonic chicken skeletal, cardiac, and smooth muscle express a common embryo-specific myosin light chain. J. Cell Biol., *100:*2025, 1985.

129. Whalen, R.G., and Sell, S.M.: Myosin from fibroblasts contains the skeletal muscle embryonic light chain. Nature, *286:*731, 1980.

130. Cummins, P., Price, K.M., and Littler, W.A.: Feotal myosin light chain in human ventricle. J. Muscle Res. Cell Motil., *1:*356, 1980.

131. Strohman, R.C., Micou-Eastwood, J., Glass, C.A., and Matsuda, R.: Human fetal muscle and cultured myotubes derived from it contain a fetal specific myosin light chain. Science, *221:*955, 1983.

132. Kawashima, M., Nabeshima, Y.-I., Obinata, T., and Fujii-Kuriyama, Y.: A common myosin light chain is expressed in chicken embryonic skeletal, cardiac, and smooth muscles and in brain continuously from embryo to adult. J. Biol. Chem., *262:*14408, 1987.

133. Whalen, R.G., Sell, S.M., Eriksson, A., and Thornell, L.-E.: Myosin subunit types in skeletal and cardiac tissues and their developmental distribution. Dev. Biol., *91:*478, 1982.

134. Price, K.M., Littler, W.A., and Cummins, P.: Human atrial and ventricular myosin light chain subunits in the adult and during development. Biochem. J., *191:*571, 1980.

135. Barton, P.J.R. et al.: The same myosin alkali light chain gene is expressed in adult cardiac atria and fetal skeletal muscle. J. Muscle Res. Cell Motil., *6:*461, 1985.

136. Conti, M.A., and Adelstein, R.S.: Relationship between calmodulin binding and phosphorylation of smooth muscle myosin kinase by the catalytic subunit of cyclic AMP dependent protein kinase. J. Biol. Chem., *256:*3178, 1981.

137. Small, J.V., and Sobieszek, A.: The contractile apparatus of smooth muscle. Int. Rev. Cytol., *64:*241, 1980.
138. Nairn, A.C., and Perry, S.V.: Calmodulin and myosin light chain kinase of rabbit fast skeletal muscle. Biochem. J., *179:*89, 1979.
139. Yagi, K., et al.: Identification of an activator protein for myosin light chain kinase as the calcium ion dependent modulator protein. J. Biol. Chem., *253:*1338, 1978.
140. Frearson, N., and Perry, S.V.: Phosphorylation of the light-chain components of myosin from cardiac and red skeletal muscles. Biochem., J., *151:*99, 1975.
141. Walsh, M., Vallet, B., Autric, F., and Demaille, J.G.: Purification and characterization of bovine calmodulin-dependent myosin light chain kinase. J. Biol. Chem., *254:*12136, 1979.
142. Hathaway, D.R., and Adelstein, R.S.: Human platelet myosin light chain kinase requires the calcium binding protein calmodulin for activity. Proc. Natl. Acad. Sci. USA, *76:*1653, 1979.
143. Dabrowska, R., and Hartshorne, D.: A calcium dependent and modulator dependent myosin light chain kinase from nonmuscle cells. Biochem. Biophys. Res. Commun. *85:*1352, 1978.
144. Yerna, M.J., Dabrowska, R., Hartshorne, D. J., and Goldman, R.D.: Calcium sensitive regulation of actin myosin interactions in baby hamster kidney BHK-21 cells. Proc. Natl. Acad. Sci. USA, *76:*184, 1979.
145. Nishikawa, M., de Lanerolle, P., Lincoln, T.M., and Adelstein, R.S.: Phosphorylation of mammalian myosin light chain kinase by the catalytic subunit of cyclic AMP dependent protein kinase and by the cyclic GMP dependent protein kinase. J. Biol. Chem., *259:*8429, 1984.
146. Scordilis, S.P., and Adelstein, R.S.: A comparative study of the myosin light chain kinases from myoblast and muscle sources. J. Biol. Chem., *253:*9041, 1978.
147. Wolf, H., and Hofmann, F.: Purification of myosin light chain kinase from bovine cardiac muscle. Proc. Natl. Acad. Sci. USA, *77:*5852, 1980.
148. Edelman, A.M., and Krebs, E.G.: Phosphorylation of skeletal muscle myosin light chain kinase by the catalytic subunit of cyclic AMP dependent protein kinase. FEBS Lett., *138:*293, 1982.
149. Guerriero, V., Jr., et al.: Domain organization of chicken gizzard myosin light chain kinase deduced from a cloned cDNA. Biochemistry, *25:*8372, 1986.
150. Bagchi, I.C., and Means, A.R.: Structure function studies on smooth muscle myosin light chain kinase. J. Cell Biol., *105:*185A, 1987.

Chapter 7

MOLECULAR BIOLOGY OF ENDOTHELIUM

Tucker Collins • Michael A. Gimbrone, Jr.

ENDOTHELIAL CELLS AND CARDIOVASCULAR FUNCTION: AN OVERVIEW

Endothelial cells constitute the first barrier between the components of the blood and the extravascular tissues. In this location, the endothelial cell can interact with both circulating cells and soluble blood products, as well as with smooth muscle cells, pericytes, and extracellular matrix components of the blood vessel wall. Although structurally simple, the endothelium performs many vital functions relevant to cardiovascular physiology and pathophysiology. For example, endothelial cells contribute to the local control of hemostasis and thrombosis, help to determine vascular tone, influence the growth of other vascular cells, and participate in local immune and inflammatory responses (reviewed in refs. 1 and 2). Additionally, increasing evidence shows that the endothelium is not a static entity but can undergo dramatic functional alterations in response to various pathophysiologic stimuli.[3-6]

Until recently, the experimental study of endothelial cells has focused primarily on the biology of these cells in culture. Since the mid-1980s, advances in molecular biology have facilitated the structural analysis of many important soluble and cell-associated endothelial components and have led to a better understanding of how these endothelial constituents function. This, in turn, has stimulated the development of new models and working hypotheses for the molecular interactions that can occur among endothelial cells, circulating blood elements, and vascular smooth muscle cells in normal and disease states.

After introducing some basic molecular biologic strategies useful for structural analysis, this chapter will outline the role of the endothelium in two pathophysiologically important processes: first, immune and inflammatory responses basic to host defense, and second, the regulation of vascular smooth muscle growth in atherogenesis and vascular repair. In each of these sections, selected molecular interactions will be examined to illustrate how the recent advances in molecular biology have contributed to a more detailed understanding of endothelial biology. Finally, the chapter will conclude with a "future directions" section, which will consider the potential application of certain powerful new molecular biologic tools to vascular disease problems.

BASIC STRATEGIES IN MOLECULAR CLONING

Much of the recent excitement in endothelial biology has come from the elucidation of the structure of certain functionally important endothelial proteins through molecular

Original research in the authors' laboratory was supported by NIH grants RO1 HL 35716 and PO1 HL 36028; T.C. is a fellow of the Pew Scholars Program.

cloning. In this section, we will briefly review the techniques commonly used to clone proteins of interest.

Certainly, the most straightforward way to obtain structural data about a new protein is to clone a message or gene, sequence the nucleic acid, and then predict the protein sequence of the primary translation product from the deduced nucleic acid sequence. Subsequent confirmation of a predicted protein sequence, while still necessary, is far easier to obtain than a priori determination of the same sequence. Additionally, the primary translation product may provide insights into the processing events that led to the mature form of the protein.

Two basic cloning approaches have been used to obtain structural information about endothelial proteins (as well as a variety of other proteins). In both approaches, an endothelial complementary DNA (cDNA) library is constructed, but in the first technique, the library is inserted into a bacterial expression vector (e.g., λgt11); in the second technique, the library is transferred into a eukaryotic expression system (e.g., CDM8). The choice of expression system is dictated by the reagents available for screening the library. Each system will be briefly introduced and its merits discussed in the following section.

Endothelial Complementary DNA Library Construction

The starting material for most human endothelial cDNA libraries is cultured human umbilical vein endothelial cells expanded in culture in the presence of heparin and endothelial growth factor.[7] In preparation of the library, RNA is usually extracted from harvested cells following lysis with guanidine isothiocyanate.[8] This procedure minimizes the actions of endogenous ribonucleases (RNases) and maintains RNA integrity. The cell lysate is layered onto a CsCl gradient and spun in an ultracentrifuge. Proteins remain in the aqueous guanidine region of the gradient, DNA bands in the CsCl, and the RNA pellets at the bottom of the centrifuge tube and is readily collected.

After total cellular RNA isolation, polyadenylated RNA species (most eukaryotic messenger RNAs [mRNAs]) are purifed from the nonpolyadenylated RNAs (ribosomal RNA [rRNA] and transfer RNAs [tRNAs]) by chromatography of the RNA on oligo(dT) cellulose. The method relies on base pairing between the poly(A) residues at the 3' end of mRNAs and the oligo(dT) residues coupled to the matrix of the column. Nonpoly(A) species are not bound and are readily washed off the column. The bound poly(A) RNA is eluted by lowering the amount of salt in the buffer used to wash the column.

After poly(A) RNA is obtained, it is converted to double-strand cDNA and inserted into an expression system before screening. Full-length, double-strand cDNA is best generated from intact poly(A) RNA by the method of Okayama and Berg[9] as modifed by Gubler and Hoffman.[10] The details of these steps are reviewed elsewhere[11] and will only be summarized here (Fig. 7-1). First, single-strand poly(A) RNA is converted

---

Fig. 7–1.   General strategies for cloning molecules from endothelial cells. Top, a schematic illustration for the conversion of endothelial poly(A) RNA into double-strand cDNA (ds cDNA). First, strand synthesis is primed by an oligo(dT) primer (. . . TTT) or a random hexanucleotide primer (N6). After cDNA is synthesized, it can be cloned in either a bacterial expression vector, λgt11, or the eukaryotic expression system, CDM8 (center of the figure). The genomic maps of wild-type phage λ, the cloning vector λgt11, and the map of the plasmid pCDM8 illustrate the structural features of each cloning vehicle discussed in the text (modified from refs. 11, 15, and 16). Once constructed, the libraries are screened, as discussed in the text, depending on the cloning strategy employed. HUVE: human umbilical vein endothelial cells. CMV: cytomegalovirus. Sup F: A gene containing a tyrosine tRNA amber-suppressor; the tRNA will provide genetic selection by suppressing the ampicillin and tetracycline amber mutations carried in certain bacterial strains (e.g., MC1061/P3).

GENERAL CLONING STRATEGY

CONSTRUCTION OF cDNA LIBRARIES

CULTURED HUVE POLY(A) RNA

FIRST STRAND cDNA SYNTHESIS

OLIGO(dT)

RANDOM

SECOND STRAND SYNTHESIS

INSERTION OF ds cDNA

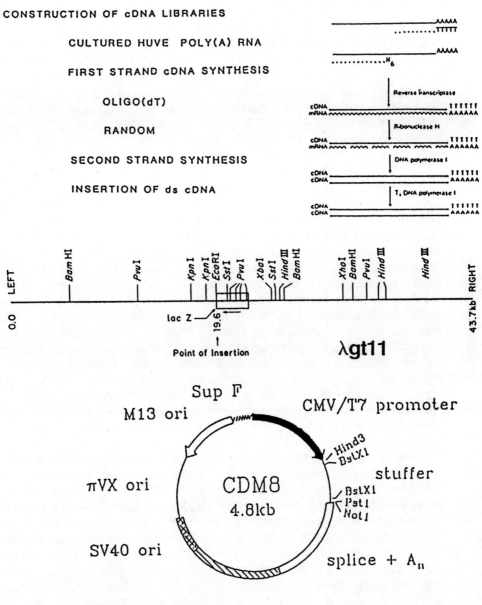

SCREEN HUVE cDNA LIBRARIES

SUBCLONE LARGEST + INSERTS

RESTRICTION MAP

to an RNA–cDNA hybrid by reverse transcriptase using either oligo(dT) or random oligomer priming. Second, the RNA–cDNA hybrid is converted into double-stranded cDNA using RNase H, which only cleaves RNA while in hybrid RNA–DNA form, in combination with DNA polymerase I, which then uses the fragments of nicked RNA as primers to synthesize the second strand of cDNA. Third, any nicks in the double-stranded cDNA are repaired by *Escherichia coli* DNA ligase. Finally, T4 DNA polymerase is used to fully convert the cDNA to blunt-ended form. Following synthesis, linkers are added to the double-strand cDNA before ligation into an expression vector. Two technical advances have greatly facilitated cDNA library construction: First a variety of manufacturers sell kits that provide detailed instructions and tested reagents; and second, several companies will prepare cDNA libraries as a custom synthesis. (The days of having to dedicate oneself to becoming a "librarian" have come to an end.)

λgt11 Vectors

One of the most frequently used cloning vectors are the bacteria phage λ-vehicles. One of these, λgt11, has been used extensively to clone a variety of molecules (Fig. 7-1). The vector is a linear double-stranded derivative of λ that is designed exclusively for cloning small *Eco*RI fragments (less than 6 kilobases [kb]) and in particular cDNA fragments with *Eco*RI linkers. Following double-strand cDNA synthesis (see preceding discussion), the internal *Eco*RI sites are protected from digestion by methylation, and *Eco*RI linkers are added by blunt-end ligation. The cDNAs are then cut with *Eco*RI and separated according to size. The pool containing cDNAs greater than about 1 kb are ligated into *Eco*RI-cut λgt11 phage arms, and viable phage are assembled. The unique *Eco*RI site used for inserting foreign DNA is located near the carboxy-terminus (C-terminus) of the β-galactosidase gene. Foreign DNA sequences in this cloning vehicle can be expressed as β-galactosidase fusion proteins. Thus, recombinant libraries generated from λgt11 can be screened with antibodies, because the antibody may recognize a fusion protein expressed by the recombinant λgt11 bacteriophage.

In a typical λgt11 screening, bacterial host cells are infected with the library and plated onto agar plates. After a short growth period at 42°C (λgt11 has a temperature-sensitive repressor of lytic growth that is inactive at 42°C, allowing selection against lysogenic growth), the plates are overlaid with nitrocellulose filters saturated with isopropyl-β-D-thiogalactopyranoside (IPTG) (an inducer of β-galactosidase gene expression) and incubated for another 3 to 4 hours at 37°C. During this time, the phage plaques are transferred to the filters, along with the β-galactosidase fusion proteins released from lytically infected cells. The filters are removed from the plates and blocked by incubation in a solution containing excess proteins, such as calf serum, to saturate nonspecific binding on the nitrocellulose filters. The primary antibody reacting with the molecule to be cloned is then incubated with the filters, followed by a washing step and another incubation with the appropriate (antimouse or antirabbit) second antibody. These second antibodies can be commercially obtained as alkaline phosphatase, horseradish peroxidase, or iodinated reagents. The positive phage are then located and removed from the original plates, and the recombinants are "plaque purified" by repeating the same screening procedure until all plaques are recognized by the primary antibody. This cloning strategy works well if a high-titer, high-affinity antibody is available. Unfortunately, a few antibodies, particularly some monoclonals, are conformation dependent and do not recognize antigens immobilized on a surface. Others can stick nonspecifically to surfaces such as nitrocellulose and are not suitable for immunoscreening. These techniques, however, were used to clone a wide variety of proteins expressed by endothelial cells. These include several proteins involved in maintaining the hemostatic–thrombotic balance. For example, cDNA clones for von Willebrand factor,[12] thrombomodulin,[13] and plasminogen activator–inhibitor[14] have been obtained by using immunologic reagents to screen endothelial expression libraries.

A Eukaryotic Expression Vector

A powerful new technique has recently been developed that is based on transient expression in COS cells and physical selection of expressing cells by adhesion to antibody-coated dishes.[15] This procedure has allowed many cell surface antigen cDNAs to be cloned quite rapidly. The technique has two major advantages: First cDNAs of interest are recovered in a form containing the necessary sequences for surface expression. In most cases, this means that the cDNAs are almost full length, thus eliminating the need for repeated rounds of library screening to obtain full-length of overlapping clones. Second, monoclonal antibodies can be used effectively, unlike the screening methods based on fusion protein expression.

The pCDM8 expression vector used in these procedures contains both SV40 and polyoma replication origins, a cytomegalovirus/T7 RNA polymerase promoter, and an M13 origin of replication (Fig. 7-1). Thus, the vector has several important features[16]. First the eukaryotic transcription unit allows high-level expression in COS cells of coding sequences placed under its control; second, the small size and arrangement of sequences permit high-level replication of the plasmid in COS cells; and third, the presence of two identical BstX I sites in an inverted orientation can be used to promote efficient cDNA insertion into the vector. In this procedure, double-strand cDNA is synthesized as described previously. Following synthesis, BstX I nonpalindromic linkers are added to the double-stranded, blunt-end cDNA. The linkers are ligated to sized cDNA pools to produce a "CACA" 3' overhang end that will subsequently ligate to the BstX I-digested pCDM8 vector (Fig. 7-1). In contrast to the assembly of λgt11 vectors, subsequent restriction endonuclease digestion after linker addition is not necessary, because the BstX I overhang does not allow more than one linker to add to each end of the cDNA strand. Following ligation into the eukaryotic expression vector, the plasmid is transformed into competent *E. coli* and the library expanded.

In a typical eukaryotic expression library screening, surface cDNAs are isolated from the library using an antibody-enrichment method. In this method, the library in pCDM8 is introduced into COS cells by transfection. The COS cells replicate and express the inserts contained in the expression vector. The cells are harvested (without trypsin) and treated with monoclonal antibodies specific for the surface antigens to be cloned. The cells are then distributed in dishes coated with affinity-purified antibody to mouse immunoglobulin. Cells expressing the surface antigen adhere, and the remaining cells can be removed by gentle washing. From the adherent cells, plasmid is prepared by "Hirt" extraction technique and transformed back into *E. coli* for further rounds of transfection and selection. Using these techniques, cDNAs have been obtained for several molecules expressed by endothelial cells (e.g., lymphocyte function-associated antigen-3 (LFA-3), intracellular adhesion molecule-2 (ICAM-2), as well as endothelial-leukocyte adhesion molecule-1 (ELAM-1), which are involved in the association of circulating blood elements with the vascular wall during immune and inflammatory responses. Additionally, the eukaryotic expression vector system has several new applications in endothelial molecular biology, which will be discussed subsequently.

ENDOTHELIAL CELLS IN IMMUNE AND INFLAMMATORY RESPONSES

Observations over a century old established the importance of the interaction of circulating blood elements with blood vessel wall cells in host inflammatory and immune responses.[4] Work during the late 1980s, however, has demonstrated that vascular endothelial cells are active players rather than passive bystanders in these interactions.[1,3,6] Using the cloning techniques just outlined, exciting new structural data have recently been obtained characterizing several molecules expressed on the surface of endothelial cells that appear to play important roles in mediating interactions with circulating lymphocytes and blood leukocytes.

Cell Adhesion Receptors of the Immune System

The study of antigen-specific interactions of T lymphocytes has yielded a variety of adhesion molecules (reviewed in ref. 17). These molecules regulate lymphocyte adhesion in response to signals from the T-cell antigen receptor. After briefly defining the molecules regulating lymphocyte adhesion, the role of these proteins in the interactions of lymphocytes with endothelial cells will be discussed. Endothelial cells interact with lymphocytes in three general contexts: First, endothelial cells can serve as targets of immune-effector cells; second, endothelial cells can function as antigen-presenting cells in initiating immune responses; and third, lymphocytes can selectively interact with high endothelial venules (HEV) during recirculation from the blood to secondary lymphoid organs. With the information gained from the study of lymphocyte interactions, we will generate models for the adhesive interactions between lymphocytes and endothelial cells.

Antigen-specific cytolytic T lymphocytes (CTL) are elicited in response to major histocompatibility complex disparities or to virally infected cells (e.g., reviewed in ref. 18). The specificity of the cellular immune responses is mediated by the antigen-specific T-cell receptor (TcR)–CD3 complex. Cells with the appropriate antigen receptor clonally proliferate in response to the stimulus. When CTL bind to target cells bearing the specific antigen, they are activated to lyse the target cell. This system has provided a model for defining the surface structures required for cell-mediated killing. Some of these structures were identified by screening for monoclonal antibodies that would inhibit T-cell mediated killing (e.g., LFA-1 and LFA-3). Others such as CD2, CD3, CD4, CD8, and TcR were initially defined as T-cell specific and are members of the immunoglobulin supergene family.[19] These molecules must interact in a cooperative manner for successful T-cell function. All of these molecules are involved in adhesion of the CTL to the target, but several are also involved in the signaling events that follow antigen recognition.

Circulating T cells can be divided into two subsets based on their expression of CD4 and CD8. It appears that the CD4/CD8 phenotype of the T cell correlates with the class of major histocompatibility complex determinants that the T cell recognizes.[20] CD8 cells recognize major histocompatibility complex class I determinants and CD4-positive cells recognize class II major histocompatibility complex determinants. CD4 and CD8 are believed to be receptors for nonpolymorphic determinants of the major histocompatibility complex molecules, in contrast to the TcR–CD3 complex, which interacts with the polymorphic regions of these molecules. In both instances, increased expression of either CD4 or CD8 augments the responsiveness of T cells to antigen stimulation.[21] These observations, together with the recent demonstration of direct binding of CD4 to major histocompatibility complex class II molecules,[22] supports the suggestion that CD4 or CD8 expression increases the strength with which the T cell binds to the antigen-presenting cell or the target cell. In addition to strengthening the specific interaction of the TcR with the antigen–major histocompatibility complex, CD4 and CD8 may transduce signals to the lymphocyte involved in the initiation of effector function.

Antigen stimulation causes resting T cells to enlarge and divide. These changes are associated with increased adhesiveness for many cell types. This antigen-independent adhesion is due to increased expression of LFA-1 and CD2 molecules on the T-cell surface and the presence of ICAM and LFA-3 molecules on the target cell.[23] These molecules appear to function as multipurpose adhesion receptors stabilizing cell–cell interactions.

LFA-1 is part of a family of three related protein molecules found on lymphoid and myeloid cells and consists of a unique α-subunit (relative molecular mass [M_r] 180,000) noncovalently associated with a common β-subunit (M_r 95,000; reviewed in ref. 23). The β-subunit of LFA-1 has been cloned,[24] and analysis of the predicted protein reveals homology to the integrin family of adhesion proteins.[25] Expression of LFA-1 is broadly distributed on almost all peripheral blood cells, including monocytes and neutrophils. It is not expressed on endothelial cells or fibroblasts. The other members of the LFA-1 family include CR3 (M_r 170,000) and p150,95 (M_r 150,000). CR3 is the complement receptor for the inactivated fragment of C3b (iC3b). It may also be involved in the recognition of opsonins stimulating phagocytosis. The function of p150,95 is not known. Both CR3 and p150,95 are stored in intracellular storage granules in leukocytes and are induced on treatment of cells with cytokines.[23]

The biologic relevance of this family of adhesion molecules is illustrated in congenital "leukocyte adhesion deficiency."[26,27] In this disorder, all three members of the family of cell-surface molecules including LFA-1 are not expressed, because of a mutation of the common subunit. Patients with this disorder have recurrent life-threatening bacterial infections. Monocytes and neutrophils from these patients cannot bind to and cross the endothelium at sites of infection, leading to diminished pus formation. Interestingly, lymphocytes can leave the circulation in the deficiency, suggesting the functional importance of other adhesion molecules.

cDNAs for the ligands of these T-cell molecules have recently been described. Intracellular adhesion molecule-1 (ICAM-1, CD54) is a ligand for LFA-1. Monoclonal antibody-blocking studies of LFA-1-dependent adhesion have defined a glycoprotein (M_r 90 to 114,000), designated ICAM-1.[28] This protein is expressed on endothelial cells as well as monocytes, lymphocytes, and fibroblasts. The cDNA sequence of ICAM-1 indicates that the glycoprotein is composed of five extracellular constant region (C)-like immunoglobulin domains, making it a member of the immunoglobulin superfamily.[29] The protein has extensive sequence similarity to other cell-adhesion molecules such as neural cell adhesion molecule (N-CAM) and myelin-associated glycoprotein. ICAM-1 on epithelial cells, endothelial cells, and fibroblasts mediates LFA-1-dependent adhesion of lymphocytes.[28] Several lines of evidence, however, suggest the existence of at least one other LFA-1 ligand[30]: First, a monoclonal antibody reacting with LFA-1 but not ICAM-1, inhibits homotypic adhesion of some but not all cell lines; second, an LFA-1-dependent, ICAM-1-independent pathway of adhesion to endothelial cells has been reported; and third, in some types of target cells, LFA-1-dependent T-lymphocyte adhesion and lysis are independent of ICAM-1. The recent cloning of ICAM-2 establishes a second LFA-1 ligand. Because it had not been possible to raise monoclonal antibodies reacting with ICAM-2, the eukaryotic expression system described earlier was modified to a "ligand"-type cloning strategy.[30] COS cells transfected with an endothelial cell expression library in CDM8 were incubated in LFA-1-coated dishes with ICAM-1 monoclonal antibodies to prevent isolation of ICAM-1 cDNA. Adherent COS cells were then eluted with ethylenediamine tetraacetic acid (EDTA), and plasmids were isolated and amplified in *E. coli.* After three cycles, a plasmid was obtained, which conferred LFA-1 dependent, ICAM-1 independent adherence to LFA-1 in a high percentage of the transfected COS cells. The plasmid so isolated contained an insert designated ICAM-2. Analysis of the cDNA reveals that the protein is an integral membrane protein with two immunoglobulin-like domains, in contrast to ICAM-1, which has five. Surprisingly, ICAM-2 is much more closely related to the two most N-terminal domains of ICAM-1 than to other members of the immunoglobulin superfamily. This raises the possibility that the ICAMs are part of a subfamily of immunoglobulin-like ligands that bind the same integrin receptor.[30]

In addition to the TcR recognition of antigen–major histocompatibility complex and the LFA-1–ICAM-1 association, a third association between CD2 and LFA-3 is important in stimulating T-cell activation during cell–cell contacts. Monoclonal antibody-blocking experiments have defined LFA-3 as a ligand for CD2 in the process of lymphocyte adherence to thymic epithelial cells, T-cell recognition of antigen presenting cells, and cytotoxic cell recognition of target cells (e.g., reviewed in ref. 23). Expression of LFA-3 is widely distributed, being found on erythrocytes, epithelial cells, endothelial cells, fibroblasts, and most cells of hematopoietic origin. Research has demonstrated that purified CD2 can adhere to LFA-3 cells and that this adhesion can be blocked by either anti-LFA-3 or CD2 monoclonal antibodies.[31] Similarly, purified LFA-3 incorporated into synthetic membranes binds to CD2-bearing cells.[32,33] Structural analysis of LFA-3 indicates that the molecule exists in two alternative forms, differing only at the extreme carboxy-terminal end. cDNAs have been obtained that encode a phosphatidylinositol (PI)-linked form of LFA-3[34] as well as a transmembrane form.[35] The deduced amino acid sequences for the PI-linked form and the transmembrane form are identical from the amino terminus to the point of divergence in the carboxy terminus at the membrane-attachment site. The effect of this small variation in the carboxy-terminal structure on the functional properties of the molecule is not known. LFA-3 is homologous to its ligand CD2 and also is a member of the immunoglobulin superfamily.

Based on this information, instructive models have been generated for the binding of T cells to either antigen-presenting cells or target cells (Fig. 7-2). In these models, the TcR–CD3 complex interacts with antigen in association with major histocompatibility complex molecules on the antigen-presenting cell. LFA-1 interacts with ICAM-1 (or ICAM-2) and may have other as-yet-unidentified ligands; CD2 interacts with LFA-3. The specific interaction between the TcR–CD3 complex and antigen–major histocompatibility complex is "strengthened" by the associations of the various accessory molecules. This pattern of association (i.e., the presence of a ligand pair conferring specificity and a series of accessory interactions that stabilize the specific association) may occur in other cell–cell associations. The validity of this paradigm, however, remains to be established.

Endothelial Cells and Immune Recognition

Some of the molecules used in the interactions between CTL and lymphoid targets are also involved in the interactions of lymphocytes and endothelial cells. Endothelial cells constitutively express class I major histocompatibility complex antigens and can be induced to express class II antigens by products of activated T cells or immune interferon.[36] Endothelial cells can also perform all of the functions of a traditional antigen-processing cell. For example, human umbilical vein endothelium can present nominal antigen to resting or activated memory T cells[37,38] and can stimulate naive T cells in a primary allogeneic response.[39,40] Additionally, endothelial cells can serve as targets for CTL.[40,41] In fact, monoclonal antibody-blocking experiments suggest that the array of accessory molecules used by the CTL in the recognition of class II major histocompatibility antigen–positive (Ia-positive) human endothelial cells is qualitatively similar to that used in the recognition of lymphoid target cells[41] (Fig. 7-2). This inter-

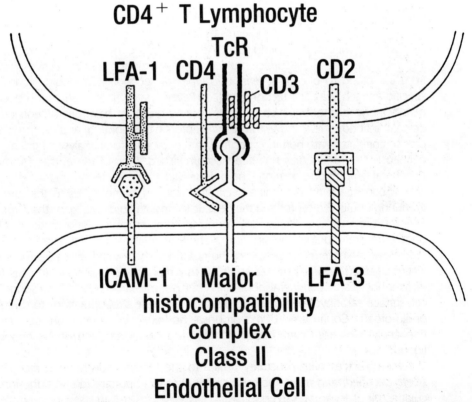

Fig. 7–2. A model for lymphocyte–endothelial interaction. Binding of a CD4⁺ T lymphocyte to the surface of an endothelial cell induced to express class II major histocompatibility complex (MHC) antigens by immune interferon. The T-cell receptor (TcR)–CD3 complex recognizes the class II major histocompatibility complex molecule expressed on the endothelial cell; CD4 also interacts with the major histocompatibility complex molecule. This specific interaction may be strengthened by the lymphocyte function-associated antigen-1 (LFA-1)–intracellular adhesion molecule-1 (ICAM-1) and LFA-3–CD2 associations. (Modified with permission from Bierer, B., and Burakoff, S.J.: T cell adhesion molecules. *FASEB J.*, *2*:2584, 1988.)

action between a CTL and an endothelial cell induced to express class II antigens may occur in vivo during allograft rejection. In this setting, the presence of class II major histocompatibility complex antigens on the luminal surface of vessels in an allograft may make the donor endothelium a prime target for host T cells. Whether all of these assessory molecules are important in other immune cell–endothelial interactions (e.g., in endothelial cell antigen presentation or in an allogeneic response) is not established. The display of accessory molecules may determine the type of signals delivered to the T cell and may determine the type of effector cell generated by the interaction with endothelium (i.e., T helper or suppressor).

In addition to serving as targets of immune effector cells and initiating immune responses, specific endothelial cells found within lymph nodes (HEVs) express cell-surface molecules that selectively recruit specific lymphocyte subsets to a peripheral immune response (reviewed in refs. 42 and 43). Lymphocytes emigrating from the thymus are believed to be "naive." After recognition of a specific antigen outside the thymus, they become "memory" lymphocytes. Memory T lymphocytes differ from naive T lymphocytes in the expression of a variety of adhesion molecules. For example, shortly after antigen stimulation, the levels of LFA-1, LFA-3, and CD2 are increased. The increased expression of these molecules probably facilitates interactions of memory T cells with an antigen-presenting cell.[44] Additionally, the memory T cells express "homing" receptors. These cell-surface molecules allow lymphocytes to selectively recirculate from the blood to one of at least two types of secondary lymphoid organs where specific antigen was first recognized (i.e., Peyer's patch [intestinal] lymph nodes and peripheral lymph nodes). Two types of homing receptors have been identified. Monoclonal antibodies reacting with these receptors block homing, and expression of the protein correlates with homing specificity. Recently, structural analysis of the molecules defined by the antibodies has been accomplished. One of these, Mel-14, mediates peripheral lymph node homing. The murine Mel-14 molecule has an interesting structure. It consists of an N-terminal lectin-like domain homologous to the low-affinity immunoglobulin E receptor and the hepatic lectins, an epidermal growth-factor repeat, two short repeats characteristically found in complement-binding proteins, a hydrophobic transmembrane domain, and a short cytoplasmic tail.[45,46] The present of a prominent aminoterminal sequence homologous to a variety of calcium-dependent lectins supports previous studies showing that mannose 6-phosphate inhibits binding of lymphocytes to peripheral node HEV but not intestinal HEV.[47] This suggests that cell-surface carbohydrate is used to direct lymphocyte subpopulations to diverse lymphoid organs. Confirmaton of this proposal, however, will depend on structure–function analysis of each extracellular domain and characterization of the endogenous ligand.

Two molecules with structures similar to the lymph node homing receptor have been identified; and these molecules may play an important role in adhesion of circulating blood leukocytes to endothelial cells: First, ELAM-1 and, second, granule membrane protein-140 (GMP-140). Activation of vascular endothelium by interleukin-1 (IL-1), tumor necrosis factor (TNF), or endotoxin increases the adhesion of blood leukocytes to vascular endothelium.[48] Monoclonal antibodies have been generated that react with an inducible cell-surface protein present on cytokine-stimulated, but not control, endothelial cells.[49] These monoclonal antibodies block the adhesion of neutrophils and related cell lines. Tthe eukaryotic expression vector system discussed previously was used in conjunction with one of the antibodies to clone ELAM-1.[50] Structural analysis of the cDNA reveals that the inducible adhesion molecule has an N-terminal lectin-like domain, an epidermal growth-factor domain, and six complement receptor repeats. GMP-140 is a putative adhesion molecule present in platelets and endothelial cells that is stored in granules and is mobilized to the cell surface after activation of the cells. Like the Mel-14 and ELAM-1 molecules, GMP-140 has single lectin and epidermal growth-factor domains, but it has nine complement-receptor domains.[51] The function of the various domains of these molecules is not yet known.

The second type of lymphocyte homing receptor is the CD44, or Hermes, molecule.[52,53] This molecule is broadly expressed on hematopoietic as well as nonhematopoietic cells. The lack of significant biochemical differences in the CD44 antigens isolated from cell lines binding exclusively to either lymph node or mucosal HEVs, together with the broad inhibitor specificity of the antibodies to CD44, suggests that the protein functions as a multipurpose adhesion receptor. Structural analysis of the molecule revealed that the molecule is a transmembrane molecule unrelated to Mel-14. A potentially significant N-terminal homology (30%) with cartilage link proteins may provide a structural basis for interactions with extracellular matrix components. The role that recognition of matrix components might play in homing specificity is not known.

ENDOTHELIAL CELLS AND THE REGULATION OF VASCULAR SMOOTH MUSCLE PROLIFERATION

An abnormal proliferation of cells has long been recognized as characteristic of atherogenesis. Not until the early 1970s, however, was a model proposed to formalize these concepts. This "response to injury" hypothesis[54] proposed that the smooth muscle cells in the arterial media are quiescent under normal condition. Abnormal proliferation and migration of these smooth muscle cells would occur in response to intimal "injury." The proliferating smooth muscle cells would increase their production of extracellular matrix materials and form the basis of the characteristic atherosclerotic plaque.

A major stimulus to recent experimental work on the cellular aspects of atherogenesis was the observation that, when blood platelets degranulate, they release protein growth factors that stimulate the proliferation of vascular smooth muscle cells as well as fibroblasts.[55,56] The most well characterized of these mitogens was originally designated platelet-derived growth factor (PDGF). During the 1970s several groups used standard protein-purification techniques to characterize PDGF. The platelet-derived mitogen was a small cationic protein, stable to heat and acidic pH (reviewed in ref. 57). As purified from platelets, the protein was a heterodimer composed of two chains (A and B in one nomenclature). That the mitogen has assumed major conceptual significance in understanding of atherosclerosis can be appreciated by examining the current model for atherosclerosis.

The working concept[58] for the development of atherosclerosis is that both nondenuding and denuding injury to endothelium leads to adherence of monocytes, which enter the vessel wall and secrete PDGF-like substances. Injured endothelial cells can also secrete PDGF-like mitogens. If the injury is sufficiently extensive, platelets can aggregate, degranulate, and release PDGF. The growth factor deposited in these sites is chemotactic and mitogenic for vascular smooth muscle cells. The recruited smooth muscle cells may be activated to secrete PDGF-like substances that may act in an autocrine fashion, inducing the intimal proliferative lesion characteristic of atherosclerosis (reviewed in refs. 57 and 58).

Definition of the PDGF A- and B-Chain Genes

The results of recent studies on the molecular biology of PDGF have shed new light on possible sources of PDGF-like molecules as well as on the potential diversity of control mechanisms. In the spring of 1983, a remarkable convergence of research fields occurred. The acutely transforming simian sarcoma virus had been isolated from a sarcoma found in the woolly monkey, and the transforming gene (v-*sis*) had been sequenced.[59] Additionally, a normal human gene (c-*sis*) was noted to be highly homologous to the v-*sis* gene and was located on chromosome 22.[60] When the partial sequence analysis of both chains of human PDGF were reported, it was immediately clear that one chain (designated the B chain in one nomenclature) was highly homologous to the transforming gene v-*sis*.[61,62] By inference, therefore, the normal cellular gene c-*sis* encoded the B chain of PDGF. Several groups then worked to determine the structural differences between the transforming gene v-*sis* and the product of the normal c-*sis* gene, as well as to complete the architecture of the B-chain gene.[63,64] For example, structural analysis of a cDNA clone from endothelial cells indicated that the predicted protein made by a normal cultured cell was identical to that made in transformed cells and revealed the presence of an unusually long 5' untranslated region.[65]

The biochemical data, as well as the reported partial protein sequence information, could be interpreted to suggest the existence of another protein chain in PDGF. By screening a glioma cDNA library with synthetic oligonucleotides corresponding to the partial protein sequence, a putative A chain cDNA was isolated, and the gene was mapped to chromosome 7.[66] Comparison of the amino acid sequences of the mature products of the PDGF A- and B-chain genes shows a high degree of homology (56%). Interestingly, when cDNAs were obtained from human endothelial cells, the structures were different from the reported A-chain sequence.[67,68] The endothelial clones lacked an internal 69-basepair (bp) region near the C-terminus. The change in transcript structure shortens the endothelial transcript by 15 amino acids, removing a highly basic carboxy-terminal region found in the glioma-derived precursor. Identification of an A-chain species in *Xenopus* oocytes containing these 15 amino acids indicates that the A-chain splicing process may be conserved during evolution, although the functional significance of this region of the A chain is uncertain.

To determine the basis for the structural differences in the A-chain cDNAs, the endothelial cDNA was used to clone the human A-chain gene.[69,70] Examination of the structure of the gene reveals that the two different A-chain precursors, which differ by the presence or absence of a basic C-terminus, are generated as a result of alternative mRNA splicing events, which include or exclude exon 6 (Fig. 7-3A). The positions of intervening sequences closely match those of the homologous B-chain gene.

In parallel with the definition of the genes for the A and B chains, it has become clear that the subunit composition of the mitogen varies, depending on the cell type. Analysis of purified PDGF obtained using classic protein purification techniques has revealed that all three dimeric isotypes of PDGF—PDGF-AA, -AB, and -BB—are biologically active. For example, the predominant form of PDGF derived from human platelets consists of a heterodimer containing equivalent amounts of A and B chains.[71] The homodimer form of PDGF encoded by the v-*sis* transforming gene is closely related to the BB homodimer in its amino acid sequence. When expressed in cells that have PDGF receptors, this oncogene can transform cells by interacting with receptors inside

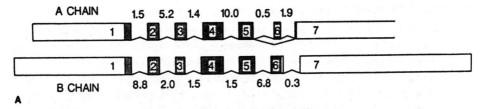

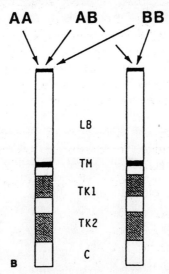

Fig. 7–3. The structure of the A- and B-chain genes of PDGF and the interaction of the mitogen with the platelet-derived growth-factor (PDGF) receptors. A. Exon–intron structure and splicing patterns of the PDGF A- and B-chain genes. Sizes of the exons are drawn to scale; lengths of the introns are indicated (kilobase). B. Schematic representation of the binding of the various PDGF isoforms to the two different PDGF-receptor types. The functional domains of the receptors are indicated as follows: LB, ligand-binding domain; TM, transmembrane domain; TK1 and TK2, the two regions of the split tyrosine kinase domain; and C, cytoplasmic domain. (Fig. 7-3A reprinted with permission from Bonthron, D.T., Morton, C.C., Orkin, S.H., and Collins, T.:Platelet-derived growth factor A chain: gene structure, chromosomal location, and basis for alternative mRNA splicing. Proc. Natl. Acad. Sci. USA, *85*:1492, 1988.)

the cells.[72,73] Interestingly, porcine platelets contain a mitogen that is a B-chain homodimer.[74] Growth factor produced by an osteosarcoma-derived cell line and certain glioma cells is a homodimer of PDGF A.[75] With such a variety of PDGF species, it became important to determine whether the mitogens were functionally equivalent.

Recent studies have been interpreted to suggest that the various isoforms of PDGF may have different functional activities. Interestingly, a glioma-derived PDGF A-chain homodimer, which is structurally similar to PDGF-AA, is less potent in terms of its mitogenic activity and its ability to stimulate receptor autophosphorylation and actin reorganization when compared with PDGF purified from human platelets.[76] Additionally, glioma-derived PDGF-AA does not exhibit chemotactic activity, but it causes transmodulation of epidermal growth-factor receptors. Recent results suggest that bacterially expressed and assembled PDGF-BB effectively stimulates cell growth, calcium flux, inositol triphosphate (IP_3) metabolism, diacylglycerol release, and vasoconstriction of aortic strips, while similarly produced PDGF-AA is less effective.[77] These preliminary results suggest that these mitogens transmit different signals to the responsive cell.

The functional differences in the isoforms of PDGF indicate that it will be important to define the nature of the mitogen being produced in a particular setting in which PDGF is implicated in a pathophysiologic process. For example, the smooth muscle cell replication in normal development, the medial thickening in hypertension, and the focal proliferations in the arterial intima, characteristic of early atherosclerosis, may be the result of growth factors made within the arterial wall. With the availability of PDGF cDNA clones, it became possible to determine which vascular cells express the PDGF A- and B-chain genes. Both endothelial cells (as noted previously) and macrophages express both chains of PDGF. Additionally, these cells secrete material that resembles PDGF immunologically or in a radioreceptor assay.[78–80] Arterial smooth muscle cells can also secrete a mitogen that structurally and functionally resembles PDGF.[81–83] Aortic smooth muscle cells from newborn rats, were shown to release PDGF-like activity into conditioned medium and to have greatly reduced numbers of PDGF receptors on their surface.[82] The smooth muscle cells from the rat pups expressed both the A and B chains of PDGF.[84] In contrast, cells from the adult expressed only the A chain of PDGF.[83,84] Because production of PDGF in the artery may play a role in the formation and maintenance of the vessel wall, the independent regulation of the A and B chains suggests that these species of PDGF may contribute in different ways to these activities.

If this mitogen plays an important role in the development of vascular proliferative lesions, then smooth muscle cell proliferation should correlate with PDGF expression. In fact, several groups have reported such observations. Shortly after balloon injury to the rat carotid artery, an intimal hyperplastic lesion develops, composed of actively proliferating smooth muscle cells. Cells isolated from such hyperplastic lesions secrete PDGF-like mitogens.[85] Analysis of RNA content in these cells reveals that smooth muscle cells express the A chain of PDGF and putatively assemble an A-chain homodimer. Additionally, vascular cells isolated from human atheromatous lesions secrete a PDGF-like mitogen and selectively express the A chain of PDGF.[86] The PDGF produced by these cells may act locally to cause an autocrine or paracrine stimulation of cell proliferation, which may initiate the process of intimal thickening in atherosclerosis.

PDGF Receptors and the Specificity of Isotype Binding

A PDGF receptor was first identified as a 180-kilodalton (kD) membrane glycoprotein by the covalent cross-linking of labeled PDGF to intact cells (reviewed in ref. 87). High-affinity binding sites of PDGF are found on cells such as fibroblasts, vascular smooth muscle cells, and a variety of tumor cells but not endothelial cells. Like many growth-factor receptors, the PDGF receptor has intrinsic tyrosine kinase activity. In fact, when quiescent fibroblasts are stimulated by PDGF, the predominant tyrosine-phosphorylated species is the receptor itself. This property allowed purification of the PDGF receptor by affinity chromatography with antibodies that recognize phosphotyrosine.[88] Oligonucleotide probes designed from the partial amino acid sequence were used to isolate the cDNAs from the mouse[88] and closely related human[89] PDGF receptors. Analysis of the protein predicted by the cDNAs reveals a ligand-binding domain consisting of five immunoglobulin-like domains, a single membrane-spanning segment, and a tyrosine kinase region that is split into two parts. This receptor preferentially binds the

BB form of PDGF but will also recognize the AB species; it binds the AA form of PDGF only at high concentrations (ref. 90 and Fig. 7-3B).

Shortly after binding to its 180-kD receptor, PDGF stimulates a variety of responses, including tyrosine kinase activity, increased phosphoinositide hydrolysis, activation of protein kinase C, enhanced expression of a group of genes including c-*myc* and c-fos, activation of phospholipase A2, changes in cell shape, increased intracellular calcium concentration, characteristic alterations in intracellular pH, and internalization of PDGF (reviewed in ref. 87). The interrelationships of these diverse responses to PDGF binding and eventual DNA synthesis are not fully understood. However, site-directed mutagenesis of the PDGF receptor has provided important insights into the mechanisms of signal transduction. That is, most responses to PDGF (e.g., phosphoinositol turnover, calcium changes, alterations in cellular pH, and DNA synthesis) could not be elicited in cells expressing a tyrosine kinase-deficient PDGF receptor.[91] Tyrosine kinase activity, however, is not required for ligand-induced receptor down-regulation. This type of detailed structure–function analysis is sure to reveal additional important insights into the mechanisms of intracellular pathway activation in the near future.

Evidence now exists that a second PDGF receptor is present on many cells and that its specificity for PDGF isoforms differs from the initially described PDGF receptor. Two groups have shown that labeled AA and BB forms of PDGF bind to cells with at least two different patterns of specificity.[92] Additionally, a monoclonal antibody that selectively binds to the type B receptor was produced.[93] Finally, a cDNA encoding a glycoprotein that efficiently binds PDGF A-chain homodimers but is structurally distinct from the PDGF-B type receptor was isolated recently.[94] This protein contained all of the characteristics of a membrane-spanning tyrosine kinase receptor. The ligand-binding domain is about 30% homologous to the existing PDGF receptor. The cytoplasmic domain contains a split tyrosine kinase domain, which is strikingly conserved. These findings convincingly demonstrate that the PDGF receptor has at least two forms. Whether the two receptors function differently or whether the relative expression on the cell surface dictates cellular responsiveness to the three PDGF isoforms is presently unclear. The PDGF receptor has been proposed to exist as either a homo- or heterodimer on the target cell or to form a dimer on stimulation by PDGF (reviewed in ref. 87). The high-affinity PDGF receptor consists of dimers of two types of subunit: an α-subunit, which can bind via either an A or a B chain of a dimeric PDGF molecule, and a β-subunit, which can bind only a B chain of PDGF. The relative quantity of each receptor type expressed on the cell surface may dictate which dimeric receptor complex is generated on the responsive cell.

Although the PDGF receptors are structurally similar, the fact that the functional effects of the various isoforms of PDGF on cells are different suggests that these receptors may transmit different intracellular signals. How could two structurally similar tyrosine kinases transmit different signals? One possibility is that the PDGF-AA receptor is coupled to a different G protein that transduces PDGF-AA activity to a distinct protein kinase C isoform. Thus, PDGF-AB or BB may stimulate phosphatidylinositol turnover via classic pathways, while the AA isoform may activate the cell via a novel mechanism. Such a proposal is not unprecedented. Interleukin-3 stimulates proliferation via a protein kinase C activation without detectable accumulation of IP_3.[95]

With a growth factor composed of two chains and at least two functionally different tyrosine kinase receptors, the potential levels of complexity involved in the control of PDGF expression increase dramatically. Control of mitogen expression may be exerted at the level of PDGF A- and B-chain gene regulation or growth-factor assembly; alternatively, the levels of PDGF receptors could be tightly coupled to growtth-factor effects. How this receptor–ligand complex functions to govern smooth muscle proliferation in the vasculature is unknown.

FUTURE DIRECTIONS IN ENDOTHELIAL MOLECULAR BIOLOGY

Several new and exciting avenues of investigation are being explored in endothelial molecular biology. The first approach is to use modifications of existing technology to try to define new endothelial molecules that may have an important role in endothelial function. The second approach is to design expression constructs to test models of endothelial function in vivo, either by overexpressing a specific endothelial molecule or by deliberately deleting an endothelial-specific gene product from an animal.

Expression Cloning of Endothelial Genes in Eukaryotic Systems

The eukaryotic expression vector system described earlier has at least three potential new applications in endothelial molecular biology: first, ligand cloning; second, cloning of secreted proteins; and third, cloning of proteins regulating endothelial gene expression. In some situations, generating monoclonal antibodies against a particularly important receptor may be difficult. Rather than using the monoclonal antibody-panning system, as described previously, a cloning strategy could be developed that exploits the ability of the cell-surface receptor to bind immobilized ligand. In brief, the expression library in pCDM8 would be constructed from a cell line that expresses large quantities of the receptor. The library would be transfected into COS cells, as before, but the cells would be screened on plates coated with ligand rather than antibody. Adherent COS cells would be eluted with EDTA and plasmids isolated and amplified in *E. coli*. After several cycles of transfection, adherence, and plasmid isolation, a construct would be obtained that confers receptor binding on the COS cells. Characterization of the cDNA contained within the expression vector would define the structure of the receptor. Such a strategy has recently been used to clone the IL-1 receptor.[96] and ICAM-2[30] and may be useful in obtaining structural information on a variety of receptors that play a role in endothelial physiology (e.g., the endothelin, thrombin, adenosine diphosphate, and angiotensin receptors).

The second novel application of eukaryotic expression vector technology would be to define new soluble mediators produced by endothelial cells. Endothelial cells produce a variety of soluble substances that are important in vascular biology. These cells, however, may produce as yet uncharacterized mediators that could alter the functional properties of other vascular cells or circulating blood elements. In the past, identification of these mediators has involved standard protein-purification procedures, often using cumbersome biologic assays for activity. In addition, this approach requires vast quantities of endothelial-conditioned medium, which can be prohibitive. To circumvent these problems, an endothelial cDNA library in a eukaryotic expression vector (e.g., pCDM8) is divided into pools of several thousand cDNA clones and used to transfect COS cells in individual dishes. Conditioned medium is collected from the transfected cells, and the desired activity is measured (e.g., positive and negative regulators of smooth muscle cell growth or vasoconstriction). Pools of cDNAs that yield the activity of interest are subject to sib selection to obtain the appropriate cDNA clone. Once a cDNA is identified, the open reading frame contained within the expression vector is further characterized by DNA sequence analysis.

A third application of the eukaryotic expression vector system would be to clone genes involved in the regulation of endothelial gene expression. Endothelial cells, like many other highly differentiated cells, have a unique pattern of gene expression. This cellular commitment depends on the establishment of specific patterns of gene expression achieved through a network of regulatory factors. Some of these factors bind in a sequence-specific manner, to promoter or enhancer elements of the expressed genes, while others interact with DNA-binding proteins to modulate transcription. Cell-specific gene expression probably involves the interaction of a limited array of tissue-restricted factors with multiple ubiquitous factors. Although a substantial body of information is accumulating on the characterization of proteins made by endothelial cells, almost nothing is known about the factors governing endothelial gene expression. In part this is due to technical difficulties associated with cloning DNA-binding proteins.

In the past, two methods have been used to obtain structural information about transcriptional regulatory factors: First, cDNAs have been identified by conventional protein purification; and second, expression libraries in λgt11 have been screened with short oligonucleotides containing a cognate binding site for the putative transcription factor. These methods have significant limitations. Standard protein purification of a DNA-binding protein for antibody production or microsequence analysis requires large amounts of nuclear extract; obtaining the number of endothelial cells required for such a purification would be difficult. Additionally, whether the majority of DNA-binding domains will function as parts of β-galactosidase fusion proteins is unclear. A technique has recently been described that takes advantage of the eukaryotic expression vector system to avoid these limitations.

The new strategy for the cloning of DNA-binding proteins employs a variation of the eukaryotic expression system previously described. To clone endothelial-specific transcription factors, an endothelial expression library would be divided into pools of several thousand cDNA clones and used to transfect COS-1 cells. Lysates of the transfected cells would be prepared and examined by gel-shift assay for the presence of factors that bind to a defined region (e.g., promoter) of a gene whose expression is limited to endothelial cells. As with the cloning of the soluble mediators from endothelial cells, pools of cDNAs that yield binding activity in the gel-shift assay are subjected to sib selection to obtain the appropriate cDNA. This technique has recently been used to obtain a DNA-binding protein that is a new member of the zinc-finger family of transcriptional regulators whose expression is limited to cells of the hematopoietic lineage.[97]

Structure–Function Analysis of Endothelial Products

Structural analysis of endothelial-derived cDNAs encoding cell-surface molecules or growth factors has suggested possible modes of action for these proteins. To further characterize these molecules, a three-step process will be employed: First, the existing molecules will be altered in defined locations; second, the altered forms of the molecule will be expressed in appropriate cells; and third, the specific effects of the alterations will be analyzed in functional assays.

Acquisition of full-length cDNA clones permits a more detailed analysis of the functional properties of the molecules. For example, deletions in the cDNA can be generated that remove large domains of a cell-surface molecule, or site-directed mutagenesis techniques can be used to alter the cDNA clones in precisely defined ways. The effect of these structural alterations can then be assessed in two types of systems. First, COS cells can be transiently transfected with an expression construct containing the modified molecule. The second system involves generating cell lines that express the modified protein. Frequently, 3T3 cells are co-transfected with the expression construct and a selectable marker (e.g., aminoglycoside resistance), and clonal transformants are selected that express large quantities of the protein of interest. The functional capacity of the protein can then be assessed by a variety of techniques (e.g., the ability to bind monoclonal antibodies, mediate leukocyte adhesion, etc.). Using either approach, structural modifications in the cDNA may be correlated with the changes observed in the functional properties of the recombinant protein. From these data, maps of the functional domains of the endothelial protein may be assembled, and possible modes of association with other proteins may be suggested.

Expression In Vivo

Although the models discussed in the previous sections are generally accepted, they leave open the question of whether the identified cell-surface molecules and growth factors are epiphenomena or major signals for the events occurring in vivo. Testing these models will ultimately involve analysis in whole organisms. At least two general types of systems are currently available for such experiments: first, seeding grafts or natural blood vessels with endothelial cells that have been genetically modified in vitro and, second, analysis of transgenic animals.

The first approach was recently described simultaneously by two groups.[98,99] Both groups used murine amphotropic retrovirus vectors to insert a marker gene (β-galactosidase) into culture endothelial cells. These modified endothelial cells were then transferred to test animals. In one set of experiments, porcine endothelial cells expressing recombinant galactosidase from a murine amphotrophic retroviral vector were introduced with a catheter into denuded iliofemoral arteries of putatively syngeneic Yucatan minipigs. Arterial segments explanted 2 to 4 weeks after seeding contained endothelial cells expressing the marker gene. In the second set of experiments, prosthetic vascular grafts seeded with endothelial cells, also expressing recombinant β-galactosidase from a retrovirus vector, were implanted as carotid interposition grafts into the dogs from which the endothelial cells were originally harvested. Analysis of the graft 5 weeks after implantation revealed genetically modified endothelial cells lining the luminal surface of the graft.

These experiments demonstrate that genetically altered autologous endothelial cells can be introduced into the vascular wall and survive. This technology potentially holds promise as a drug-delivery system, as well as an experimental tool for exploring mechanisms of pathobiology. As a drug-delivery system, this approach might be used to treat systemic or inherited diseases requiring secretion of gene products directly into the circulation. Because of their proximity to the blood stream, endothelial cells are a good candidate for systemic delivery of therapeutic agents. Additionally, grafts could contain endothelial cells capable of secreting vasodilators and thrombolytic or angiogenic factors to ischemic myocardium or capable of producing antineoplastic agents to primary or metastatic tumors. As a tool to test models of how endothelial cells interact with smooth muscle cells or with circulating blood elements, the approach also has much to offer. For example, to test the role of endothelial cells in atherogenesis, endothelial cells could be altered to overexpress specific growth factors. Similarly endothelial cells could be engineered to secrete proteins capable of preventing thrombosis or inhibiting smooth muscle cell hyperplasia. The impact of these changes on neointima formation could be assessed in appropriate animal model systems.

The second general type of model for examining the role of a cloned endothelial molecule in vivo is the transgenic animal (reviewed in refs. 100 to 102). *Transgenic* refers to the integration of foreign DNA into an animal's germ line as a result of experimental manipulation. The exogenous or foreign DNA placed into the animal is designated the *transgene*. The production of transgenic animals generally involves several basic steps: (1) The gene of interest is isolated and placed into an expression construct. (2) The gene is microinjected into fertilized egg pronuclei and implanted into pseudopregnant female animals. (3) Those animals that carry the transgene (founders) are identified. (4) Offspring expressing the transgene are generated from the founder animals. The technical aspects of this procedure have been reviewed extensively elsewhere[103] and will not be discussed here.

Transgenic mice have been useful in studying a variety of disease processes (reviewed in ref. 102), and all evidence suggests that this system will be helpful in examining vascular pathophysiology. Several issues need to be considered, however, when applying this technology to vascular questions. The first is specificity of expression. Expressing gene products normally restricted to vascular cells inappropriately in all cells of the body is not particularly meaningful. In designing a transgene to test questions relevant to vascular pathophysiology, it will be important to selectively direct production of the protein under scrutiny to the cellular components of the blood vessel, either to smooth muscle cells or endothelial cells. Expression of several genes (e.g., von Willebrand factor, endothelin) is limited to endothelial cells. Analysis of the 5' flanking sequences of these genes may provide insights into the mechanisms controlling this unusual tissue specificity. By coupling these regulatory regions to the structural gene being examined, expression of this gene may be directed selectively to endothelial cells. Similarly, promoter regions of genes specifically expressed in macrophages or smooth muscle cells could be identified and exploited to confer tissue specificity in the transgenic setting. The second issue to be considered in designing a transgene to examine vascular questions is level of gene expression. A construct that is constitutively exprressed at high levels may be lethal early in development. Ideally, a construct would be inducible (e.g., under the transcriptional control of an inducible promoter system like the steroid-responsive murine mammary tumor virus promoter or metallothionine) by a relatively innocuous agent (such as steroids or heavy metals, respectively). Thus, by inducing the production of a transgene specifically expressed in vascular cells, it may be possible to correlate overexpression of the product with interpretable pathology.

The new advances in molecular biology have facilited the structural analysis of both soluble and cell-associated endothelial proteins. This has stimulated the generation of models illustrating how the components interact. In the new future, these models will be tested in vivo using the recombinant-DNA methodologies, and a detailed understanding of the function of the proteins should become possible. These insights, in turn, will provide a greater understanding of how the vascular endothelium interacts with the circulating blood elements and vascular smooth muscle cells in both normal and pathologic settings.

REFERENCES

1. Gimbrone, M.A., Jr.: Vascular Endothelium in Hemostasis and Thrombosis. New York, Churchill Livingstone, 1986.

2. Simionescu, N., and Simionescu, M.: Endothelial Cell Biology in Health and Disease. New York, Plenum Press, 1988.
3. Pober, J.S.: Cytokine-mediated activation of vascular endothelium. Am. J. Pathol., *133*:426, 1988.
4. Wallis, M.W., and Harlan, J.M.: Effector functions of endothelium in inflammatory and immunologic reactions. Pathol. Immunopathol. Res., *5*:73, 1986.
5. Davies, P.F.: Vascular cell interactions with special reference to the pathogenesis of atherosclerosis. Lab. Invest., *55*:5, 1986.
6. Cotran, R.C.: New roles for the endothelium in inflammation and immunity. Am. J. Pathol., *129*:407, 1988.
7. Thornton, S.C., Mueller, S.N., and Levine, E.M.: Human endothelial cells: Use of heparin in cloning and long-term serial cultivation. Science, *222*:623, 1981.
8. Chirgwin, J.M., Przybyula, A.E., MacDonald, R.J., and Rutter, W.J.: Isolation of biologically active ribonucleic acid from sources enriched in ribonuclease. Biochemistry, *18*:5294, 1979.
9. Okayama, H., and Berg, P.: High-efficiency cloning of full-length cDNA. Mol. Cell. Biol., *2*:161, 1982.
10. Gubler, U., and Hoffman, B.J.: A simple and efficient method for generating cDNA libraries. Gene, *25*:263, 1983.
11. Huyuh, T.V., Young, R.A., and Davis, R.W.: Constructing and screening cDNA libraries in gt10 and gt11. *In* DNA Cloning. Vol. I: A Practical Approach. Edited by D.M. Clover, Washington, DC, IRL Press, 1985.
12. Ginsburg, D., et al.: Human von Willebrand factor (vWF): Isolation of complementary DNA (cDNA) clones and chromosomal localization. Science, *228*:1401, 1985.
13. Jackman, R.W., Beeler, D.L., Van De Water, L., and Rosenberg, R.D.: Characterization of a thrombomodulin cDNA reveals structural similarity to the low density lipoprotein receptor. Proc. Natl. Acad. Sci., *83*:8834, 1986.
14. Ginsburg, D. et al.: cDNA cloning of human plasminogen activator-inhibitor from endothelial cells. J. Clin. Invest., *78*:1673, 1986.
15. Seed, B., and Aruffo, A.: Molecular cloning of the CD2 antigen, the T-cell erythrocyte receptor, by a rapid immunoselection procedure. Proc. Natl. Acad. Sci., *84*:3365, 1987.
16. Aruffo, A., and Seed, B.: Molecular cloning of a CD28 cDNA by a high-efficiency COS cell expression system. Proc. Natl. Acad. Sci., *84*:8573, 1987.
17. Bierer, B., and Burakoff, S.J.: T cell adhesion molecules. FASEB J.,*2*:2584, 1988.
18. Marrack, P., and Kappler, J.: The T cell receptor. Science, *238*:1073, 1987.
19. Hood, L., Kronenberg, M., and Hunkapiller, T.: T cell antigen receptors and the immunoglobulin supergene family. *Cell, 40*:225, 1985.
20. Swain, S.L.: Significance of Lyt phenotypes: Lyt 2 antibodies block activities of T cells that recognize class I major histocompatibility complex antigens regardless of their function. Proc. Natl. Acad. Sci., *78*:7101, 1981.
21. Sleckman, B.P., et al.: Expression and function of CD4 in a murine T-cell hybridoma. Nature, *328*:351, 1987.
22. Ratnofsky, S.E., Peterson, A., Greenstein, J.L., and Burakoff, S.J.: Expression and function of CD8 in a murine T-cell hybrioma. J. Exp. Med., *166*:1747, 1987.
23. Springer, T.A., Dustin, M.L., Kishimoto, T.K., and Marlin, S.D.: The lymphocyte function-associated LFA-1, CD2, and LFA-3 molecules: Cell adhesion receptors of the immune system. Ann Rev. Immunol., *5*:223, 1987.
24. Kishimoto, T.K., O'Connor, K., Lee, A., Roberts, T.M., and Springer, T.A.: Cloning of the beta subunit of the leukocyte adhesion proteins: Homology to an extracellular matrix receptor defines a novel supergene family. Cell, *48*:681, 1985.
25. Hynes, R.O.: Integrins: A family of cell surface receptors. Cell, *48*:549, 1987.
26. Anderson, D.C. et al.: The severe and moderate phenotypes of heritable Mac-1, LFA-1 deficiency: Their quantitative definition and relation to leukocyte dysfunction and clinical features. J. Infect. Dis., *152*:668, 1985.
27. Springer, T.A.: Inherited deficiency of the Mac-1, LFA-1, p150,95 glycoprotein family and its molecular basis. J. Exp. Med., *160*:1901, 1984.
28. Makgoba, M.W. et al: ICAM-1 a ligand for LFA-1-dependent adhesion of B, T and myeloid cells. Nature, *331*:86, 1988.
29. Simmons, D., Makgoba, M.W., and Seed, B.: ICAM, an adhesion ligand of LFA-1, is homologous to the neural cell adhesion molecule NCAM. Nature, *331*:624, 1988.
30. Staunton, D.E., Dustin, M.I., and Springer, T.A.: Functional cloning of ICAM-2, a cell adhesion ligand for LFA-1 homologous to ICAM-1. Nature, *339*:61, 1989.
31. Selvaraj, P., et al.: The T lymphocyte glycoprotein CD2 binds the cell surface ligand LFA-3. Nature, *326*:400, 1987.
32. Takai, Y., Reed, M., Burakoff, S.J., and Herrmann, S.: Direct evidence for a physical interaction between CD2 and LFA-3. Proc. Natl. Acad. Sci., *84*:6864, 1987.
33. Dustin, M.L., Sanders, M.E., Shaw, S., and Springer, T.A. Purified lymphocyte function-associated antigen 3 binds to CD2 and mediates T lymphocyte adhesion. J. Exp. Med., *165*:677, 1987.

34. Seed, B.: An LFA-3 cDNA encodes a phospholipid-linked membrane protein homologous to its receptor CD2. Nature, *329:*840, 1987.
35. Wallner, B.P. et al.: Primary structure of lymphocye function-associated antigen 3 (LFA-3). J. Exp. Med., *166:*923, 1987.
36. Pober, J.S. et al.: Ia expression by vascular endothelium is inducible by activated T cells and by human immune interferon. J. Exp. Med., *157:*1339, 1983.
37. Hirschberg, H., Bergh, O.J., and Thorsby, E.: Antigen-presenting properties of human vascular endothelial cells. J. Exp. Med., *152:*249, 1980.
38. Wagner, C.R., Veto, R.M., and Burger, D.R.: The mechanism of antigen presentation by endothelial cells. Immunobiology, *168:*453, 1984.
39. Hirschberg, H., Eversen, S.A., Henriksen, T., and Thorsby, E.: The human mixed lymphocyte-endothelium culture interaction. Transplantation, *19:*495, 1975.
40. Pober, J.S. et al.: Lymphocytes recognize human vascular endothelial and dermal fibroblast Ia antigens induced by recombinant immune interferon. Nature, *305:*726, 1983.
41. Collins, T. et al.: Human cytolytic T lymphocyte interactions with vascular endothelium and fibroblasts: Role of effector and target cell molecules. J. Immunol., *133:*1878, 1984.
42. Stoolman, L: Adhesion molecules controlling lymphocyte migration. Cell, *56:*907, 1989.
43. Berg, E.L. et al.: Homing receptors and vascular addressins: Cell adhesion molecules that direct lymphocyte traffic. Immunol. Rev., *108:*1, 1989.
44. Dustin, M.L., and Springer, T.A.: Lymphocyte function-associated antigen-1 (LFA-1) interaction with intercellular adhesion molecule-1 (ICAM-1) is one of at least three mechanisms for lymphocyte adhesion to cultures endothelial cells. J. Cell Biol., *107:*321, 1988.
45. Laskey, L.A.: Cloning of a lymphocyte homing receptor reveals a lectin domain. Cell, *56:*1045, 1989.
46. Siegelman, M.H., Rijn, M., and Weissman, I.L.: Mouse lymph node homing receptor cDNA encodes a glycoprotein revealing tandem interaction domains. Science, *243:*1165, 1989.
47. Yednock, T.A., and Rosen, S.D.: Lymphocyte homing. Adv. Immunol., *44:*313, 1989.
48. Bevilacqua, M.P. et al.: Interleukin 1 acts on cultured human vascular endothelial cells to increase the adhesion of polymorphonuclear leukocytes, monocytes, and related cell lines. J. Clin. Invest., *76:*2003, 1985.
49. Bevilacqua, M.P. et al.: Identification of an inducible endothelial-leukocyte adhesion molecule, E-LAM 1. Proc. Natl. Acad. Sci., *84:*9328, 1987.
50. Bevilacqua, M.P., Stengelin, S., Gimbrone, M.A. Jr., and Seed, B.: Endothelial leukocyte adhesion molecule 1: An inducible receptor for neutrophils related to complement regulatory proteins and lectins. Science, *243:*1160, 1989.
51. Johnston, G.I., Cook, R.G., and McEver, R.P.: Cloning of GMP-140, a granule membrane protein of platelets and endothelium: Sequence similarity to proteins involved in cell adhesion and inflammation. Cell, *56:*1033, 1989.
52. Goldstein, L.A. et al.: A human lymphocyte homing receptor, the Hermes antigen, is related to cartilage proteoglycan core and link proteins. Cell, *56:*1063, 1989.
53. Stamenkovic, I., Amiot, M., Pesando, J.M., and Seed, B.: A lymphocyte molecule implicated in lymph node homing is a member of the cartilage link protein family. Cell, *56:*1057, 1989.
54. Ross, R., and Glomset, J.A.: The pathogenesis of atherosclerosis. N. Engl. J. Med., *295:*369, 1976.
55. Kohler, N., and Lipton, A.: Platelets as a source of fibroblast growth-promoting activity. Exp. Cell Res., *87:*297, 1974.
56. Ross, R., Glomset, J., Kariya, B., and Harker, L.: A platelet-dependent serum factor that stimulates the proliferation of arterial smooth muscle cells in vitro. Proc. Natl. Acad. Sci., *71:*1207, 1974.
57. Ross, R., Raines, E.W., and Bowen-Pope, D.F.: The biology of platelet-derived growth factor. Cell, *46:*155, 1986.
58. Ross, R.: The pathogenesis of atherosclerosis. N. Engl. J. Med., *134:*488, 1986.
59. Devare, S.G. et al: Nucleotide sequence of the simian sarcoma virus genome: Demonstration that its acquired cellular sequences encoded the transforming growth factor product p28sis. Proc. Natl. Acad. Sci., *80:*731, 1982.
60. Dalla Favera, R., Gallo, R.C., Giallongo, A., and Croce, C.M.: Chromosomal localization of the human homolog of the (c-sis) of the similar sarcoma virus onc gene. Nature, *292:*31, 1982.
61. Waterfield, M.D. et al.: Platelet-derived growth factor is structurally related to the putative transforming protein p28sis of similar sarcoma virus. Nature, *304:*35, 1983.
62. Doolittle, R.F. et al.: Simian sarcoma virus onc gene, v-sis, is derived from the gene (or genes) encoding a platelet-derived growth factor. Science, *221:*275, 1983.
63. Johnson, A. et al.: The c-sis gene encodes a precursor of the B chain of platelet-derived growth factor. EMBO J., *3:*921, 1984.
64. Josephs, S.F., Guo, C., Ratner, L., and Wong-Staal, F.: Human protooncogene nucleotide sequences corresponding to the transforming region of simian sarcoma virus. Science, *223:*487, 1984.
65. Collins, T. et al.: Cultured human endothelial cells express platelet-derived growth factor chain 2: cDNA cloning and structural analysis. Nature, *316:*748, 1985.

66. Betsholtz, C. et al.: cDNA sequence and chromosomal localization of human platelet-derived growth factor A-chain and its expression in tumor cell lines. Nature, *320:*695, 1986.
67. Collins, T., Bonthron, D.T., and Orkin, S.H.: Alternative RNA splicing affects function of encoded platelet-derived growth factor A chain. Nature, *328:*621, 1987.
68. Tong, B.D. et al.: cDNA clones reveal differences between human glial and endothelial cell platelet-derived growth factor A chains. Nature, *328:*619, 1987.
69. Bonthron, D.T., Morton, C.C., Orkin, S.H., and Collins, T.: Platelet-derived growth factor A chain: Gene structure, chromosomal location, and basis for alternative splicing. Proc. Natl. Acad. Sci., *85:*1492, 1988.
70. Rorsman, F. et al: Structural characterization of the human platelet-derived growth factor A-chain cDNA and gene: Alternative exon useage predicts two different precursor proteins. Mol. Cell. Biol., *8:*571, 1988.
71. Hammacher, A. et al.: A major part of platelet-derived growth factor purified from human platelets is a heterodimer of one A and one B chain. J. Biol. Chem., *263:*16493, 1988.
72. Deuel, T.F. et al.: Expression of a platelet-derived growth factor-like protein in simian sarcoma virus transformed cells. Science, *221:*1348, 1983.
73. Keating, M.T., and Williams, L.T.: Autocrine stimulation of intracellular PDGF receptors in v-sis-transformed cells. Science, *239:*914, 1988.
74. Stroobant, P., and Waterfield, M.: Purification and properties of porcine platelet-derived growth factor. EMBO J., *3:*2963, 1984.
75. Heldin, C-H. et al.: A human osteosarcoma cell line secretes a growth factor structurally related to a homodimer of PDGF A-chains. Nature, *319:*511, 1986.
76. Nister, M. et al.: A glioma-derived PDGF A chain homodimer has different functional activities from a PDGF AB heterodimer purified from human platelets. Cell, *52:*791, 1988.
77. Block, L.H. et al.: Calcium-channel blockers inhibit the action of recombinant platelet-derived growth factor in vascular smooth muscle cells. Proc. Natl. Acad. Sci., *86:*2388, 1989.
78. Collins, T. et al.: Cultured human endothelial cells express platelet-derived growth factor A chain. Am. J. Pathol., *126:*7, 1987.
79. Shimokado, K. et al.: A significant part of macrophage-derived growth factor consists of at least two forms of PDGF. Cell, *43:*277, 1985.
80. Martinet, Y. et al.: Activated human monocytes express the c-sis protooncogene and release a mediator showing PDGF-like activity. Nature, *319:*158, 1986.
81. Nilsson, J. et al.: Arterial smooth muscle cells in primary culture produce a platelet-derived growth factor-like protein. Proc. Natl. Acad. Sci., *83:*4418, 1985.
82. Seifert, R.A., Schwartz, S.M., and Bowen-Pope, D.F.: Developmentally regulated production of platelet-derived growth factor-like molucules. Nature, *311:*669, 1984.
83. Sjolund, M. et al.: Arterial smooth muscle cells express platelet-derived growth factor (PDGF) A chain mRNA, secrete, PDGF-like mitogen, and bind exogenous PDGF in a phenotype- and growth state-dependent manner. J. Cell Biol., *106:*403, 1988.
84. Majesky, M.W., Benditt, E.P., and Schwartz, S.M.: Expression and developmental control of platelet-derived growth factor A-chain and B-chain/Sis genes in rat aortic smooth muscle cells. Proc. Natl. Acad. Sci., *85:*1524, 1988.
85. Walker, L.N., Bowen-Pope, D.F., Ross, R., and Reidy, M.A.: Production of platelet-derived growth factor-like molecules by cultured arterial smooth muscle cells accompanies proliferation after arterial injury. Proc. Natl. Acad. Sci., *83:*7311, 1986.
86. Libby, P., Warner, S.J.C., Salomon, R.N., and Birinyi, L.K.: Production of platelet-derived growth factor-like mitogen by smooth-muscle cells from human atheroma. Proc. Natl. Acad. Sci., *318:*1493, 1988.
87. Williams, L.T.: Signal transduction by the platelet-derived growth factor receptor. Science, *243:*1564, 1989.
88. Yarden, Y. et al.: Structure of the receptor for platelet-derived growth factor helps define a family of closely related growth factor receptors. Nature, *323:*226, 1986.
89. Gronwald, R. et al.: Cloning and expression of a cDNA for the human platelet-derived growth factor receptor: Evidence for more than one receptor class. Proc. Natl. Acad. Sci., *85:*3435, 1988.
90. Escobedo, J.A. et al.: A common PDGF receptor is activated by homodimeric A and B forms of PDGF. Science, *240:*1532, 1988.
91. Fantl, W.J., Escobedo, J.A., and Williams, L.T. Mutations of the platelet-derived growth factor receptor that cause a loss of ligand-induced conformational change, subtle changes in kinase activity, and impaired ability to stimulate DNA synthesis. (1989): Mol. Cell. Biol., *9:*4473.
92. Heldin, C-H et al.: Binding of different dimeric forms of PDGF to human fibroblasts: Evidence for two separate receptor types. EMBO J., *7:*1387, 1988.
93. Hart, C.E. et al.: Two classes of PDGF receptor recognize different isoforms of PDGF. Science, *240:*1529, 1988.
94. Matsui, T. et al.: Isolation of a novel receptor cDNA establishes the existence of two PDGF receptor genes. Science, *243:*800, 1989.

95. Whetton, A.D. et al.: Interleukin 3 stimulates proliferation via protein kinase C activation without increasing inositol lipid turnover. Proc. Natl. Acad. Sci., *85*:3284, 1988.

96. Sims, J.E. et al.: cDNA expression cloning of the IL-1 receptor, a member of the immunoglobulin superfamily. Science, *241*:585, 1988.

97. Tsai, S-F. et al.: Cloning of cDNA for the major DNA-binding protein of the erythroid lineage through expression in mammalian cells. Nature, *339*:446, 1989.

98. Wilson, J.M. et al.: Implantation of vascular grafts lined with genetically modified endothelial cells. Science, *244*:1344, 1989.

99. Nable, E.G. et al.: Recombinant gene expression in vivo within endothelial cells of the arterial wall. Science, *244*:1342, 1989.

100. Jaenisch, R.: Transgenic animals. Science, *240*:1468, 1988.

101. Palmiter, R.D., and Brinster, R.L.: Germ-liner transformation of mice. Ann. Rev. Genet., *20*:465, 1986.

102. Cuthbertson, R.A., and Klintworth, G.K.: Transgenic mice—a gold mine for furthering knowledge in pathobiology. Lab. Invest., *58*:484, 1988.

103. Hogan, B., Constantini, F., and Lacy, E.: Manipulating the Mouse Embryo. New York, Cold Spring Harbor Laboratory, 1986.

Chapter 8

MOLECULAR BIOLOGY OF ANGIOGENIN

Edward A. Fox • James F. Riordan

Unraveling the molecular events of angiogenesis holds promise for new therapeutic strategies that would involve the inhibition or augmentation of vascularization in diseased or injured tissues. Diseases associated with vascular proliferation such as growth of solid tumors and neovascular glaucoma[1] are likely candidates for this approach using inhibitors of angiogenic molecules. Similarly, wound healing may be improved by administration of angiogenic molecules themselves. Compelled by such practical interests, along with the realization that the study of angiogenesis affords an opportunity to dissect the complex process of organ development, this laboratory initiated a program to purify and characterize an angiogenic molecule produced by human cells. The techniques of molecular biology have played an indispensable role in this research. In this chapter, we emphasize the reasoning behind and rationale for employing molecular biology in our studies of the molecular basis of angiogenesis.

A cultured human cell line, HT-29, derived from a colon adenocarcinoma, was chosen as the biologic starting material because tumor cells were purported to be the most appropriate source of an angiogenesis factor.[2] In addition, this cell line can be maintained in a defined protein-free medium.[3] The chick embryo chorioallantoic membrane assay[4–6] served to detect the presence of angiogenic molecules in the conditioned medium of the HT-29 cells (Fig. 8-1). Fractionation of the medium by cation exchange and reversed phase high-performance liquid chromatography (HPLC) gave a homogeneous preparation of a potent angiogenic protein.[5] This protein, named angiogenin, is a single-chain polypeptide with a molecular mass of 14.1 kilodaltons (kD), and an isoelectric point >9.5. Sequence analysis of the gene and the protein yielded a surprising result: Angiogenin displays significant (~35%) identity to pancreatic ribonuclease (RNase) A.[7,8] This finding provided an important basis for subsequent efforts to understand how angiogenin induces neovascularization and to obtain or design therapeutic angiogenin inhibitors. It also was critical to the discovery of angiogenin's ribonucleolytic activity[9] and to identifying a tight-binding protein inhibitor of angiogenin's ribonucleolytic and angiogenic properties.[10] In addition, the RNase A homology has given clues regarding residues that may be important in catalysis.

Initially, structural analysis of angiogenin aimed at elucidating the relationship between its enzymatic and angiogenic activities was not possible. The amounts of protein required to carry out such structure–function studies were much larger than could be obtained from the cultured tumor cells, which yielded only 0.5 μg of angiogenin/L of medium. Two new sources of angiogenin therefore were developed.

First, angiogenin was discovered to be present in normal human plasma and could be isolated with a yield of 60 to 150 μg/L.[6] This new biologic source increased the

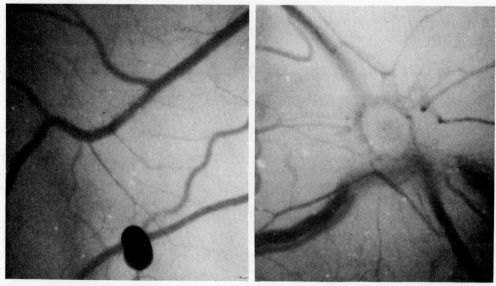

A   B

Fig. 8–1.   Chick embryo chorioallantoic membrane assay depicting (A) a negative and (B) a positive angiogenic response. The positive response was induced by 25 ng of angiogenin. A. A control disk contained 5 μL of
water applied near the black spot. (Reprinted with permission from Fett, J.W., Bethune, J.L., and Vallee,
B.L.: Induction of angiogenesis by mixtures of two angiogenic proteins, angiogenin and acidic fibroblast
growth factor, in the chick chorioallantoic membrane. Biochem. Biophys. Res. Commun., *146*:1122, 1987.)

amount of accessible angiogenin by more than one hundredfold per liter. The plasma-
derived protein was found to be chemically and biologically identical to the tumor-
derived material and, hence, suitable for further studies.

The work with HT-29 cells began as a search for a tumor-specific angiogenesis
factor, but the presence of angiogenin in normal human plasma suggested that angiogenin might play a role in non-neoplastic processes. This inference was foreshad-
owed by studies that led to the development of the second new source of angiogenin,
a recombinant expression system.

THE HUMAN ANGIOGENIN GENE

To obtain even greater amounts of angiogenin than could be obtained from HT-29
cells, a recombinant expression system was seen to be required, and this in turn ne-
cessitated that the gene for angiogenin first had to be cloned. This was done by
screening a normal human liver complementary, DNA (cDNA) library with a mixture
of oligonucleotides that could encode amino acids 12 to 20 of angiogenin. Of 350,000
clones screened, seven were found to be positive. The sequence of the largest clone
was determined and found to encode angiogenin as well as a signal peptide, which
is 22 or 24 amino acids long. The presence of the signal peptide, along with the absence
of a detectable level of angiogenin in cells, indicates that angiogenin is likely a secreted
protein and does not get into the culture medium as a result of cell death. The cDNA
also contained 7 nucleotides complementary to untranslated RNA at its 5' end and
175 nucleotides complementary to untranslated RNA at its 3' end along with a poly(A)
tail of 36 nucleotides.[8]

The presence of the cDNA in a normal liver library was actually the first evidence
that angiogenin might be produced by non-neoplastic tissues. Furthermore, the an-

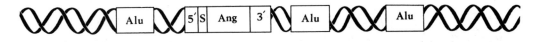

├── 500 bp ──┤

Fig. 8-2.   Structure of human angiogenin gene. The angiogenin transcription unit, including a 5' untranslated region (5'), the coding region for the signal peptide (S), and the mature protein (Ang), and a 3' untranslated region (3'), with flanking *alu* sequences.

giogenin gene appears to be expressed in all cells and tissues examined.[11,12] Therefore, together with the presence of the protein in normal serum, these results indicate that the protein probably has a normal physiologic role in addition to its possible function in neoplasia.

The cDNA was used as a probe to isolate the angiogenin gene from a human genomic library (Fig. 8-2). The gene is unusual in that it apparently contains no introns, although the presence of an intron in the 5' untranslated region of the gene has not been completely ruled out. It is of interest that the bovine pancreatic RNase A gene contains a single intron in its untranslated 5' region.[13] Another noteworthy feature of the gene is the presence of two *alu* sequences[14] flanking the gene, beginning approximately 400 base pairs (bp) upstream and 300 bp downstream from the coding region. A third *alu* sequence is present beginning approximately 1100 bp downstream from the gene. The significance of the apparent absence of introns in the gene and the presence of *alu* sequences flanking the gene is unknown.

ANGIOGENIN FROM RECOMBINANT SOURCES

Mammalian Expression System

A recombinant expression plasmid was synthesized that incorporates the angiogenin gene and the mouse metallothionein I promoter along with the dihydrofolate reductase amplifiable selectable marker under the control of the simian virus (SV) 40 late promoter[15] (Fig. 8-3). This construct was used to transfect baby hamster kidney (BHK) cells. Transfectants were cultured in the presence of increasing concentrations of methotrexate, beginning with 1 μM up to a final concentration of 1 mM to amplify the dihydrofolate reductase gene and the contiguous angiogenin gene. Cells that were viable in 1 mM methotrexate were induced to secrete angiogenin into the medium by addition of 80 μM $ZnSO_4$ and 2 μM $CdSO_4$. Approximately 400 μg of angiogenin is routinely purified per liter of BHK cell medium. The recombinant product was found to be chemically indistinguishable from that purified from HT-29 cells based on amino acid composition, tryptic peptide mapping, and partial sequence analysis of eight tryptic peptides. The immunologic reactivity as well as the ribonucleolytic (see following discussion) and angiogenic activities of the recombinant protein were also comparable to angiogenin isolated from HT-29 cells. Therefore, the mammalian cell expression system provided a somewhat higher level of pure angiogenin than was routinely obtained from plasma and constituted a more convenient and less costly source.

Angiogenin purified from the three sources described previously was used in experiments to explore its ribonucleolytic activity. Angiogenin was found to be 10^5- to 10^6-fold less active than RNase A with 8 commonly employed mono-, di-, and polynucleotide substrates, as well as with 24 other substrates. Because assays with polynucleotide substrates detect the reaction product by spectrophotometric monitoring of the acid-soluble fraction of the reaction mixture, limited cleavage of the substrate might

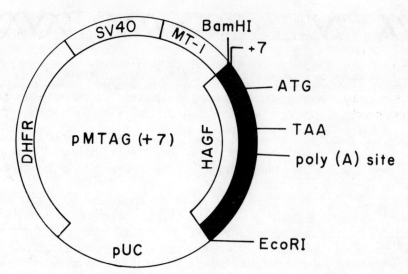

Fig. 8–3. Recombinant plasmid for expression of angiogenin by baby hamster kidney cells. The expression plasmid is composed of the human angiogenin gene (HAGF), the inducible metallothionein-I promoter (MT-I), the dehydrofolate reductase gene (DHFR), the SV-40 late promoter (SV40), and the plasmid pUC 13 (pUC). (Reprinted, in part, with permission from Kurachi, K. et al.: Expression of human angiogenin in cultured baby hamster kidney cells. Biochemistry, 27:6557, 1988. Copyright 1988 American Chemical Society.)

go undetected; larger products would not be acid soluble. Therefore, the action of angiogenin on 28S and 18S ribosomal RNA was visualized by ethidium bromide staining following separation of the reaction products by agarose gel electrophoresis. Angiogenin thereby was found to catalyze the limited cleavage of these RNA substrates to give products that ranged mainly from 100 to 500 nucleotides in length, a pattern that was unchanged on prolonged incubation.[9] Subsequently angiogenin was also found to cleave yeast transfer RNA (tRNA) to acid-soluble products.[16] These critical results established that angiogenin, like RNase A, can cleave RNA but that the two proteins display strikingly dissimilar catalytic action.

Indeed, RNase A degrades 28S and 18S ribosomal RNA (rRNA) to small oligonucleotides under conditions during which angiogenin generates the limited cleavage described previously. Moreover, in experiments that measure production of acid-soluble RNA fragments from tRNA, RNase A was about 55,000 times more active than angiogenin (R. Shapiro, unpublished data). A comparison of the cleavage specificities toward 5S RNA by RNA sequence analysis established that angiogenin significantly cleaves particular bonds that are not appreciably cleaved by RNase A, and vice versa.[17] Because angiogenin can induce neovascularization and RNase A cannot, these results raised the possibility that the catalytic specificity of angiogenin may be the basis of its angiogenic activity.

Even before experimental demonstration, ribonucleolytic activity for angiogenin was predicted based on its primary and calculated tertiary structural similarity with RNase A.[18] The finding of detectable ribonucleolytic activity made it possible to establish the identity of the critical catalytic residues in angiogenin and to confirm their presumable correspondence to those in RNase A. Chemical modification experiments demonstrated that angiogenin contains one or more each of histidine, lysine, and arginine residues, which are essential for its ribonucleolytic activity.[16] Importantly, modification of active-site histidine residues of angiogenin with bromoacetate also led to a significantly diminished angiogenic activity.[9]

Extensive chemical and physical data were available on the identity of functional residues in RNase A, and these data were used to guide our search for the corresponding catalytic residues in angiogenin. The results of the chemical modification experiments clearly indicated that the association between ribonucleolytic and angiogenic activities could be readily tested at the hands of angiogenin modified by oligonucleotide-directed mutagenesis at specific residues critical for catalysis.

The availability of purified angiogenin also permitted investigations into its effect on vascular endothelial cells in culture. Efforts were made to determine whether angiogenin exerted its biologic activity through induction of second messenger molecules. It indeed stimulates a transient but significant increase in the intracellular concentration of diacylglycerol, resulting from the activation of phospholipase C.[19] This response may be relevant to the angiogenic activity of angiogenin, because modification of one or both active-site histidines abolished both the diacylglycerol response in calf pulmonary artery endothelial cells and the ability of angiogenin to induce neovascularization. A discussion of the effect of other, mutagenic changes in the structure of angiogenin on diacylglycerol formation and neovascularization follows.

Escherichia coli Expression System

A second recombinant expression system, this one in *Escherichia coli*, has recently been developed by both our laboratory[20] and another laboratory.[21] Our objectives were to obtain an even higher level of protein than had been obtained from the initial sources and to permit site-directed mutagenesis studies. To optimize synthesis of angiogenin in bacterial cells, an angiogenin gene was chemically synthesized based on *E. coli* codon usage. *E. coli* cells were transformed with an expression plasmid containing a synthetic angiogenin gene under the transcriptional control of the *trp* promoter and translational enhancement of a modified ribosome-binding sequence (Fig. 8-4). The transformed bacteria could be induced to synthesize high levels of angiogenin by addition of indoleacrylic acid to a mid-log phase culture. On centrifugation and lysis of the bacteria, angiogenin was found in the insoluble fraction. Following denaturation in guanidine hydrochloride, reduction with β-mercaptoethanol and reoxidation by dilution, fully active angiogenin can be purified by cation exchange HPLC followed by reversed phase HPLC. In our laboratory, approximately 2 mg of purified protein is routinely obtained from each liter of bacterial culture. This represents a fivefold increase in yield over the mammalian cell expression system.

Amino-terminal sequence analysis showed that the bacterially synthesized angiogenin differed from angiogenin purified from the three other sources described previously. The HT-29, plasma, and BHK angiogenin each has pyroglutamic acid at its amino terminus formed by spontaneous cyclization of glutamine, whereas angiogenin from *E. coli* contains an amino-terminal methionine preceding that glutamine. Therefore, the angiogenin from *E. coli* was treated with *Aeromonas* aminopeptidase under conditions in which the new N-terminal glutamine cyclizes nonenzymatically, yielding a protein that is chemically and biologically indistinguishable from angiogenin purified from the other sources. A comparison of the yields of angiogenin from the four sources is summarized in Table 8-1.

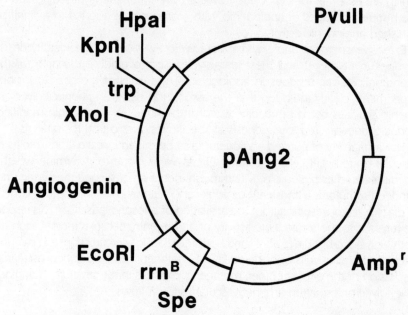

Fig. 8–4. Recombinant plasmid for expression of human angiogenin in *Escherichia coli*. The expression plasmid is a pUC-derived construct containing the *E. coli trp* promoter (*trp*), a chemically syn-thesized angiogenin-coding sequence (angiogenin), a transcription termination sequence (rrn[B]), and an ampicillinase gene (Amp[r]). Cleavage sites for the restriction endonucleases PruII, Hpal, Kpnl, Xhol, and Spe are indicated. (Reprinted with permission from Shapiro, R., et al.: Expression of Met-(-1) Angiogenin in *E. coli*: conversion to the authentic <Glu-1 protein. Anal. Biochem., *175*:450, 1988.)

MUTAGENESIS

Site-Specific Mutagenesis

Site-specific and regional mutagenesis of the angiogenin gene and subsequent expression of the altered protein in *E. coli* have permitted us to test the relationship between the in vitro activities of angiogenin and angiogenesis. Three different mutant proteins, each with strikingly different ribonucleolytic activities, have been produced. The results indicate a critical role for angiogenin's active site in its biologic activity. (A representation of active site residues in the homologous protein, RNase A, interacting with the dinucleotide CpA is shown in Fig. 8-5.) Furthermore, they support the idea

Table 8–1. Yield of Angiogenin From Natural and Recombinant Sources

Source	Yield (µg/L)
HT-29 cell medium	0.5
Human plasma	100.0
BHK cell medium	400.0
Escherichia coli	2000.0

BHK, baby hamster kidney.

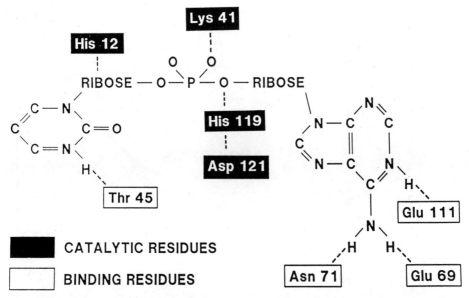

Fig. 8–5. Interaction of ribonuclease A with the dinucleotide, cytidylyl (3′ → 5′) adenosine (CpA).

that angiogenin's ability to induce the release of diacylglycerol from membrane inositol phospholipids is also associated with its angiogenic activity.

The most unexpected results were obtained with an angiogenin altered by mutagenesis in which aspartic acid-116 (D-116, using the single-letter amino acid code) has been changed.[22] This residue was chosen because it is conserved in all pancreatic-type RNases,[23] and because replacement of the corresponding aspartic acid-121 by asparagine in a semi-synthetic RNase diminished catalysis 20-fold with cytidine 2′,3′ cyclic monophosphate as substrate but did not appreciably affect binding of the substrate.[24]

Further evidence that the aspartic acid residue may be important for catalysis was derived from crystal structures of RNase A,[25–27] which indicated a hydrogen bond between this residue and the catalytic histidine at position 119. Together, these data implied, but did not clearly establish, that D-121 of RNase A does play a functional role in ribonucleolytic activity. Mutagenesis of D-116 in angiogenin could test whether this was true for angiogenin, and by implication, could confirm the role of D-121 in RNase A. In addition, if the mutant exhibited decreased ribonucleolytic activity but not decreased substrate binding, it might act as a competitive inhibitor of angiogenin.

Mutagenesis was carried out employing a mixture of oligonucleotides that code for nine possible alternative amino acids at the position of aspartic acid 116.[22] Mutants were screened by DNA sequence analysis, and three were chosen for expression. The gene for each mutant was sequenced in its entirety to ensure that no unintended mutations had occurred.

The D116N mutant, in which the aspartic acid (D) was changed to an asparagine (N), was chosen because it removed the β-carboxylate but still possessed hydrogen-bonding potential with little change in size. The D116H mutant should also be capable of hydrogen bonding and in addition the histidine would acquire a positive charge at low pH. In the third mutant, D116A, the alanine has no potential for hydrogen bonding through its side chain. Altered proteins encoded by each of the three mutants were

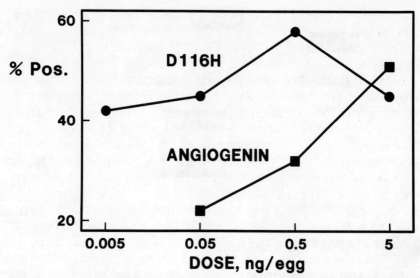

Fig. 8–6. Comparison of angiogenin and D116H activities in chick embryo chorioallantoic membrane assay. In D116H, the aspartic acid (D) in amino acid 116 of angiogenin is replaced by histidine (H).

expressed in *E. coli* and purified. Amino acid analyses were obtained for the intact protein as well as for the purified peptides from a tryptic digest of each protein. This established that only the expected change in the composition had occurred and that disulfide bonds had formed correctly.

Surprisingly, all were found to possess *increased* ribonucleolytic activity with tRNA as a substrate. The D116N, D116A, and D116H angiogenins are 8-, 15-, and 18-fold more active than unaltered angiogenin. Moreover, while catalytically more active toward tRNA, their ribonucleolytic specificity was not changed: The mutant proteins still give angiogenin's characteristic pattern of rRNA cleavage. The appearance of this pattern, in which most of the 28S and 18S RNA is not degraded beyond 100 to 500 nucleotides, is typical of angiogenin but not of RNase A. Furthermore, this pattern could be produced by D116H employing an enzyme concentration that is only $\frac{1}{15}$ of that required with unaltered angiogenin.

Induction of neovascularization by D116H angiogenin was tested extensively in the chick embryo chorioallantoic membrane assay. It was found that the mutant protein induced blood vessel formation at significantly lower doses than were needed for angiogenin. We estimate that the angiogenic potency of angiogenin has been increased one to two orders of magnitude by alteration of D116 (Fig. 8-6). These data provided a strong indication that the characteristic ribonucleolytic activity of angiogenin and its ability to induce neovascularization may be closely related.

Because increased characteristic ribonucleolytic activity was clearly related to increased angiogenic potency in this one case, we sought to establish whether the converse is true: Is a loss of ribonucleolytic activity accompanied by a loss of biologic activity? This question has been addressed by site-directed mutagenesis of the angiogenin DNA sequence encoding lysine 40.[28]

This particular residue was chosen based on data obtained from both RNase A and angiogenin. With RNase A, chemical modification of the corresponding lysine 41 residue can be carried out specifically with several different reagents, resulting in inactivation of the enzyme.[29–32] Crystallographic and neutron-diffraction studies,[33–35] confirm that this lysine is in the active site of the enzyme. Angiogenin is also inactivated by lysine-specific reagents, albeit three- to ten-fold more slowly. It was not possible, however, to obtain an inactive angiogenin modified at a unique lysine.

Based on structural homology with RNase A, residue 40 was chosen as the one lysine most likely to be catalytically important in angiogenin. It was, therefore, changed by site-directed mutagenesis to a glutamine (K40Q) in one case to retain the approximate size of the side chain while changing the charge from positive to neutral. A second mutant protein was produced in which the lysine was replaced by arginine (K40R) to examine the effect of a seemingly very conservative change. Amino acid analysis, tryptic mapping, chromatographic analysis, and circular dichroism spectroscopy all established that the altered angiogenins do not differ significantly from the unaltered protein except at the site of replacement. K40Q angiogenin was devoid of measurable ribonucleolytic activity with tRNA as a substrate, but K40R angiogenin possessed 2.2% of the activity of unaltered angiogenin in the same assay. These results established that lysine 40 is critical for the enzymatic activity of angiogenin; its role in catalysis can only be marginally fulfilled by a residue of very similar charge and size.

Because one major objective of this work was to test whether a loss of ribonucleolytic activity is accompanied by a loss of angiogenic potency, K40Q angiogenin was tested in the chick embryo chorioallantoic membrane assay. In this assay, K40Q showed little, if any, angiogenic activity at concentrations at which the unaltered molecule induces neovascularization. These results establish that loss of enzymatic activity of K40Q angiogenin coincides with a loss of blood vessel-inducing activity, but they do not establish a causal relationship. Clearly, an intact active site is critical for angiogenic activity; whether the site must be catalytically competent or whether it must only retain the ability to bind to the appropriate molecule is not clear. In addition, while an intact active site is critical for angiogenic activity, it cannot be ruled out that other regions may be critical as well.

Because these data indicate that the structural integrity of K40Q is preserved, except at position 40, and because K40Q is essentially devoid of angiogenic activity, this mutant can be used to test whether any in-vitro activity of angiogenin may reflect the mechanism by which angiogenin induces neovascularization. Several in-vitro activities of angiogenin have been reported. These include ribonucleolytic activity, placental ribonuclease-inhibitor binding, inhibition of protein synthesis,[36,37] and stimulation of diacylglycerol production in cultured cells. If any of these activities is involved in the induction of angiogenesis by angiogenin, the activity of K40Q in the assay should be significantly diminished. K40Q therefore could be a powerful tool for distinguishing whether an in-vitro activity is relevant to the angiogenic activity of angiogenin.

Lys-40 has been shown to be essential for the ribonucleolytic activity of angiogenin. Furthermore, as was true for its ribonucleolytic and angiogenic activities, the ability of K40Q to stimulate the release of diacylglycerol from the membranes of calf pulmonary artery endothelial cells was also greatly attenuated. In this case, the altered protein was completely inactive at six concentrations between 0.01 and 1000 ng/mL (R. Bicknell, unpublished data). Angiogenin itself is optimally active with these cells between 0.1 and 10 ng/mL. Because K40Q is not angiogenic, these data are consistent with the possibility that the induced production of diacylglycerol by angiogenin is one of the steps leading to neovascularization.

Regional Mutagenesis

Similarities between the structures of angiogenin and RNase A have led to an understanding that the common active site is critical for the ribonucleolytic activity of angiogenin. Conversely, because RNase A is not angiogenic, positions in angiogenin that differ in RNase A must contribute to the angiogenic activity of angiogenin.

One of the most salient differences between the primary structures of angiogenin and RNase A lies in the region between residues 62 and 71 of angiogenin, which correspond to residues 63 to 74 of bovine RNase A. In this region, 9 of the 10 residues differ, and RNase A contains two additional residues. Although very different from RNase A, this region of human angiogenin is very similar to the corresponding region in bovine angiogenin (ref. 38 and M. D. Bond and D. J. Strydom, unpublished data). Six of the ten bovine residues are conserved when compared to human angiogenin, indicating that this region may be important for angiogenesis. Moreover, angiogenin lacks the two cysteines (positions 65 and 72 in RNase A) that form a disulfide bond in RNase A: angiogenin contains three disulfide bonds, whereas RNase A possesses four. Crystallographic studies of RNase A show that the disulfide loop from Cys-65 to Cys-72 is involved in purine binding.[25,33] If this region of angiogenin is critical for its biologic activity, then replacement of the region with the corresponding region of RNase A should diminish that activity.

Replacement of a small segment of a protein by a new segment, termed *regional mutagenesis*, has an advantage over site-directed mutagenesis because a larger number of residues can be evaluated for each mutant. Then, if the region is found to be important for an activity, it can be dissected by step-wise, site-directed mutagenesis to identify the specific critical residues. One potential pitfall of regional mutagenesis is that the new segment may interfere with the proper folding and resulting tertiary structure of the remainder of the molecule. Because the desired disulfide loop occurs on the surface of RNase A and because the computed angiogenin tertiary structure indicates that the angiogenin and RNase A backbones do not differ significantly in the segments bordering the loop, the chance that the hybrid protein would fold properly was thought to be good.

Construction of the hybrid gene[39] was carried out as shown in Figure 8-7. Briefly, the segments of the gene encoding residues 1 to 57 and 73 to 123 were prepared by restriction endonuclease cleavage followed by agarose gel electrophoretic separation and purification of the fragments. The two fragments, A and C, along with an *E. coli* high-copy-number plasmid and an oligonucleotide duplex, fragment B, encoding the RNase A replacement region, were ligated together, and the ligation mix was used to transform *E. coli*. The angiogenin–RNase hybrid gene (ARH-I) was then cleaved from the plasmid and ligated into an expression vector. Subsequent transformation, culture, and expression in *E. coli* yielded the hybrid protein.

Structural characterization confirmed that the desired protein had been obtained and the ARH-I contained a fourth disulfide bond not present in angiogenin. The ribonucleolytic activity of ARH-I was found to be increased dramatically over that of angiogenin. In the rRNA assay during which angiogenin catalyzes the limited cleavage of 18S and 28S RNA, ARH-I is approximately 20 times more active than native angiogenin. However, while nanomolar concentrations of ARH-I generate the same characteristic product pattern as angiogenin, an increase of ARH-I concentration into the micromolar range completely degrades fragments that are not degraded by angiogenin at a similar

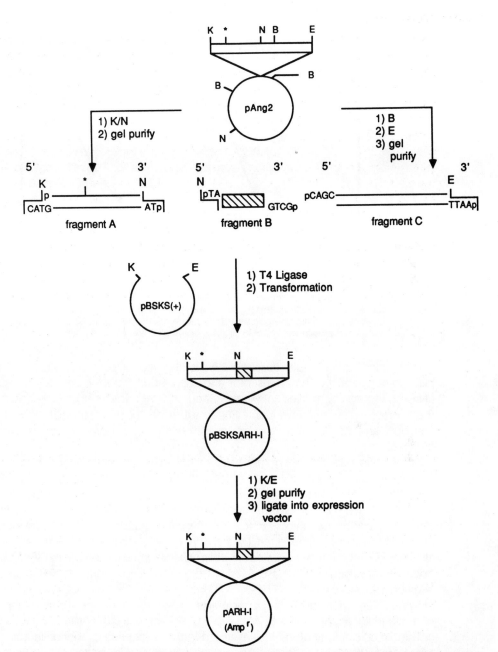

Fig. 8–7. Regional mutagenesis of angiogenin. A plasmid, pAng2, for the expression of angiogenin in *Escherichia coli* was cleaved with Kpn I (K) and Nde I (N) or with Bst M I (B) and *Eco*RI (E), and appropriate fragments (A and C) were purified and ligated with a synthetic duplex (fragment B) coding for Cys-58 to Tyr-73 of RNase A, and an *E. coli* cloning vector, pBSKS(+) to yield plasmid BSKSARH-I. The hybrid-coding region was cleaved from pBSKSARH-I, purified, and ligated into the *E. coli* expression vector to yield the expression plasmid pARH-I. (Reprinted with permission from Harper, J.W., and Vallee, B.L.: A covalent angiogenin/ribonuclease hybrid with a fourth disulfide bond generated by regional mutagenesis. Biochemistry, *28*:1875, 1989. Copyright 1989 American Chemical Society.)

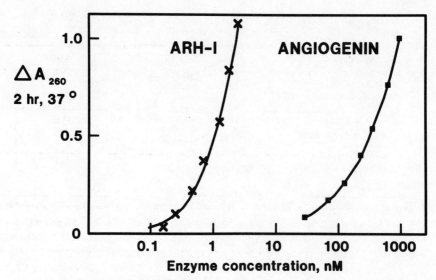

Fig. 8–8. Cleavage of transfer RNA (tRNA) by ARH-I and angiogenin. The A_{260} of perchloric acid-soluble tRNA fragments generated by ARH-I and angiogenin is shown as a function of the increasing concentration of each protein.(Reprinted with permission from Harper, J.W., and Vallee, B.L.: A covalent angiogenin/ribonuclease hybrid with a fourth disulfide bond generated by regional mutagenesis. Biochemistry, *28*:1875, 1989. Copyright 1989 American Chemical Society.)

concentration. These results demonstrate that the ribonucleolytic activity of ARH-I retains some specificity characteristic of angiogenin but that it possesses a substantial RNase A-like activity. Indeed, ARH-I has about 200-fold greater ribonucleolytic activity than angiogenin with conventional RNase substrates including poly(U), poly(C), CpA, and UpA, and it is 300-fold more active with tRNA as a substrate than is angiogenin (Fig. 8-8).

Angiogenin has also been shown to inhibit in-vitro protein synthesis in a rabbit reticulocyte lysate specifically by cleavage of the 18S rRNA in the intact ribosome. RNase A is somewhat less active in this assay. Experiments with ARH-I showed it to be at least 10 times more potent than angiogenin, indicating that it not only retains the cleavage specificity of angiogenin toward intact ribosomes but also acts at an increased rate. This inference is supported by electrophoretic analysis of the reticulocyte RNA cleaved by ARH-I, which exhibited a pattern characteristic of angiogenin and different from that observed with RNase A. Much higher concentrations of ARH-1 produced additional cleavage products. These results indicate that ARH-1, like its action on purified 18S and 28S rRNA, retains the ribonucleolytic activity on intact ribosomes characteristic of angiogenin, and in addition, ARH-1 possesses some RNase A-like activity.

The biologic activity of ARH-I was tested in the chick embryo chorioallantoic membrane assay. Its ability to induce neovascularization was found to be significantly diminished compared with that of angiogenin. This result leads to several important considerations. First, because ARH-I has an enhanced ability to inhibit protein synthesis—specificially by cleavage of the 18S rRNA—but a decreased angiogenic activity, it appears, prima facie, that angiogenin does not induce angiogenesis via inhibition of protein synthesis. Second, ARH-I does not induce diacylglycerol in calf pulmonary artery endothelial cells in the concentration range that is optimal for angiogenin (R. Bicknell, unpublished data); this is consistent with the possibility of a causal relationship

Table 8-2. Summary of Activiities of Three Angiogenin Mutants Compared to Angiogenin

Activity	D-116H	K40Q	ARH-I
Angiogenesis	10–100x	Decreased	Decreased
Ribonucleolytic			
rRNA	15x	n.d.	Characteristic pattern transient
tRNA	18x	Abolished	300x
Inhibitor binding	n.d.	0.0008x	~0.5x
Phospholipase C activation	n.d.	Abolished	10^{-3}–10^{-4}

n.d., not determined; rRNA, ribosomal RNA; tRNA, transfer RNA.

between diacylglycerol formation and angiogenesis. Third, the fact that ARH-I has diminished angiogenic activity yet strikingly enhanced ribonucleolytic activity with a variety of substrates requires careful analysis. It is particularly noteworthy that both ARH-I and D116H display increased ribonucleolytic activity compared to angiogenin, but ARH-I has diminished angiogenic activity while D116H has enhanced activity.

The activities of three angiogenin mutants, D116H, K40Q, and ARH-I, are compared to that of angiogenin in Table 8-2. Several points are to be noted. First, because D116H, K40Q, and ARH-I all contain alterations in at least one active site residue, it is evident that perturbations to the active site have in all cases resulted in changes in ribonu-cleolytic activity. Second, because all mutants display altered angiogenic activity, the structure of the active site must be critical for its biologic activity. Third, employing the RNA substrates now available, changes in ribonucleolytic activity do not clearly predict how angiogenic activity will be affected. The discovery of an appropriate substrate may allow more reliable predictions. We emphasize, however, that a causal relationship between ribonucleolytic activity and angiogenic activity has not been established.

The active site may have a binding function that is relevant to angiogenin's biologic activity, along with a catalytic action that is not relevant to that activity. The active site may have evolved to bind effectively to a ligand or receptor, and the residual ribo-nucleolytic activity may be an evolutionary vestige persisting by virtue of the binding requirements of the active site. Changes in the active site would not necessarily result in corresponding changes in both activities.

Binding to an RNA molecule covalently or noncovalently associated with a com-ponent of the cell membrane could conceivably trigger the activation of phospholipase C and the consequent generation of diacylglycerol. A protein receptor in the cell mem-brane is another possible target to which angiogenin could bind.

PLACENTAL RIBONUCLEASE INHIBITOR

When angiogenin induces neovascularization via catalysis or binding, the active site is clearly involved in its biologic activity. Therefore one would expect that the angiogenic activity of angiogenin would be inhibited by a competitive inhibitor. An inhibitor of ribonuclease A, at first, reported to be a noncompetitive inhibitor, had been purified to homogeneity from human placenta[40] well before the discovery of angiogenin. This ribonuclease inhibitor subsequently was found to inhibit both the angiogenic and ribonucleolytic activities of angiogenin. The inhibitor has been shown to be a tight binding ($K_i = 4.3 \times 10^{-14}$ M), competitive (rather than noncompetitive) inhibitor of RNase A.[41] By inference from the homology between RNase A and angiogenin, the placental ribonuclease inhibitor (PRI) is also a competitive inhibitor of angiogenin (K_i = 7.1 × 10^{-16} M). This is supported by the fact that the K40Q mutant, which contains a change in the active site, binds to PRI with a K_i that is 1200-fold higher than that with unaltered angiogenin.[42]

These data indicate that angiogenin is capable of strong protein interaction through contact with its active site. The evidence obtained to date does not clearly indicate whether PRI acts in vivo with angiogenin to modulate the process of neovascularization. Thus, while the K40Q and ARH-I mutants display diminished angiogenic activity, and the binding of K40Q is greatly decreased, the binding of ARH-I to PRI is unchanged.

Whether or not PRI acts in vivo to regulate angiogenesis. it must be considered a potential therapeutic agent for inhibiting undesired vascular development. In addition, structural information on PRI could provide a basis for developing more stable and less costly inhibitors of angiogenin.

With this goal in mind, we undertook a project to determine the primary sequence of PRI. Because cloning and sequencing of a PRI cDNA would be more rapid than direct amino acid sequence analysis of a 50-kD protein, the former techniques were employed. Furthermore, the availability of the cDNA would permit overexpression in a heterologous host such as *E. coli*. As with angiogenin, an expression system would afford a convenient source of large amounts of material as well as the potential to carry out mutagenesis.

Seven cDNA clones were isolated from a human placental library in λgt11 employing affinity-purified anti-PRI antibodies, and the nucleotide sequences of two were determined completely.[43] Amino acid sequence data from 30 tryptic PRI peptides confirmed that the correct clones had been isolated and that both contained the entire PRI coding region. Subsequently, human PRI has been cloned by another group.[44]

The inferred amino acid sequence is noteworthy for the presence of seven direct, internal repeats, each 57 amino acids in length, which make up 87% of the molecule. Because the stoichiometry of binding between angiogenin and PRI is 1:1, it will be of interest to understand the significance of the 7-fold repeated structure. Knowledge of the primary structure may now permit us to locate the region in PRI that is involved in binding to angiogenin employing chemical modification, cross-linking, and mutagenesis studies and may ultimately lead to the design of new RNase and angiogenin inhibitors.

COMMENTS

Angiogenin, a potent inducer of neovascularization, displays significant homology to RNase A. The techniques of molecular biology have facilitated biochemical and biologic studies by allowing production of large amounts of material for further studies. Moreover, mutagenesis experiments have permitted us to test whether each of the several in-vitro activities of angiogenin is associated with its angiogenic activity (Table 8-2). The results of these experiments indicate that perturbations to the active site of angiogenin affect its biologic activity, but the results obtained thus far have not established whether catalysis by or simple binding to the active site is critical for neovascularization. The capability to dissect the structural basis of angiogenin's biologic activity based on rational predictions has provided an incisive and fruitful approach toward understanding the molecular basis of angiogenesis.

Acknowledgments

We thank F. Lee and Drs. R. Bicknell, J. W. Harper, R. Shapiro, D. J. Strydom, and B. L. Vallee for helpful discussions. This work was supported by funds from Hoechst, A. G., under an agreement with Harvard University. E.A.F. was supported by National Research Service Award HL-07582 from the National Heart, Lung and Blood Institute.

REFERENCES

1. Folkman, J., and Klagsbrun, M.: Angiogenic factors. Science, *235*:442, 1987.
2. Folkman, J., and Cotran, R.S.: Relation of vascular proliferation to tumor growth. Int. Rev. Exp. Pathol., *16*:207, 1986.
3. Alderman, E.M., Lobb, R.R., and Fett, J.W.: Isolation of tumor-secreted products from human carcinoma cells maintained in a defined protein-free medium. Proc. Natl. Acad. Sci. USA, *82*:5771, 1985.
4. Knighton, D., Ausprunk, D., Tapper, D., and Folkman, J.: Avascular and vascular phases of tumor growth in the chick embryo. Br. J. Cancer, *35*:347, 1977.
5. Fett, J.W. et al.: Isolation and characterization of angiogenin, an angiogenic protein from human carcinoma cells. Biochemistry, *24*:5480, 1985.
6. Shapiro, R., Strydom, D.J., Olson, K.A., and Vallee, B.L.: Isolation of angiogenin from normal human plasma. Biochemistry, *26*:5141, 1987.
7. Strydom, D.J. et al.: Amino acid sequence of human tumor-derived angiogenin. Biochemistry, *24*:5486, 1985.
8. Kurachi, K. et al.: Sequence of the cDNA and gene for angiogenin, a human angiogenesis factor. Biochemistry, *24*:5494, 1985.
9. Shapiro, R., Riordan, J.F., and Vallee, B.L.: Characteristic ribonucleolytic activity of human angiogenin. Biochemistry, *25*:3527, 1986.
10. Shapiro, R., and Vallee, B.L.: Human placental ribonuclease inhibitor abolishes both angiogenic and ribonucleolytic activities of angiogenin. Proc. Natl. Acad. Sci. USA, *84*:2238, 1987.
11. Rybak, S.M., Fett, J.W., Yao, Q.Z., and Vallee, B.L.: Angiogenin mRNA in human tumor and normal cells. Biochem. Biophys. Res. Commun., *146*:1240, 1987.
12. Weiner, H.L., Weiner, L.H., and Swain, J.L.: Tissue distribution and developmental expression of the messenger RNA encoding angiogenin. Science. *237*:280, 1987.
13. Carsana, A. et al.: Structure of the bovine pancreatic ribonuclease gene: the unique intervening sequence in the 5' untranslated region contains a promoter-like element. Nucleic Acids Res., *16*:5491, 1988.
14. Kariya, Y. et al.: Revision of consensus sequence of human *alu* repeats—a review. Gene, *53*:1, 1987.
15. Kurachi, K. et al.: Expression of human angiogenin in cultured baby hamster kidney cells. Biochemistry, *27*:6557, 1988.
16. Shapiro, R., Weremowicz, S., Riordan, J.F., and Vallee, B.L.: Ribonucleolytic activity of angiogenin: essential histidine, lysine, and arginine residues. Proc. Natl. Acad. Sci. USA, *84*:8783, 1987.
17. Rybak, S.M., and Vallee, B.L.: Base cleavage specificity of angiogenin with *Saccharomyces cerevisiae* and *E. coli* 5S RNAs. Biochemistry, *27*:2288, 1988.
18. Palmer, K.A., Scheraga, H.A., Riordan, J.F., and Vallee, B.L.: A preliminary three-dimensional structure of angiogenin. Proc. Natl. Acad. Sci. USA, *83*:1965, 1986.
19. Bicknell, R., and Vallee, B.L.: Angiogenin activates endothelial cell phospholipase C. Proc. Natl. Acad. Sci. USA, *85*:5961, 1988.
20. Shapiro, R. et al.: Expression of Met-(-1)Angiogenin in *E. coli*: conversion to the authentic <Glu-1 protein. Anal. Biochem., *175*:450, 1988.
21. Denèfle, P. et al.: Chemical synthesis of a gene coding for human angiogenin, its expression in *Escherichia coli* and conversion of the product into its active form. Gene, *56*:61, 1987.
22. Harper, J.W., and Vallee, B.: Mutagenesis of aspartic acid-116 enhances the ribonucleolytic activity and angiogenic potency of angiogenin. Proc. Natl. Acad. Sci. USA, *85*:7139, 1988.
23. Beintema, J.J., Fitch, W.M., and Carsana, A.: Molecular evolution of pancreatic-type ribonucleases. Mol. Biol. Evol., *3*:262, 1986.
24. Stern, M.S., and Doscher, M.S.: Aspartic acid-121 functions at the active site of bovine pancreatic ribonuclease. FEBS Lett., *171*:253, 1984.
25. Wodak, S.Y., Lin, M.Y., and Wyckoff, H.W.: The structure of cytidylyl(2',5')adenosine when bound to pancreatic ribonuclease. J. Mol. Biol., *116*:855, 1977.
26. Campbell, R.L., and Petsko, G.A.: Ribonuclease structure and catalysis: crystal structure of sulfate-free native ribonuclease A at 1.5 Å resolution. Biochemistry, *26*:8579, 1987.
27. Wlodawer, A., Svensson, L.A., Sjolin, L., and Gilliland, G.L.: Structure of phosphate-free ribonuclease A refined at 1.26 Å. Biochemistry, *27*:2705, 1988.
28. Shapiro, R., Fox, E.A., and Riordan, J.F.: The role of lysines in human angiogenin: chemical modification and site-directed mutagenesis. Biochemistry, *28*:1726, 1989.

29. Hirs, C.H.W., Halman, M., and Kycia, J.H.: Dinitrophenylation and inactivation of bovine pancreatic ribonuclease A. Arch. Biochem. Biophys., *111*:209, 1965.
30. Henrickson, R.L.: On the alkylation of amino acid residues at the active site of ribonuclease. J. Biol. Chem., *241*:1393, 1966.
31. Raetz, C.R.H., and Auld, D.S.: Schiff bases of pyridoxal phosphate with active center lysines in ribonuclease A. Biochemistry, *11*:2229, 1972.
32. Bello, J., Iijima, H., and Kartha, G.: A new arylating agent, 2-carboxy-4,6-dinitrochlorobenzene: reaction with model compounds and bovine pancreatic ribonuclease. Int. J. Pept. Protein Res., *14*:199, 1979.
33. Richards, F.J., and Wyckoff, H.W.: Atlas of molecular structures in biology: 1. Ribonuclease S. Edited by D.C. Phillips and F.M. Richards. Oxford, Clarendon Press, 1973.
34. Borkakoti, N., Palmer, R.A., Heneef, I., and Moss, D.S.: Specificity of pancreatic ribonuclease A: an X-ray study of a protein-nucleotide complex. J. Mol. Biol., *169*:743, 1983.
35. Wlodawer, A., Miller, M., and Sjolin, L.: Active site of RNase: neutron diffraction study of a complex with uridine vanadate, a transition-state analog. Proc. Natl. Acad. Sci. USA, *80*:3628, 1983.
36. St. Clair, D.K., Rybak, S.M., Riordan, J.F., and Vallee, B.L.: Angiogenin abolishes cell-free protein synthesis by specific ribonucleolytic inactivation of ribosomes. Proc. Natl. Acad. Sci. USA, *84*:8330, 1987.
37. St. Clair, D.K., Rybak, S.M., Riordan, J.F., and Vallee, B.L.: Angiogenin abolishes cell-free protein synthesis by specific ribonucleolytic inactivation of 40S ribosomes. Biochemistry, *27*:7263, 1988.
38. Maes, P. et al.: The complete amino acid sequence of bovine milk angiogen. FEBS Lett., *241*:41, 1988.
39. Harper, J.W., and Vallee, B.L.: A covalent angiogenin/ribonuclease hybrid with a fourth disulfide bond generated by regional mutagenesis. Biochemistry, *28*:1875, 1989.
40. Blackburn, P., Wilson, G., and Moore, S.: Ribonuclease inhibitor from human placenta: purification and properties. J. Biol. Chem., *252*:5904, 1977.
41. Lee, F.S., Shapiro, R.S., and Vallee, B.L.: Tight-binding inhibition of angiogenin and ribonuclease A by placental ribonuclease inhibitor. Biochemistry, *28*:225, 1989.
42. Lee, F.S., and Vallee, B.L.: Binding of placental ribonuclease inhibitor to the active site of angiogenin. Biochemistry, *28*:3556, 1989.
43. Lee, F.S. et al.: Primary structure of human placental ribonuclease inhibitor. Biochemistry, *27*:8545, 1988.
44. Schneider, R. et al.: The primary structure of human ribonuclease/angiogenin inhibitor (RAI) discloses a novel highly diversified protein superfamily with a common repetitive module. EMBO J., *7*:4151, 1988.

Chapter 9

MOLECULAR BIOLOGY OF THE RED BLOOD CELL MEMBRANE PROTEINS

Pamela S. Becker • Edward J. Benz, Jr.

The red blood cell is charged with the vital task of oxygen delivery to all tissues. To perform this function, it must withstand the shear stresses of the circulation and must deform to negotiate narrow passages in organs such as the spleen. To maintain durability, flexibility, and cell shape, the red blood cell membrane is equipped with an interlocking network of proteins that lines its inner surface. This structure is termed the *membrane skeleton.*

The major protein elements of the membrane skeleton have been identified and their interactions defined. Proteins analogous in structure and function to the mammalian erythrocyte membrane proteins have been described in many species and tissues. Several of the genes encoding the erythrocyte and nonerythrocyte proteins have been isolated and their chromosome origins assigned. Extensive homology at both the amino acid and nucleotide sequence level has been observed. In some cases, intricate pathways of gene expression have been shown to be involved in the generation of the proper types and quantity of the tissue-specific skeletal proteins.

In both mouse and man, several hereditary defects of the erythrocyte membrane cause varying degrees of hemolytic anemia and abnormal cell shape. These include hereditary spherocytosis, elliptocytosis, and pyropoikilocytosis. In certain kindreds, the primary molecular defects of these disorders have been discovered and in some cases, confirmed at the gene level.

This chapter will first describe the structural details of the normal erythrocyte membrane skeleton, then define the molecular biology of these proteins, and finally, discuss the known molecular defects responsible for the hereditary disorders.

THE MEMBRANE SKELETON: COMPONENT PROTEINS

The erythrocyte membrane is approximately 50% protein by weight. The proteins are of two types, designated integral and peripheral. The integral membrane proteins are typically transmembrane glycoproteins that span the lipid bilayer via peptide segments consisting of particularly hydrophobic amimo acids. The peripheral membrane proteins associate with the transmembrane proteins or polar part of the lipid bilayer by chemical interactions.

The erythrocyte membrane proteins derived from the hemoglobin-depleted red blood cell "ghost" are designated numerically according to their relative mobility during sodium dodecyl sulfate polyacrylamide gel electrophoresis (SDS-PAGE) (Table 9-1). For a schematic diagram of the erythrocyte membrane, refer to Figure 9-1.

Table 9–1.   The Human Erythrocyte Membrane Proteins

Band Name	Molecular Weight	Chromosomal Location
1 α-Spectrin	240,000	1q22–q25
2 β-Spectrin	220,000	14
2.1 Ankyrin	210,000	8p11–p21
3	95,000	17
4.1	80–78,000	1p36.2–p34
5 Actin	42,000	7pter–q22
Glycophorin A	36,000	4q28–q31
Glycophorin B	20,000	4q28–q31
Glycophorin C	32,000	2q14–21
Glycophorin D*	23,000	2q14–21
Nonerythroid α-spectrin		9q33–q34

* Chromosomal assignment is based on analysis of abnormal "fusion" glycophorins thought to have arisen by unequal crossover between linked loci.

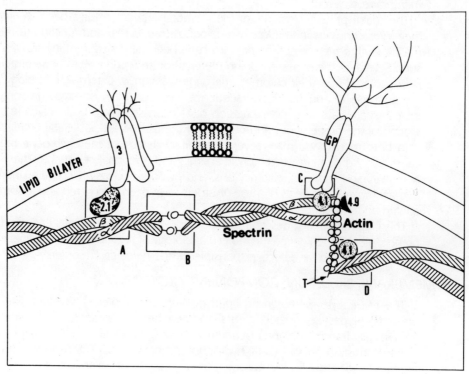

Fig. 9–1.   The erythrocyte membrane skeleton. Spectrin dimers composed of intertwined α- and β-chains self-associate at one end to form tetramers or higher oligomers and bind to short filaments of F-actin and protein 4.1 at the opposite end. The membrane skeleton is attached to the lipid bilayer via interactions with ankyrin to the transmembrane protein 3 and with protein 4.1 to the transmembrane protein, glycophorin (A or C) or protein 3. Tropomyosin (T) lies in the groove of the actin protofilament. (GP, glycophorin.)

The major erythrocyte skeletal protein, spectrin, is composed of proteins 1 (α-subunit) and 2 (β-subunit). Protein 2.1 is ankyrin, the protein that mediates the attachment of spectrin to the lipid bilayer via protein 3. Protein 3, also known as band 3, is the transmembrane anion-transport channel. Protein 4.1 is an 80-kilodalton (kD) globular protein that mediates spectrin's binding to protein 5, erythrocyte actin, and also binds to glycophorin and protein 3. Protein 4.9 acts as an actin-bundling protein in vitro. Protein 6 is glyceraldehyde-3-phosphate dehydrogenase. One of the proteins contained in the band 7 region is erythrocyte tropomyosin.

The erythrocyte membrane skeleton is composed primarily of spectrin, short filaments of actin, and proteins 4.1 and 4.9.[1] The interlocking network of proteins has been visualized by negative staining electron microscopy (Fig. 9-2).[2] The membrane skeleton maintains the shape of the erythrocyte, as evidenced by the fact that skeletons isolated from elliptocytes[3] or irreversibly sickled cells[4] retain the shape of the cell from which they were derived. It follows that the skeletal proteins must be major determinants of cell shape.

Spectrin

Spectrin is the major constituent of the erythrocyte membrane skeleton, accounting for 75% of its weight. It consists of two nonidentical chains, α and β, of considerable size, 240 and 220 kD, respectively, arranged in antiparallel alignment with respect to their aminotermini. The long chains are variably twisted with irregular points of inter-chain attachment. The protein is composed of many (20 on the α-chain and 18 on the β-chain) repeating homologous subunits, each 106 amino acids long, arranged in triple α-helices (Fig. 9-3).[5] Limited proteolytic digestion reproducibly results in produc-tion of large polypeptides of 28 to 80 kD called *domains*.[6,7] Digestion with trypsin at 0°C results in production of five domains from the α-chain, designated α-I through V, and four from the β-chain, designated β-I through IV (Fig. 9-4). Abnormal domain maps have been demonstrated in several individuals with hereditary hemolytic anemias. A single phosphorylated site occurs on the β-I polypeptide.[8]

Spectrin dimers participate in several binding interactions, including self-association to tetramers[9] and higher oligomers,[10] attachment of ankyrin,[11] and involvement in a ternary complex with short filaments of actin and protein 4.1.[12–14] The interactions occur through specific sites: The oligomerization occurs at the "head" end of the molecule[15] comprised of the N-terminus of the α-I and the C-terminus of the β-I domains, ankyrin binding occurs in the region of the β-I–B-II junction,[15] and the in-teraction with actin and protein 4.1 takes place at the "tail" end[16] opposite to the self-association site.

Proteins immunologically cross-reactive with erythrocyte spectrin have been iden-tified in nearly all tissues (for review, see ref. 17). The nonerythroid spectrins share the linear morphology and functional properties of cross-linking of actin filaments and binding to membrane receptors characteristic of erythroid spectrin. In addition, there is a strong calmodulin-binding site on the α-chain of nonerythroid spectrin. The major types are fodrin, which is widespread and is the type found in the brain, and TW260/240, which is the type found in the intestinal brush border.

Protein 4.1

Protein 4.1 is a globular phosphoprotein of approximately 80 kD. On Laemmli poly-acrylamide gels, it appears to be a doublet of proteins 4.1a and b, which are sequence-

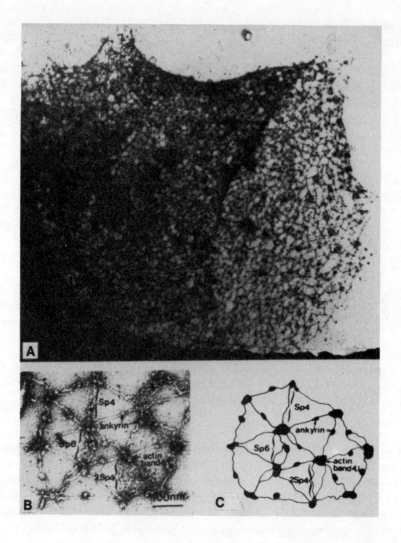

Fig. 9–2. Electron microscopy of negative-stained erythrocyte membrane skeleton. A. Low-power view of spread membrane skeleton, with marginal region of exposed bottom layer. B. High-power view, with schematic diagram as described in C. C. Hexagonal lattice of junctional complexes, composed of F-actin protofilaments and protein 4.1 molecules, cross-linked by spectrin tetramers (Sp4), spectrin hexamers (Sp6), and double-spectrin tetramers (2Sp4). Globular ankyrin molecules are attached to spectrin near the junction of spectrin dimers. (Reprinted with permission from Liu, S.-C., Derick, L. H., and Palek, J.: Visualization of the hexagonal lattice in the erythrocyte membrane skeleton. J. Cell Biol., *104*:527, 1987.)

related.[18,19] It binds to spectrin at the tail end and greatly amplifies the binding of spectrin to actin. A 10-kD central peptide is responsible for this function.[20] In addition, the binding of protein 4.1 to glycophorins A[21] and C[22] and protein 3[23] has been described. It is thought that the binding to glycophorin serves as a second major attachment site of the membrane skeleton to the lipid bilayer. The protein 4.1–glycophorin A interaction is apparently regulated by polyphosphoinositides.[24]

Fig. 9–3. Model of spectrin dimer as a series of homologous repeating subunits. Both the α- and β-chains are composed of a series of triple α-helical segments joined by short segments of nonhelical sequence. Each triple helical segment is composed of 106 amino acids of homologous sequence. The α-chain has 20 repeats, and the β-chain 18. The tenth repeat of the α-chain and the carboxy terminus of the β-chain are thought to be nonhomologous. In nonerythroid spectrin, repeats 10, 11, 21, and 22 are nonhomologous. (Reprinted by permission from *Nature* Vol. 311 p. 177. Copyright © 1984 Macmillan Magazines Ltd.)

Actin

Erythrocyte actin is similar to that in other cells. It is of the β-type and is organized into double helical filaments (termed *F-actin*) of 12 to 16 monomers.[25] Spectrin dimers bind to the side of actin filaments with an association constant of only 10^3 M^{-1}, compared to 10^{12} M^{-2} for the ternary spectrin–actin–protein 4.1 interaction.[14]

Protein 4.9

Protein 4.9 is isolated with the other skeletal proteins by high salt or detergent. It is a 48-kD phosphoprotein that exists as a trimer and bundles actin filaments in vitro.[26] Its function has not been further elucidated.

Tropomyosin

The erythrocyte form of tropomyosin consists of a heterodimer of 27- and 29-kD subunits.[27] Enough copies are in the red cell to cover all the actin protofilaments. It is capable of limiting the amount of spectrin that co-sediments with F-actin.

Adducin

Adducin is a heterodimer of 103- to 105-kD α-subunits and 97- to 100-kD β-subunits that amplifies spectrin-actin binding and promotes the binding of additional spectrin molecules to actin filaments.[28,29] It binds much more tightly to spectrin–actin complexes than to either protein alone.[28] Adducin binds to actin and bundles it into groups of laterally arranged filaments.[29] Calmodulin and calcium inhibit the effect of adducin on spectrin–actin binding.[28,29] Adducin is phosphorylated by protein kinase C.[28]

ALPHA CHAIN

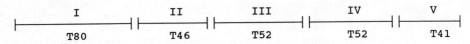

BETA CHAIN

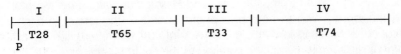

Fig. 9–4. Alignment of the tryptic domains of spectrin. The large, proteolytically resistant polypeptides (domains) are aligned 80-46-52-52-41 on the α-chain and 28-65-33-74 on the β-chain. The P indicates the peptide that contains the phosphorylation site. The T indicates that the peptides are produced by trypsin. The domains are designated by the chain of origin and the Roman numeral indicating their relative positions.

THE ASSOCIATION OF THE MEMBRANE SKELETON TO THE LIPID BILAYER VIA TRANSMEMBRANE GLYCOPROTEINS

Ankyrin

Ankyrin is a 210-kD globular peripheral membrane protein that attaches spectrin to protein 3.[11] It is highly susceptible to proteolysis, yielding lower molecular weight forms known as proteins 2.2, 2.3, and 2.6, which all originate from the native 2.1.[30,31] The binding site for spectrin is located on a 72-kD chymotryptic peptide.[32] The stoichiometry is one spectrin tetramer to one ankyrin, although each tetramer contains two binding sites.[33] By electron microscopy, the binding site for ankyrin is located about 20 nm from the head end of spectrin[16] and is on the β-chain.[15] A distinct 90-kD peptide of ankyrin binds to the cytoplasmic portion of protein 3.[34] When the ankyrin structure was analyzed by limited trypsin digestion, a peptide of 82- to 83-kD binds protein 3, and one of 55- to 65-kD binds spectrin.[35] The K_d for ankyrin–spectrin binding is 10^{-7} M[36] and for ankyrin–protein 3 binding, 8×10^{-8} M.[37]

Protein 3

Protein 3 is the red cell membrane anion exchange protein of 100 kD, of which there are a million copies per red cell. It is a transmembrane glycoprotein with heterogeneous glycosylation that accounts for the diffuse band on SDS gels. The amino-terminal 43-kD cytoplasmic peptide serves as the binding site for ankyrin,[11,37] as well as a number of other proteins, including proteins 4.1 and 4.2, hemoglobin, and a few of the glycolytic enzymes. The carboxy-terminal 52-kD peptide functions as the anion transport channel for chloride–bicarbonate exchange.[38] Protein 3 is associated noncovalently into tetramers,[39] each of which binds one ankyrin molecule. Analogs to protein 3 have been identified immunologically in many nonerythroid cells.[39a]

Glycophorins

Four species of transmembrane sialoglycoproteins are known as the glycophorins (for review, see ref. 40), designated A, B, C, and D, which account for 85%, 15%, 4%, and 1%, respectively, of the periodic acid-Schiff (PAS) staining intensity on gels. They all consist of an external glycosylated segment that can have a receptor function, a hydrophobic segment that traverses the lipid bilayer, and an inner segment that can bind to other cellular or membrane proteins. Glycophorin A (molecular weight 36 kD) is also known as PAS-2 and glycoprotein α and contains the determinants for the MN blood group antigens. Glycophorin B (molecular weight 20 kD) is also known as PAS-3 and glycoprotein δ and contains the Ss antigens. Glycophorin C (molecular weight 32 kD) is also known as PAS-2' and glycoprotein β and carries the blood group Gerbich antigens. Glycophorin D (molecular weight 23 kD) is also known as glycoprotein γ and is the only glycophorin to remain in organic phase after extraction with hot phenol-saline. Glycophorin A has the influenza and *Plasmodium falciparum* receptors. Protein 4.1 is thought to bind to both glycophorin A and C, the former in a phosphatidyl inositol-dependent fashion. This protein 4.1–glycophorin interaction serves as an attachment site for the membrane skeleton to the lipid bilayer.

MOLECULAR BIOLOGY

The genes encoding portions of nearly all of the erythrocyte membrane proteins have been cloned. In some cases, gene families have been identified that encode the erythrocyte and nonerythrocyte forms of the proteins, while in others, a complex mechanism of mRNA processing allows generation of the various tissue isoforms. One of the latter mechanisms is alternate splicing of various exons to construct mRNA species for different tissue types, which occurs for protein 4.1 and β-spectrin.

The methods of antibody and oligonucleotide screening of cDNA libraries have been used to isolate the cDNAs for the membrane proteins. The cDNA has then been used to determine chromosome location and to identify the structural gene.

Erythroid and Nonerythroid Spectrin

Study of the molecular biology of erythroid and nonerythroid spectrin in several species has revealed remarkable preservation of the structure of nonerythroid spectrin, with relative divergence in the structure of erythroid spectrin. Various nonerythroid tissues appear to share a common α-subunit, and the β-subunit differs. The most widely distributed nonerythroid spectrin is fodrin, and the type in avian intestinal tissue found in the terminal web is designated TW260/240. All spectrins have the repetitive structure comprised of homologous subunits of 106 amino acids. Furthermore, spectrin may be part of a larger gene family of similar proteins, including α-actinin and dystrophin. All these proteins have rod-like morphology, with an actin-binding domain and a series of spectrin-like, α-helical repeats.

ERYTHROCYTE α-SPECTRIN

Both murine and human erythrocyte α-spectrin have been cloned using the techniques of antibody and oligonucleotide screening of several cDNA libraries. Antibody screening of a mouse anemic spleen cDNA library enabled the isolation of a 750-base pair (bp) insert that directed synthesis of spectrin-like proteins.[41] The cDNA hybridized to an 8-kilobase (kb) mRNA.

The human cDNA homolog was isolated from a cDNA library derived from human erythroleukemic cell line K562.[42] The mouse and human cDNAs were sequenced and confirmed the amino acid sequence of the α-14 to α-17 repeat units of human spectrin and α-15 to α-18 repeat units of murine spectrin. The mouse and human sequences were completely identical over 360 bp, and overall were 82% homologous. Restriction enzyme analysis of total mouse and human DNA demonstrated the existence of a single erythrocyte α-spectrin.

A sizable part of the 5' end of α-spectrin was cloned by using a synthetic oligonucleotide of 90 bp derived from the amino-terminal end of the α-I domain to screen a human genomic library.[43] The clone contained a 16.8-kb insert, and of 3074 bp sequenced, 12.8% code for amino acids in three exons separated by introns of various sizes. The sequence obtained correlates with amino acid residues 314 to 444, or the fourth 106 amino acid repeat.

The α-spectrin gene was localized to chromosome 1 in both mouse and man[44] by Southern blotting of somatic cell hybrid DNAs using cDNA probes. In-situ hybridization demonstrated the location at human chromosome 1q22 to 1q25.

β-SPECTRIN

The erythrocyte β-spectrin gene was cloned by antibody screening of a human fetal liver cDNA library. Clones totaling 3.3 kb that encode over 50% of the erythrocyte β-spectrin were isolated by Winkelmann and co-workers.[45] A smaller fragment, 822 bases, apparently representing the same mRNA species, was isolated from a reticulocyte library by another group.[46] The size of the β-spectrin mRNA in human erythroleukemia cells was found to be 7.5 kb. Both groups localized the gene for β-spectrin to chromosome 14. The longer cDNA extends from repeat 10 to the 3' noncoding sequence and demonstrated marked sequence divergence after amino acid 94 of repeat 18.

The mouse β-spectrin gene was cloned by antibody screening of an anemic spleen cDNA expression library.[47] This cDNA probe identified mRNA transcripts of 6 and 8 kb in the anemic spleen, 6 and 10 kb (faintly) in the mouse brain, and none in kidney, liver or normal spleen. A Southern blot of mouse genomic DNA demonstrated a single band with several restriction endonucleases, suggesting a single gene copy. The cDNA insert encoded the β-8 to β-10 repeats, with maintenance of 60 to 70% of the conserved amino acids. The gene for mouse β-spectrin was localized to chromosome 12.[48]

The presence of multiple mRNA species corresponding to a single gene in the mouse suggests possible alternative splicing in production of the various β-spectrin mRNAs in different tissues. This idea was confirmed by the finding that human erythroid and skeletal muscle cDNAs were identical over much of their lengths, until a sharp divergence 300 bp from the 3' end of the erythroid cDNA and 1 kb from the 3' end of the muscle cDNA.[49] The point of divergence occurs 22 amino acids from the carboxy terminus of the erythroid form, at a potential splice site.

NONERYTHROID SPECTRIN

Isolation of the cDNAs for nonerythroid spectrin has revealed that the amino acid sequence of nonerythroid spectrin is more tightly conserved across species than is erythroid spectrin (Table 9-2).[50]

cDNAs for nonerythroid α-spectrin were obtained by antibody screening of rat brain and human neuroblastoma cDNA libraries.[50] Comparison of chicken and rat cDNA sequences revealed 81% homology, leading to greater than 96% identical amino acid sequence. Comparison of rat and human demonstrated 89% homology at the nucleotide level, with 97% homology of the peptide sequences. Similarly, the human vs. chicken comparison revealed 82% nucleotide homology, but 95% amino acid identity.[50] The near identity of the peptide sequence is due to the fact that the majority of base changes occurred at the third position of degenerate codons.

Table 9–2. Homologies* Between Nonerythroid and Erythroid α-Spectrins

Origin	Homology (%)	
	Nucleotide	Amino Acid
Nonerythroid		
Rat vs. human	89	97
Rat vs. chicken	81	96
Human vs. chicken	82	95
Erythroid		
Human vs. mouse	82	83
Nonerythroid vs. erythroid		
Human repeat 16	—	52
Human repeat 19	—	54
Rat vs. mouse	—	66

* Homology is expressed as percent identity.
Modified from Leto, T. L. et al.: Comparison of nonerythroid alpha spectrin genes reveals strict homology among diverse species. Mol. Cell. Biol., *8*:1, 1988.

This striking homology of the nonerythroid spectrin at the protein level contrasts with the divergence of the erythroid genes, for which nucleotide homology of mouse and human was 82%, but amino acid sequence identity was only 83%.[42] These findings imply that erythroid spectrin is highly adapted to erythroid cells and that erythrocyte spectrins of different species are drifting rapidly. Nonerythroid cells tolerate almost no variation in the α-chain of nonerythroid spectrin; hence, there is tremendous amino acid sequence identity. The chromosomal location of the nonerythroid α-spectrin gene is at human chromosome 9q33–34.[50]

AVIAN SPECTRIN AND ANKYRIN

Moon et al.[51] studied gene expression of spectrin and ankyrin in the chicken. They found a single species of α-spectrin transcript in myogenic and erythroid cells and distinct forms of β-spectrin and ankyrin in the two tissue types of 7.8 kb and 3.4 kb for β-spectrin and ankyrin in the myotubes, and 7.1 kb and 8.8 kb in the erythroid cells. Moon and co-workers found 5 to 10 times more RNA for the three polypeptides in the erythroid cells than in cultured myotubes, while chicken embryo fibroblasts had only the RNA species for α-spectrin but not for β-spectrin or ankyrin. The major problem with these results is that the myotube type of ankyrin is 235 kD (compared with 260 kD for the erythroid type), but the 3.4-kb transcript can only encode for a protein of 125 kD.

Birkenmeier et al.[52] isolated a cDNA clone for chicken smooth muscle α-spectrin by immunologic screening of a cDNA library. It hybridized to an 8-kb mRNA, and a 1419-base open reading frame was identified that encoded two partial and three complete 106 amino acid repeats homologous to the repeats of erythroid spectrin. The gene was present in a single copy in the chicken genome, and the homology of the chicken nonerythroid to human erythroid α-spectrin was 20 to 36%.

Wasenius et al.[53] identified two clones for nonerythroid spectrin by antibody screening of a chicken smooth muscle cDNA expression library. They sequenced a 1.5-kb insert and found that the chicken nonerythroid and human erythroid spectrin species were highly homologous. The 478 amino acids encoded by the chicken cDNA demonstrated subunit structure of 106 amino acid repeats, with predicted secondary structure of α-helices flanked by β-turns.

STRUCTURAL SIMILARITIES OF NONERYTHROID α-SPECTRIN, α-ACTININ, AND DYSTROPHIN

Nonerythroid spectrin demonstrates striking structural similarity to α-actinin and dystrophin. Spectrin, α-actinin, and dystrophin all are believed to be composed of long, rod-shaped dimers with antiparallel alignment of chains, with multiple repeating subunits, and actin-binding domains.

The entire coding region for nonerythroid chicken α-spectrin has now been sequenced.[54] Nonrepetitive structure was identified in the center and at the carboxy-terminal part of the protein, in subunit repeats α-10, -11, -21, and -22. The carboxy terminus shows homology to α-actinin, with 38% identity. Both α-actinin and α-spectrin have two EF-hands (calcium-binding loops) and a potential calmodulin-binding site composed of clusters of basic and hydrophobic residues. The site for calmodulin binding on nonerythroid mammalian α-spectrin has been localized to a site just distal to the C-terminal region of the eleventh repeat by two groups.[55,56] The α-10 subunit shows homology to the src family of protein kinases and to phospholipase C, which are types of proteins that like spectrin, associate with the inner surface of the plasma membrane.

α-Actinin, a cytoplasmic actin-binding protein, has an aminoterminal actin-binding domain, four spectrin-like repeats of 120 amino acids, and a carboxy-terminal domain of two EF-hand calcium-binding sites.[57] The repeats show internal conservation of sequence in 45 out of 120 possible positions and conservation of sequence in 23 out of 106 positions when compared with α-spectrin. Sequencing of the cDNA for dystrophin, the protein encoded by the gene found to be defective in Duchenne or Becker muscular dystrophy, enabled discovery of four structural domains.[58] The N-terminal 240 amino acids appear to constitute an actin-binding domain similar to α-actinin.[59,60] The second domain is composed of 25 triple helical segments analogous to the spectrin repeats. The third domain is cysteine rich and similar to the C-terminal domain of *Dictyostelium* α-actinin, whereas the fourth domain of 420 amino acids is divergent. The sequence predicts a rod structure, 150 nm in length.

INVERTEBRATE SPECTRIN

Drosophila spectrin was purified from *Drosophila* S3 tissue culture cells.[61] It consisted of 234- and 236-kD subunits that appeared as intertwined elongated strands by electron microscopy. The *Drosophila* protein bound actin and calmodulin, and was immunologically recognized by rabbit antibody to chicken erythroid α-spectrin. *Drosophila* α-spectrin cDNA was isolated from a λgt11 expression library by antibody screening.[62] The cDNA clones encoded fusion proteins that were immunologically related to vertebrate α-spectrin and functionally similar in their ability to bind calmodulin and β-spectrin.

Protein 4.1: Isoforms and Motifs

Protein 4.1 is present in several tissues in various isoforms. cDNA cloning has revealed that several nucleotide sequence "motifs" are inserted or deleted by alternate splicing to produce tissue-specific mRNAs (Fig.9-5). For example, the motif that encodes the spectrin-actin binding domain of protein 4.1 is present in committed erythrocytes but not in lymphoid cells. Protein 4.1 cDNA clones were isolated by both antibody screening of a human reticulocyte cDNA library[63] and by screening of a human bone marrow cDNA library with a synthetic 63-mer derived from the 16-kD peptide.[64] Tang et al.[64] then used the cDNA to screen a MOLT-4 T-cell leukemia cell line cDNA library, enabling the cloning of lymphoid protein 4.1. A comparison of the lymphoid cDNA sequence to the published reticulocyte sequence of Conboy et al.[63] revealed close homology except for five nucleotide sequence "motifs" that were thought to be inserted or deleted by alternative splicing. Motif I encodes a 21-amino acid segment of the 10-kD spectrin–actin-binding domain. It is present in erythroid and absent in lymphoid cDNA. Motif II encodes 34 amino acids near the carboxy terminus of the 22/24-kD domain and is present in the lymphoid cDNA but absent in the erythroid cDNA. Motif III encodes 35 amino acids near the amino terminus of the 30-kD domain and is present in erythroid, but not lymphoid, cDNA. Motifs IV and V are in the 5' untranslated region of the reticulocyte cDNA. Synthetic oligonucleotides and synthetic peptides corresponding to some of the motifs were constructed to study expression. For example, human reticulocytes contained two mRNA species, one with and one without motif I, whereas other tissues only had mRNA without it. Motif I was found to be expressed only in dimethyl sulfoxide-induced and not -uninduced mouse erythroleukemia cells. Sequence complementary to motif III was found in MOLT-4 mRNA, despite its absence in the cDNA. These results suggest heterogeneity in the protein 4.1 isoforms. The mRNA

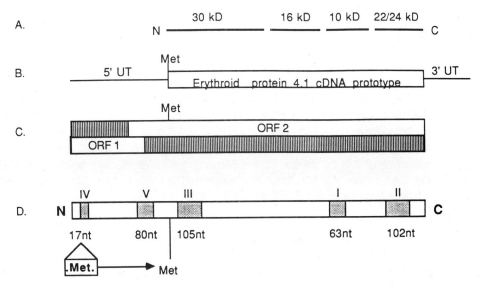

Fig. 9–5. Structural model of protein 4.1 and alignment of complementary DNA (cDNA) sequences containing motifs. A. Structural domains of human erythrocyte protein 4.1.[19] B. Structure of erythroid protein 4.1 cDNA prototype. C. Positions of two potential translational open reading frames (ORF). If motif IV is spliced in and motif V is spliced out, ORF 1 is shifted to ORF 2, and is "open." It can then be read continuously to the end of ORF 2. A protein of 135-kD will be produced instead of the 80-kD protein produced by translation of ORF 2. D. Location of the five identified sequence-motif regions alternatively spliced in or out of the protein 4.1 cDNA. The initiator methionine (Met) enclosed by the box is present in motif IV. It corresponds to the initiation site for the 135-kD protein, while the other Met site corresponds to the start site for the 80-kD protein. (nt, nucleotides.) (Modified with permission from Tang, T. K., et al.: Heterogeneity of mRNA and protein products arising from the protein 4.1 gene in erythroid and non-erythroid tissues. J. Cell Biol., in press, 1990.)

data were confirmed by immunologic studies: Antibody to motify I peptide reacted only with erythrocyte proteins, while antibody to motif II peptide reacted with both erythrocytes and MOLT-4 cells. Thus, motif I appears to be limited to erythroid cells, which is consistent with its location in the spectrin-actin binding domain.

Conboy et al.[63] identified a single, long open reading frame at nucleotides 799 to 2562 out of a total of 2867 nucleotides sequenced. The amino acid sequence of the various protein 4.1 domains were identified, including the hydrophobic 30-kD transmembrane protein-binding domain with predicted β-sheet structure, the 16-kD domain with high proline content and predicted α-helical structure, the 10-kD spectrin–actin-binding domain, which is highly charged with predicted α-helical structure, and the carboxy-terminal acidic 22/24-kD peptide, which has a calculated molecular weight of only 12.6 kD. The mRNA species was 5.6 kb in the reticulocyte and pancreas, slightly larger in brain, and two species existed in the intestine. A study of individuals deficient in protein 4.1 enabled the assignment of the chromosomal location of protein 4.1 to 1p32-pter.[65]

Tang et al.[66,67] identified patterns of mRNA splicing that open along a translatable reading frame in the 5' untranslated region. mRNAs spliced in this fashion can be translated into either a 135-kD isoform of the 80-kD prototypic form. The 135-kD form contains 209 new amino-terminal residues attached to the 80-kD form and may localize preferentially in the nucleus.

After amplification of all protein 4.1 mRNA sequences in the reticulocyte by use of the polymerase chain reaction, at least six potential variations in structure were found in the amino-terminal end of the 30-kD domain.[68] This is the location of the glycoprotein-binding function and may predict a role in the interaction of protein 4.1 with the transmembrane glycoproteins. In keeping with the potential importance of the spectrin–actin-binding domain to erythrocyte protein 4.1 function, this 21-amino acid sequence motify was found to be absent from the liver protein.[69]

cDNA clones have also been isolated for chicken protein 4.1.[70] Seven different protein 4.1 variants occur in chicken erythrocytes, ranging from 77 to 175 kD. The nucleotide sequence of 270 bp of one cDNA demonstrated 86% nucleotide homology or 98% deduced amino acid sequence homology with human reticulocyte protein 4.1 cDNA. A single gene gives rise to multiple protein 4.1 mRNAs, which all comigrate at 6.6 kb. Different mRNA species predominated at different stages of chicken embryo development.

Ankyrin

Two preliminary reports on the cloning of erythrocyte ankyrin have been published. Immunologic screening of human reticulocyte libraries enabled isolation of 1.4-kb cDNA clones.[71,72] Lux et al.[71] isolated a cDNA that encodes for 38 kD of the C-terminal domain of ankyrin and 407 bp of the 3' untranslated region. The chromosomal location was identified as chromosome 8 by hybridization of the cDNA to cell-sorted chromosomes.[71] Lambert et al.[72] found that 108 residues of the spectrin-binding domain sequence correlated with the derived amino acid sequence from the cDNA. They describe the presence of five highly conserved 33 amino acid repeats with 27 to 45% homology to each other.[73]

Glycophorins

Several glycophorin A cDNA clones were isolated by screening a cDNA library constructed from the human erythroleukemia cell line, K562, with synthetic oligonucleotides.[74] Four mRNA species of 2.8, 1.7, 1.0, and 0.6 kb were observed by hybridization with oligonucleotide probes that encoded the N-terminal part of the protein. The 0.6-kb mRNA may represent glycophorin B mRNA because only the probes that encode the amino acid sequence shared by glycophorins A and B hybridize to it. Nucleotide sequence of the largest cDNA clone suggests the presence of a 19-amino acid leader sequence on glycophorin A. This group then used the glycophorin cDNA sequence to synthesize exact-sequence oligonucleotides.[75] These probes were used in in-situ genomic DNA hybridization in agarose gels after restriction endonuclease digestion and the probes derived from the C-terminal central amino acid sequence hybridized to a single band, suggesting the presence of a single gene for glycophorin A in the human genome. However, when a probe was used that corresponded to the N-terminal amino acid sequence shared by glycophorins A and B, two hybridizing restriction fragments were observed for three of the endonucleases, suggesting the existence of a distinct, single-copy, glycophorin B gene.[75]

Another group isolated two cDNA clones for glycophorin A from human fetal cDNA libraries.[76] In-situ hybridization was used to demonstrate the chromosomal location of glycophorin A as 4q28-q31. Northern blot analysis demonstrated the presence of three mRNA species in human fetal spleen erythroblasts, of 1, 1.7, and 2.2 kb, which share 3' noncoding region. Southern blot analysis using probes to both the coding and noncoding regions suggested a large deletion of the glycophorin A gene in individuals of the En(a-) phenotype (Finnish type), who lack the protein on their red cells. Glycophorin B also has been cloned from a K562 cDNA library.[77,78]

cDNA clones for glycophorin C have been obtained by several groups.[79-82] Colin et al.[80] isolated cDNA clones for glycophorin C from a human reticulocyte cDNA library, using a mixture of 14-mers from the amino acid sequence in an N-terminal tryptic peptide. By Northern blot analysis, an mRNA species of 1.4 kb was identified in human fetal liver, K562, and HEL lines that was absent from adult liver and lymphocyte lines. Mattei et al.[79] used the cDNA to isolate a cDNA clone of 900 bp encoding nearly all the coding sequence and the 3' untranslated region from a human fetal cDNA library. They identified the chromosomal origin by in-situ hybridization as 2q14–21. High and Tanner[82] isolated cDNA clones from a human fetal liver cDNA library that together encompassed nucleotides − 120 to the 3' poly A tail of the mRNA. The nucleotide sequence predicts an absence of a cleaved N-terminal signal peptide in glycophorin C.

Study of individuals who lack Gerbich antigens (and glycophorins C and D) but do not have elliptocytosis suggested the existence of a glycophorin C–D hybrid of 25 to 28 kD.[81] Southern blot analysis of genomic DNA from leukocytes digested with *Eco*RI demonstrated hybridization of the glycophorin C cDNA to a 15-kb fragment in Gerbich-positive individuals that was absent in Gerbich-negative individuals, who instead had a 12-kb fragment. The 3-kb size difference was observed with three other restriction endonucleases. Because the proposed C–D hybrid protein had immunologic and functional properties of glycophorin C, and the physicochemical extraction property of glycophorin D, and because the Gerbich-negative individuals did possess a restriction fragment that bound the glycophorin C cDNA, the authors concluded that glycophorin C and D were encoded by the same gene by either alternative splicing or differential processing of the polypeptides.

Use of multiple cDNA probes for glycophorins A, B, and C enabled Rahuel and colleagues[78] to propose the existence of a gene-fusion product encoding a glycoprotein composed of the N-terminal portion of glycophorin A and the C-terminal portion of glycophorin B in individuals with the En(a-) phenotype, English type. Two unrelated individuals who lack S, s, and U phenotypes and glycophorin B were shown to have a deletion in the gene for glycophorin B by comparative restriction mapping and use of several cDNA probes to glycophorins A and B.[83]

Protein 3

The 17-kb gene for murine protein 3 was isolated and found to exist in a single copy in the mouse genome, with 19 intervening sequences.[84] S_1 nuclease protection experiments revealed the existence of five transcriptional start sites for protein 3 message in the mouse anemic spleen between − 146 and − 189, and a single start site at position − 260 in the kidney, suggesting that tissue-specific transcripts can arise without alternate splicing.[85] Interestingly, the 1.7-kb of genomic DNA at the 5' end of the gene lacks the TATA and CCAAT boxes typical of eukaryotic RNA polymerase II promoter regions.[85]

Protein 3 has also been cloned in the chicken,[86] in which it exists as a doublet of 100 and 105 kD. A single genetic locus encoded for two different erythroid mRNAs, which predicted polypeptides of the correct size. The predicted amino acid sequence shows striking homology to the amino acid sequence predicted from the murine cDNA.

The cDNA for a human band 3-like protein was isolated from a K562 library by screening with a mouse erythroid cDNA for protein 3.[87,88] The predicted amino acid sequence differs from the known sequence of human protein 3: The sequence is 70% homologous to the sequence predicted by the mouse cDNA in the membrane domain and is 35% homologous in the cytoplasmic domain.[87] The gene was localized to human chromosome 7q35–7q36.[87]

The human cDNA for protein 3 has been recently cloned by two groups.[89,90] Tanner et al.[89] obtained a clone that contained the entire coding region and parts of both the 5′ and 3′ noncoding regions. It predicts a protein of 911 amino acids. The cDNA was entirely sequenced and shows homology in the coding region to the chicken band 3 and the band 3-related protein from K562 cells and homology in both the coding and noncoding regions to the mouse protein. The availability of the entire amino acid sequence allowed assignment of the previously described sites of proteolytic cleavage and chemical labeling such as the binding site for 4,4′-diisothiocyanostilbene-2,2′-disulfonate (DIDS), the anion-transport inhibitor. The gene for protein 3 is located on chromosome 17.[90]

Actin

Functional genes for β-actin have been isolated in the rat,[91] the chicken,[92] and man.[93] Approximately 25 gene copies for actin are found in the human genome, most of which are thought to represent pseudogenes related to cytoplasmic actins.[94] The functional human β-actin gene was isolated from a human genomic library by screening with a synthetic oligonucleotide.[93] The derived amino acid sequence from the nucleotide sequence matches perfectly with the known sequence of fibroblast β-actin. Interspecies comparison of nucleotide sequence demonstrated greater than 90% homology of human vs. rat and greater than 85% homology of human vs. chicken sequence, whereas the deduced amino acid sequence from the human gene is identical to the chicken gene, and that from the rat gene differs by only one amino acid. Southern blot analysis of human genomic DNA suggested the presence of a single functional gene for β-actin. Unlike the cardiac and the two smooth muscle actin genes, the β-actin gene lacks the codon for cysteine after the codon for the initial methionine. The human β-actin gene is located at chromosome 7pter-q22.[95]

MUTANT ERYTHROCYTE MEMBRANE PROTEINS

Hereditary hemolytic anemias in mice and man have been shown to arise from defects in the erythrocyte membrane skeleton. Specific molecular defects in protein structure and function have been identified in hereditary spherocytosis, elliptocytosis, and pyropoikilocytosis. Some of these defects have been confirmed at the DNA level. The first clues as to the genetic basis of these defects were obtained by linkage studies. Hereditary elliptocytosis (HE) was found to be linked to the Rh locus at chromosome 1p34–36,[96,97] and in another study, HE was found to be tightly linked to the Duffy blood group locus at 1q24.[98] In fact, the defect of HE was found in many cases to be due to a defect in α-spectrin or an abnormality of protein 4.1, both of which are on chromosome 1.

Hereditary spherocytosis (HS) was mapped by two groups to chromosome 8, by linkage to 8p11[99] and association with a translocation, t8:12, t3:8.[100] Interestingly, an abnormal ankyrin gene has recently been identified in HS,[71] and use of restriction fragment-length polymorphism (RFLP) analysis of the ankyrin gene in a large kindred with HS has shown close linkage between HS and the ankyrin gene.[101] The ankyrin gene is on chromosome 8.[71] Another linkage study showed that a type of HS mapped to chromosome 14q[99] and chromosome 14 contains the gene for β-spectrin.[45,46]

The murine hemolytic anemias and HS in humans share the feature of erythrocyte spectrin deficiency. In the mouse, deficient spectrin synthesis has been demonstrated in several types, while in one murine type and in humans, defective attachment of spectrin to the membrane may be the fundamental problem. In a few cases, HS has been linked to the ankyrin gene, while in one subclass of HS, spectrin is defective in binding to its second membrane attachment site, protein 4.1. The attachment of spectrin to the membrane can be thought of as a vertical membrane skeletal interaction.[102]

In contrast, in HE and hereditary pyropoikilocytosis (HPP), defective spectrin oligo-merization is observed in a majority of cases. The self-association of spectrin within the plane of the membrane skeleton can be thought of as a horizontal interaction.[102] Thus, defective vertical interactions of spectrin apparently lead to spherocytosis, while defective horizontal interactions lead to the HE and HPP phenotypes.

Defects in other membrane skeletal proteins have been observed, as well as other defects of spectrin, and these cases do not fit well into the horizontal versus vertical interaction model. Moreover, we do not understand how such defects lead to the specified cell shapes, or why such great variation is observed in clinical presentation, even within families.

Murine Hemolytic Anemias

Five types of hemolytic anemia have been described in mutants of the common house mouse, *Mus musculus*.[103] These anemias are characterized by spherocytosis, severe hemolytic anemia, jaundice, and hepatosplenomegaly. The five types are designated ja/ja (jaundice), sph/sph (spherocytosis), sph^{ha}/sph^{ha} (hemolytic anemia), sph^{2BC}/sph^{2BC}, and nb/nb (normoblastosis). The conditions are autosomal recessive, with drastically impaired viability in the homozygotes. The three sph types represent different alleles at the sph locus. A milder but similar condition exists in mutants of the deer mouse,[104] *Peromyscus maniculatus*, designated sp/sp.

All of the mutants are deficient in spectrin, and the degree correlates with the clinical severity. The ja/ja mutant has no detectable spectrin, the sph/sph mutant lacks spectrin α-chains and has markedly reduced β-spectrin, the sph^{ha}/sph^{ha} mutant has 20% of the normal quantity of spectrin, and the nb/nb has 70% of the normal amount.[105] Bone marrow transplantation demonstrates transfer of the disease with the erythroid tissue.[103]

Reconstitution of the sph/sph erythrocyte membranes with normal spectrin resulted in decreased membrane fragmentation in vitro.[106] Studies of spectrin synthesis in the mouse mutants revealed specific defects.[107,108] No detectable α-spectrin was synthesized in the three sph mutants, and both the sph locus and the α-spectrin gene[44] have been mapped to chromosome 1 in the mouse. The ja mutant was deficient in β-spectrin synthesis. The nb mutant has normal spectrin synthesis, but very unstable ankyrin, leading to membranes deficient in spectrin. Thus, abnormal synthesis of membrane proteins has been observed in the murine hemolytic anemias, which could serve as models for human hereditary spherocytosis.

Hereditary Spherocytosis

HS is a disorder characterized by spheroidal red cell shape and hemolytic anemia. The pathogenesis of the disease is related to decreased membrane deformability and splenic trapping. A membrane defect has been long suspected as the cause of the hereditary disorder. The majority of HS is transmitted as an autosomal-dominant condition, with variable penetrance. A minority of patients have an apparently recessively inherited disorder, with marked poikilocytosis and hemolysis, severe anemia, and jaundice. The hallmark diagnostic laboratory test is the osmotic fragility test, which reflects the degree of spherocytosis. The spherocytes are lysed in higher concentrations of sodium chloride than discocytes because less of an increase in volume is allowed before the cells attain a spherical shape, at which point the cells lyse.[109]

The hereditary spherocyte was under study for decades before the primary molecular defect was identified (for review, see ref. 110). During that time, numerous abnormalities of composition and structure were identified, including abnormal cation permeability, abnormal membrane lipids, abnormal membrane protein phosphorylation, and abnormal membrane protein composition and function. A great many of the reports were met with contradictory publications or were not confirmed.

Finally, the unifying concept of spectrin deficiency was proposed. Variable spectrin deficiency was observed in all patients with hereditary spherocytosis, and the degree of deficiency correlated well with clinical severity and the degree of spherocytosis (Fig. 9-6).[111,112] Initially, the mechanism of spectrin deficiency was unclear. The possibilities include deficient synthesis of β-spectrin, which is assembled on the membrane before α-spectrin, as appears to be the case in the mouse ja mutant, or deficient or defective synthesis of ankyrin, by which spectrin is attached to the membrane.

The recent cloning of the ankyrin gene enabled the understanding of how one such mechanism leads to HS. A study of RFLPs detected by ankyrin cDNA in a large kindred

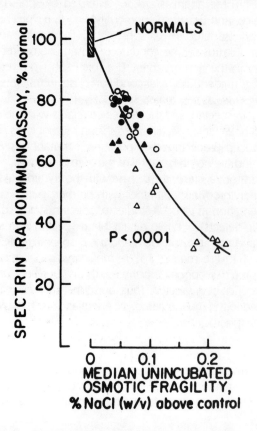

Fig. 9–6. Spectrin deficiency in hereditary spherocytosis (HS). The degree of spectrin deficiency is correlated with the unincubated osmotic fragility in all patients with HS. The *x*-axis represents the concentration of sodium chloride producing 50% hemolysis (patient minus control). The circles represent individuals with the classic, autosomal-dominant HS, and the triangles represent individuals with the recessively inherited type. Open symbols refer to those who have undergone splenectomy. The curve is a computer-generated fitting of the data to an exponential function. The degree of spectrin deficiency also correlates with clinical severity. (Excerpted from material originally appearing in Agre, P., Asimos, A., Casella, J. F. and McMillan, D.: Inheritance pattern and clinical response to splenectomy as a reflection of erythrocyte spectrin deficiency in hereditary spherocytosis. N. Engl. J. Med., *315*:1579, 1986.)

of 21 affected members with typical dominant HS enabled linkage of HS and the ankyrin gene.[101] The lod score was 3.33, indicating that the odds against linkage of HS and the ankyrin gene are less than 1 in 2000. Furthermore, hybridization of the 1.4-kb ankyrin cDNA to Southern blots of DNA from two unrelated children with severe HS and deletion of 8p11.1–p21.1 of one chromosome demonstrated that ankyrin gene dosage decreased to 53 and 65% of normal.[71]

A severe, atypical form of HS was described in one patient with a deficiency of both spectrin and ankyrin.[113] Spectrin synthesis in the patient's reticulocytes appeared to be normal, but the newly synthesized ankyrin was rapidly degraded.[114] This unstable ankyrin species is most likely responsible for the spectrin deficiency.

The recessively inherited type of HS is characterized by severe anemia, jaundice, and massive splenomegaly.[115] This disorder was first described in sisters whose parents were fourth cousins. The erythrocytes had a marked degree of spectrin deficiency, 26 to 29% of normal by radioimmunoassay. The exact mechanism by which these cells become spectrin deficient is not known but may involve either a gene for spectrin itself, or an anchoring protein.

Two kindreds with autosomal-dominant HS have been described with a defect in the binding of spectrin to protein 4.1.[116,117] Half of the patients' spectrin cannot bind protein 4.1[117] There is an abnormality in the chymotryptic digestion pattern of the spectrin β-chain in the affected family members.[118] As has thus far been observed for all patients with HS, the red cells from one of the kindreds with this type of HS are deficient in spectrin, with levels approximately 80% of normal.[118] Perhaps the deficiency is due to the defect in the ability of spectrin to interact with its second membrane attachment site, protein 4.1. The confirmation of a defect of β-spectrin in this type of HS awaits molecular genetic analysis.

Thus, the unifying concept of spectrin deficiency is fundamental in all cases of HS. It not only correlates with osmotic fragility but also with clinical severity. Furthermore, in some kindreds, HS is linked to the ankyrin gene. Defective ankyrin function could lead to spectrin deficiency by loss of binding of spectrin to the membrane.

Hereditary Elliptocytosis and Pyropoikilocytosis

HE refers to a group of heterogeneous disorders of elliptical red cell shape and variable hemolytic anemia (for review, see ref. 102). The disorder is clinically heterogeneous, and a large number of molecular defects have been identified. The clinical types of HE can be organized as follows: mild HE, HE with acute or chronic hemolysis, mild HE with poikilocytosis in infancy, mild HE with dyserythropoiesis, homozygous mild HE, HPP, spherocytic HE, and stomatocytic HE. Mild HE is autosomal dominant, and individuals have no anemia or splenomegaly but may have slight hemolysis, as evidenced by mild reticulocytosis. A similar picture can exist in certain members of such kindreds with HE, except that they have increased hemolysis and anemia. A severe presentation with hemolytic anemia, poikilocytosis, and neonatal jaundice is present in some infants from kindreds with mild HE. By the end of infancy, the disorder gradually becomes indistinguishable from the usual form. A few families have coincident HE and ineffective erythropoiesis. Patients apparently homozygous for two "mild HE" alleles can have a very severe, transfusion-dependent anemia.

HPP is a recessive, severe hemolytic anemia with extreme poikilocytosis and thermal sensitivity of the red cells. It is thought to be related to HE because in approximately 30% of cases, one parent or sibling has typical, mild HE. Spherocytic HE is an apparent hybrid of HS and HE, with both spherocytes and elliptocytes on the peripheral blood smear and mild-to-moderate hemolysis. Stomatocytic HE is described in the lowland tribes of Melanesia, possibly related to resistance of the red cells to malarial invasion. The red cells are elliptic stomatocytes or round elliptocytes with one or two transverse bars, and hemolysis is mild or absent.

Several molecular defects have been observed in HE. The first clue to a defect in the membrane skeleton was the fact that the skeletons of hereditary elliptocytes retained the shape of the original red cell.[3] The clue that spectrin was a potential culprit came from work that demonstrated a thermal sensitivity of not only the red cells from patients with HPP[119] but also of the spectrin,[3] which denatured at a lower temperature than normal. In addition, the membranes and membrane skeletons in HE fragment more easily with applied shear stress.[120] Approximately 30% of patients with HE and all patients with HPP demonstrate defective spectrin function in its ability to self-associate to oligomers.[102,121-124]

Structural defects of spectrin have been found in the individuals demonstrating abnormal spectrin oligomerization, at the head end of spectrin responsible for this function. One type of defect has been described in the β-chain of individuals with HE; a truncated polypeptide of 214 kD occurs, lacking the carboxy-terminal phosphorylated region.[125,126]

The spectrin-domain map[6] has been instrumental in allowing the identification of numerous spectrin peptide abnormalities in HE and HPP. The domain map is produced by limited proteolytic digestion of spectrin at 0°C, followed by two-dimensional electrophoresis (SDS-PAGE vs. isoelectric focusing) (Fig. 9-7). Several types of defects have been identified in the αl/80 kD, N-terminal domain, of α-spectrin. These have been recognized on two-dimensional domain maps and, in some cases, have been confirmed by amino acid and nucleotide sequences.

Several abnormalities of the αl domain have been found in individual patients with HE and HPP. The abnormal domain maps demonstrate decrease in the amount of the αl/80 kD domain and appearance of new peptides designated by the size of the abnormal peptide as follows: αl/78 kD,[127] αl/74 kD,[124,128-131] αl/68 or /65 kD,[131-135] simultaneous appearance of two peptides αl/50a or /46 and αl/21 or /17 kD,[3,136-138] αl/50 with a more basic isoelectric point (designated 50b),[139] and two peptides, αl/42 and /43 kD.[140] The amino acid sequence differences thus far identified include proline for leucine at residue 254 in αl/50a, proline for glutamine at residue 465 in αl/50b, and an extra leucine after residue 150 in αl/65 or /68.[139] These abnormal spectrin-domain maps and amino acid sequences are portrayed in Figures 9-8 and 9-9.

Another type of α-spectrin defect is a shortened chain, of 234 kD, in which the defect may reside in the αIV domain.[141] The heterozygous forms with the α-chain variants generally have been associated with mild HE. The αl/42-43 kD defect is associated with mild hemolysis but marked poikilocytosis.[140] The patients with HPP studied thus far have had either the 74 kD or 50a/46 kD αl variants. Patients with HPP have also been demonstrated to be spectrin deficient, about 70% of normal,[142] but the mechanism responsible is unknown.

Molecular genetics analysis has confirmed and explained a few of the previously described peptide defects in the spectrin αl domain in HE. For example, in the αl/65 or /68 kD defect, which is due to an inserted leucine, a duplication of the triplet codon TTG was found in five families.[143,144] Polymerase chain reaction was used to amplify specific exons in α-spectrin, and the amplified DNA was sequenced. Oligonucleotides

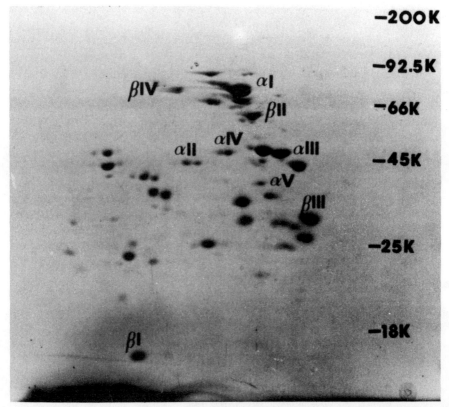

Fig. 9–7. Two-dimensional domain map of spectrin. Spectrin dimer was digested with trypsin, with a trypsin/spectrin ratio of 1:20 (w/w), for 90 min. at 0°C. The horizontal dimension is isoelectric focusing, and the vertical dimension is sodium dodecyl sulfate-polyacrylamide gel electrophoresis, by the method of Speicher et al.[6] The domains are indicated as α-I through α-V and β-I through β-IV. (Reprinted with permission from Becker, P. S.: The Spectrin–Actin–Protein 4.1 Interaction in Normal and Abnormal Red Cell Membranes. Doctoral Thesis, Harvard Graduate School of Arts and Sciences, 1984.)

corresponding to the normal and mutant sequences, and slot-blot analysis enabled demonstration of the abnormality in all affected family members.[143] The αI/50a and /50b kD abnormalities were also confirmed by sequencing of amplified DNA.[145] One individual with the 50a-peptide defect had CTG changes to CCG, corresponding to a change of leucine to proline at residue 254, while another had TCC changed to CCC, corresponding to a change in serine to proline at residue 255. Two unrelated individuals with the 50b defect had CAG changed to CCG, corresponding to a change in glutamine to proline at residue 465. Interestingly, a few individuals with the same peptide pattern had normal nucleotide sequence through the same region. Therefore, a more distal mutation or a mutation in the adjacent chain may affect the trypsin sensitivity of these regions. It was proposed that in one type of HS, abnormal kinetics of cleavage of the α-chain was caused by a defect in the β-chain, because only the latter subunit was digested abnormally if the chains were separated before proteolytic cleavage.[118]

The αI/78 kD defect was associated with a G to T transversion in the 39th codon, AGT for AGG, changing serine for arginine in four members of a family with HE.[146]

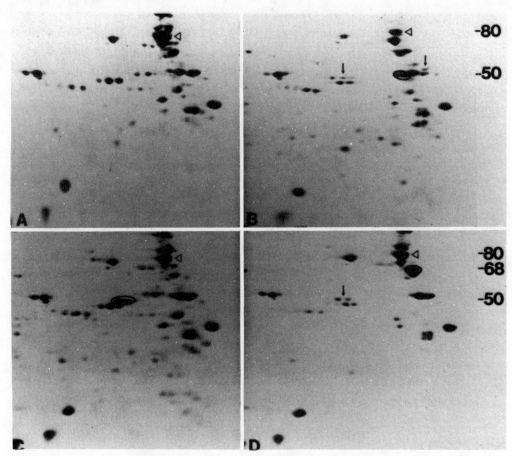

Fig. 9–8. Limited tryptic digestions (domain maps) of normal and hereditary elliptocytosis (HE) spectrin. The numbers indicate apparent molecular weight in kilodaltons (kD). The open triangles mark the position of the αI/80 kD domain. The arrows indicate polymorphisms of the α-II and α-III domains. A. Normal spectrin–normal αI/80 kD domain. B. Spectrin from individual with αI/50a HE. C. Spectrin from individual with αI/50b HE. D. Spectrin from individual with αI/68 HE. (Reprinted with permission from Marchesi, S. L. et al.: Mutant forms of spectrin alpha-subunits in hereditary elliptocytosis. J. Clin. Invest., *80*:191, 1987.)

This analysis was performed by first identifying an XbaI polymorphic site and then amplifying a 364-bp fragment containing the third exon of α-spectrin by the polymerase chain reaction. Oligonucleotide probes were prepared to the normal and mutant sequences, and the mutant probes were hybridized to DNA from the affected family members but not to DNA from a normal control.

Protein 4.1 deficiency has been demonstrated in several kindreds with HE.[65,147–151] Among the affected members, 50% deficiency occurs in heterozygotes, and no protein 4.1 occurs in the homozygotes. The homozygous patients have a severe hemolytic anemia and erythrocyte membrane fragility. The membrane stability can be restored by reconstitution of the cells with exogenous protein 4.1.[152] In one family, the mutant protein 4.1 gene has a DNA rearrangement upstream from the initiation codon for translation, and therefore the mRNA was abnormally spliced.[65] In another family, poly-

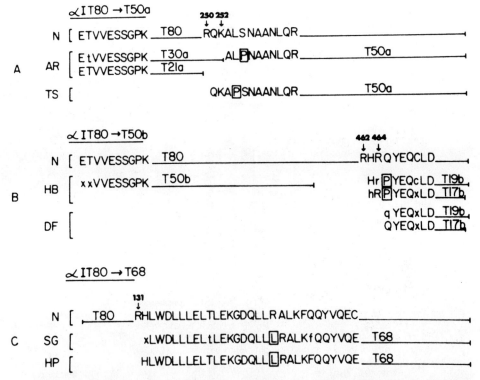

Fig. 9–9. Abnormal amino acid sequences in individuals with various types of hereditary elliptocytosis (HE). Arrows mark the tryptic cleavage sites at lysine (K) or arginine (R) residues. Unknown residues are indicated by x. Tentative residues are indicated by lower case letters. The amino acids are designated as follows: E, Glu; T, Thr; V, Val; S, Ser; G, Gly; P, Pro; K, Lys; R, Arg; Q, Gln; A, Ala; L, Leu; N, Asn; H, His; Y, Tyr; C, Cys; D, Asp; W, Try; F, Phe. A. Sequence of type αI/50a HE. B. Sequence of type αI/50b HE. C. Sequence of type αI/68 HE. (Reprinted with permission from Marchesi, S. L. et al.: Mutant forms of spectrin alpha-subunits in hereditary elliptocytosis. J. Clin. Invest., *80*:191, 1987.)

morphisms were identified in the protein 4.1 coding region by restriction endonucleases.[151] Some genetic studies demonstrated linkage of HE in some individuals to the Rh locus,[96,97] which is on chromosome 1 at 1p32–p36, close to the location of the gene for protein 4.1, 1p32–pter. These cases may have had defective protein 4.1. Other structural defects of protein 4.1 have been observed in HE: decreased protein 4.1 with new bands immunologically identified as 4.1 just below 4.2[153] and new protein 4.1 variants of 95 kD[154–156] or 68 kD.[156] Analysis of the 95-kD and 68-kD variants suggests the existence of an insertion N-terminal to the 10-kD domain in the former and deletion of the 10-kD domain in the latter.[156] Reconstitution of red cell membranes from two individuals with HE having mutant, dysfunctional protein 4.1 with normal protein 4.1 improved the mechanical stability of the membranes.[157]

Protein 4.1 deficiency may be accompanied by deficiency of protein 4.9[147] or glycophorin C.[158] Deficiency of glycophorin C occurs in homozygous protein 4.1 deficiency[158,159] or as an isolated deficiency in the Leach phenotype,[160] which has spherocytic HE and absence of Gerbich antigens. The coincident deficiency of glycophorin C and protein 4.1 lends support to the hypothesis of a binding interactions between the two proteins.

Table 9–3. Types of Molecular Defects Identified in Some Cases of HS and HE/HPP

Molecular Defect		Clinical Disorder
HS		
Spectrin deficiency		Autosomal-dominant HS
		Recessively inherited HS
		Autosomal-dominant HS
		Severe HS with partial deletion
Abnormal ankyrin gene		of chromosome 8
Combined spectrin/ankyrin deficiency		Atypical, severe HS
Spectrin defective in protein 4.1 binding		Autosomal-dominant HS
HE/HPP		
Spectrin defects		
Truncated β-chain of 214 kD, lacking phosphorylated, carboxyterminal peptide		HE
Truncated α-chain of 234 kD; defect thought to be in α-IV domain		HE
Diminished amount of αI/80 kD domain with replacement by the following:		
αI/78	arg 39 to ser: AGG to AGT	HE
αI/74		HE or HPP
αI/68	inserted leu 151: duplication	HE
(or /65)	TTG	
αI/50a-21	leu 254 to pro: CTG to CCG or	HE or HPP
(or /46-17)	ser 255 to pro: TCC to CCC	
αI/50b	gln 465 to pro: CAG to CCG	HE
αI/42-43		HE
Spectrin deficiency		HPP—feature common to 70%
Protein 4.1 defects		
Protein 4.1 deficiency		
Homozygous or heterozygous		HE
Protein 4.1 variants		
95-kD form		HE
68-kD form		HE
Other protein defects		
Protein 4.2 deficiency		HE
Concurrent 4.1 and 4.9 deficiency		HE
Concurrent 4.1 and glycophorin C deficiency		HE

HE, hereditary elliptocytosis; HPP, hereditary propoikilocytosis; HS, hereditary spherocytosis.

Defective binding of spectrin to ankyrin has been described in one kindred in which a severe HPP-like syndrome occurred when this defect was accompanied by a defect in spectrin self-association.[161]

Deficiency of protein 4.2 has been described in some Japanese families whose red cell morphology most closely resembles that of spherocytic HE.[162,163]

In summary, a number of polypeptide abnormalities have been identified in individuals with HE or HPP who have defective membrane skeletal protein function or unusual structure (Table 9-3). The majority of anomalies have been described in the head end of spectrin, at the spectrin oligomerization site, either in the N-terminal domain, the αI/80 kD, or in the carboxyterminus of the β-chain. For a few of the peptides, amino acid or nucleotide sequence data have identified the site of the mutation leading to formation of the defective protein. The other major type of molecular defect in HE is aberrant protein 4.1 characterized by either partial or total deficiency or by the presence of elongated or truncated variants.

Many of the molecular defects observed in the hemolytic anemias await confirmation at the gene level. Just as in the case of thalassemia and the hemoglobinopathies, for which a great number of genetic defects have been observed, a vast number of defects of each of the erythrocyte membrane proteins will likely be found to be responsible for the various types of hemolytic anemias.

REFERENCES

1. Yu, J., Fischman, D.A., and Steck, T.L.: Selective solubilization of proteins and phospholipids from red blood cells by non-ionic detergents. J. Supramolec. Struct., *1*:233, 1973.
2. Liu, S.-C., Derick, L.H., and Palek, J.: Visualization of the hexagonal lattice in the erythrocyte membrane skeleton. J. Cell Biol., *104*:527, 1987.
3. Tomaselli, M.B., John, K.M., and Lux, S.E.: Elliptical erythrocyte membrane skeletons and heat-sensitive spectrin in hereditary elliptocytosis. Proc. Natl. Acad. Sci. USA, *78*:1911, 1981.
4. Lux, S.E., John, K.M., and Karnovsky, M.J.: Irreversible deformation of the spectrin-actin lattice in irreversibly sickled cells. J. Clin. Invest., *58*:955, 1976.
5. Speicher, D.W., and Marchesi, V.T.: Erythrocyte spectrin is comprised of many homologous triple helical segments. Nature, *311*:177, 1984.
6. Speicher, D.W., Morrow, J.S., Knowles, W.T., and Marchesi, V.T.: Identification of proteolytically resistant domains of human erythrocyte spectrin. Proc. Natl. Acad. Sci. USA, *77*:5673, 1980.
7. Speicher, D.W., Morrow, J.S., Knowles, W.J., and Marchesi, V.T.: A structural model of human erythrocyte spectrin: Alignment of chemical and functional domains. J. Biol. Chem., *257*:9093, 1982.
8. Harris, H.W., and Lux, S.E.: Structural characterization of the phosphorylation sites of human erythrocyte spectrin. J. Biol. Chem., *255*:11512, 1980.
9. Ungewickell, E., and Gratzer, W.: Self-association of human spectrin. A thermodynamic and kinetic study. Eur. J. Biochem., *88*:379, 1978.
10. Morrow, J.S., and Marchesi, V.T.: Self-assembly of spectrin oligomers in vitro: a basis for a dynamic skeleton. J. Cell Biol., *88*:463, 1981.
11. Bennett, V., and Stenbuck, P.J.: Association between ankyrin and the cytoplasmic domain of band 3 isolated from the human erythrocyte membrane. J. Biol. Chem., *255*:6424, 1980.
12. Ungewickell, E. et al.: In vitro formation of a complex between cytoskeletal proteins of the human erythrocyte. Nature, *280*:811, 1979.
13. Cohen, C.M., and Korsgren, C.: Band 4.1 causes spectrin-actin gels to become thixotropic. Biochem. Biophys. Res. Commun., *97*:1429, 1980.
14. Ohanian, V. et al: Analysis of the ternary interaction of the red cell membrane skeletal proteins spectrin, actin and 4.1. Biochem., *23*:4416, 1984.
15. Morrow, J.S. et al.: Identification of functional domains of human erythrocyte spectrin. Proc. Natl. Acad. Sci. USA, *77*:6592, 1980.
16. Tyler, J.M., Hargreaves, W.R., and Branton, D.: Purification of two spectrin-binding proteins: biochemical and electron microscopic evidence for site-specific reassociation between spectrin and bands 2.1 and 4.1. Proc. Natl. Acad. Sci. USA, *76*:5192, 1979.
17. Lazarides, E., and Nelson, W.J.: Expression of spectrin in nonerythroid cells. Cell, *31*:505, 1982.
18. Goodman, S.R. et al.: Erythrocyte membrane skeletal protein bands 4.1a and b are sequence-related phosphoproteins. J. Biol. Chem., *257*:4564, 1982.
19. Leto, T.L., and Marchesi, V.T.: A structural model of human erythrocyte protein 4.1. J. Biol. Chem., *259*:4603, 1984.
20. Correas, I., Leto, T.L., Speicher, D.W., and Marchesi, V.T.: Identification of the functional site of erythrocyte protein 4.1 involved in spectrin-actin associations. J. Biol. Chem., *261*:3310, 1986.
21. Anderson, R.A., and Lovrien, R.E.: Glycophorin is linked by band 4.1 protein to the human erythrocyte membrane skeleton. Nature, *307*:655, 1984.
22. Mueller, T.: A band 4.1-glyconnectin (PAS 2) complex mediates the association of the red cell membrane skeleton with the bilayer domain. Fed. Proc., *43*:1849, 1984.
23. Pasternack, G.R., Anderson, R.A., Leto, T.L., and Marchesi, V.T.: Interactions between protein 4.1 and band 3. An alternative binding site for an element of the membrane skeleton. J. Biol. Chem., *260*:3676, 1985.
24. Anderson, R.A., and Marchesi, V.T.: Regulation of the association of membrane skeletal protein 4.1 with glycophorin by polyphosphoinositide. Nature, *318*:295, 1985.
25. Pinder, J.C., and Gratzer, W.B.: Structural and dynamic states of actin in the erythrocyte. J. Cell Biol., *96*:768, 1983.
26. Siegel, D.L., and Branton, D.: Partial purification and characterization of an actin-bundling protein, band 4.9, from human erythrocytes. J. Cell Biol., *100*:775, 1985.

27. Fowler, V.M., and Bennett, V.: Erythrocyte membrane tropomyosin. Purification and properties. J. Biol. Chem., *259*:5978, 1984.
28. Gardner, K., and Bennett, V.: Modulation of spectrin-actin assembly by erythrocyte adducin. Nature, *328*:359, 1987.
29. Mische, S.M., Mooseker, M.S., and Morrow, J.S.: Erythrocyte adducin: A calmodulin-regulated actin-bundling protein that stimulates spectrin-actin binding. J. Cell Biol., *105*:2387, 1987.
30. Yu, J., and Goodman, S.R.: Syndeins: the spectrin-binding protein(s) of the human erythrocyte membrane. Proc. Natl. Acad. Sci. USA, *76*:2340, 1979.
31. Luna, E.J., Kidd, G.H., and Branton, D.: Identification by peptide analysis of the spectrin binding protein in human erythrocytes. J. Biol. Chem., *254*:2526, 1979.
32. Bennett, V.: Purification of an active proteolytic fragment of the membrane attachment site for human erythrocyte spectrin. J. Biol. Chem., *253*:2292, 1978.
33. Goodman, S.R., and Weidner, S.A.: Binding of spectrin alpha$_2$-beta$_2$ tetramers to human erythrocyte membranes. J. Biol. Chem., *255*:8082, 1980.
34. Wallin, R., Culp, E.N., Coleman, D.B., and Goodman, S.R.: A structural model of human erythrocyte band 2.1: alignment of chemical and functional domains. Proc. Natl. Acad. Sci. USA, *81*:4095, 1984.
35. Weaver, D.C., and Marchesi, V.T.: The structural basis of ankyrin function: I. Identification of two structural domains. J. Biol. Chem., *259*:6165, 1984.
36. Bennett, V., and Branton, D.: Selective association of spectrin with the cytoplasmic surface of human erythrocyte plasma membranes. J. Biol. Chem., *252*:2753, 1977.
37. Hargreaves, W.R., Giedd, K. N., Verkleij, A., and Branton, D.: Reassociation of ankyrin with band 3 in erythrocyte membranes and lipid vesicles. J. Biol. Chem., *255*:11965, 1980.
38. Drickamer, L.K.: Fragmentation of the band 3 polypeptide from human erythrocyte membranes. J. Biol. Chem., *252*:6909, 1977.
39. Nigg, E.A., and Cherry, R.J.: Anchorage of a band 3 population at the erythrocyte cytoplasmic membrane surface. Protein rotational diffusion measurements. Proc. Natl. Acad. Sci. USA, *77*:4702, 1980.
39a. Kay, M.M.B. et al: Polypeptides immunologically related to band 3 are present in nucleated somatic cells. Proc. Natl. Acad. Sci. USA. *80*:6882, 1983.
40. Anstee, D.J., and Tanner, M.J.: Structure and function of the red cell membrane sialoglycoproteins. Br. J. Haematol, *64*:211, 1986.
41. Cioe, L., and Curtis, P: Detection and characterization of a mouse alpha spectrin cDNA clone by its expression in E. coli. Proc. Natl. Acad. Sci. USA, *82*:1367, 1985.
42. Curtis, P.J. et al.: Sequence comparison of human and murine erythrocyte alpha-spectrin cDNA. Gene, *36*:357, 1985.
43. Linnenbach, A.J., Speicher, D.W., Marchesi, V.T., and Forget, B.G.: Cloning of a portion of the chromosomal gene for human erythrocyte alpha spectrin by using a synthetic gene fragment. Proc. Natl. Acad. Sci. USA, *83*:2397, 1986.
44. Huebner, K. et al.: The alpha spectrin gene is on chromosome 1 in mouse and man. Proc. Natl. Acad. Sci. USA, *82*:3790, 1985.
45. Winkelmann, J.C. et al.: Molecular cloning of the cDNA for human erythrocyte beta spectrin. Blood, *72*:328, 1988.
46. Prchal, J.T. et al.: Isolation and characterization of cDNA clones for human erythrocyte beta-spectrin. Proc. Natl. Acad. Sci. USA, *84*:7468, 1987.
47. Cioe, L. et al.: Cloning and nucleotide sequence of a mouse erythrocyte beta-spectrin cDNA. Blood, *70*:915, 1987.
48. Laurila, P., Cioe, L., Kozak, C., and Curtis, P.: Assignment of mouse beta-spectrin gene to chromosome 12. Somatic Cell Mol. Genet., *13*:93, 1987.
49. Winkelmann, J.C., Costa, F.F., and Forget, B.F.: Molecular cloning of the cDNA for human skeletal muscle beta spectrin: Evidence for tissue-specific differential processing of 3' beta spectrin pre-mRNA. Blood, *72*:35a, 1988.
50. Leto, T.L. et al.: Comparison of nonerythroid alpha spectrin genes reveals strict homology among diverse species. Mol. Cell. Biol., *8*:1, 1988.
51. Moon, R.T., Ngai, J., Wold, B.J., and Lazarides, E.: Tissue-specific expression of distinct spectrin and ankyrin transcripts in erythroid and nonerythroid cells. J. Cell Biol., *100*:152, 1985.
52. Birkenmeier, C.S. et al.: Remarkable homology among the internal repeats of erythroid and non-erythroid spectrin. Proc. Natl. Acad. Sci. USA, *82*:5671, 1985.
53. Wasenius, V.-M. et al.: Sequencing of the chicken nonerythroid spectrin cDNA reveals an internal repetitive structure homologous to the human erythrocyte spectrin. EMBO J., *4*:1425, 1985.
54. Wasenius, V.-M. et al.: Primary structure of the brain alpha-spectrin. J. Cell Biol., *108*:79, 1989.
55. Harris, A.S., Croall, D.E., and Morrow, J.S.: The calmodulin-binding site in alpha-fodrin is near the calcium-dependent protease-I cleavage site. J. Biol. Chem., *263*:15754, 1988.
56. Leto, T.L. et al.: Characterization of the calmodulin binding site of nonerythroid alpha spectrin. J. Biol. Chem., *264*:5826, 1989.

57. Baron, M.D., Davison, M.D., Jones, P., and Critchley, D.R.: The sequence of chick alpha-actinin reveals homologies to spectrin and calmodulin. J. Biol. Chem., *262*:17623, 1987.
58. Koenig, M., Monaco, A.P., and Kunkel, L.M.: The complete sequence of dystrophin predicts a rod-shaped cytoskeletal protein. Cell, *53*:219, 1988.
59. Hammonds, R.G.: Protein sequence of DMD gene is related to actin-binding domain of alpha-actinin. Cell. *51*:1, 1987.
60. Davison, M.D., and Critchley, D.R.: Alpha-actinins and the DMD protein contain spectrin-like repeats. Cell, *52*:159, 1988.
61. Dubreuil, R. et al.: Drosophila spectrin: I. Characterization of the purified protein. J. Cell Biol., *105*:2095, 1987.
62. Byers, T.J. et al.: Drosophila spectrin: II. Conserved features of the alpha subunit are revealed by analysis of cDNA clones and fusion proteins. J. Cell Biol., *105*:2103, 1987.
63. Conboy, J., Kan, Y.W., Shohet, S.B., and Mohandas, N.: Molecular cloning of protein 4.1, a major structural element of the human erythrocyte membrane skeleton. Proc. Natl. Acad. Sci. USA, *83*:9512, 1986.
64. Tang, T. et al.: Selective expression of an erythroid-specific isoform of protein 4.1. Proc. Natl. Acad. Sci. USA, *85*:3713, 1988.
65. Conboy, J., Mohandas, N., Tchernia, G., and Kan, Y.W.: Molecular basis of hereditary elliptocytosis due to protein 4.1 deficiency. N. Engl. J. Med., *315*:680, 1986.
66. Tang, T.K., Mazzucco, C.E., Benz, E.J., and Marchesi, V.T.: Identification of protein 4.1 isoforms in human lymphocytes with a nuclear localization. Submitted for publication, 1989.
67. Tang, T.K. et al.: Heterogeneity of mRNA and protein products arising from the protein 4.1 gene in erythroid and non-erythroid tissues. J. Cell Biol., in press, 1990.
68. Chan, J., Mohandas, N., Kan, Y.W., and Conboy, J.G.: Detection of further transcript diversity of protein 4.1 isoforms by enzymatic amplification of reticulocyte mRNA. Blood, *72*:24a, 1988.
69. Chan, J.Y., Conboy, J.G., Kan, Y.W., and Narla, M.: Evidence for a possible liver specific form of human protein 4.1. J. Cell Biol., *107*:25a, 1988.
70. Ngai, J., Stack, J.H., Moon, R.T., and Lazarides, E.: Regulated expression of multiple chicken erythroid membrane skeletal protein 4.1 variants is governed by differential RNA processing and translational control. Proc. Natl. Acad. Sci. USA, *84*:4432, 1987.
71. Lux, S. et al.: Red Cell (RBC) ankyrin is located on chromosome 8 and is deleted or defective in some patients with hereditary spherocytosis (HS). Blood, *72*:46a, 1988.
72. Lambert, S. et al.: A cDNA for human erythrocyte ankyrin. Blood, *72*:30a, 1988.
73. Lambert, S. et al.: A conserved repeating unit within the structure of human erythrocyte ankyrin. J. Cell Biol., *107*:469a, 1988.
74. Siebert, P.D., and Fukuda, M.: Isolation and characterization of human glycophorin A cDNA clones by a synthetic oligonucleotide approach: nucleotide sequence and mRNA structure. Proc. Natl. Acad. Sci. USA, *83*:1665, 1986.
75. Siebert, P.D., and Fukuda, M.: Human glycophorin A and B are encoded by separate, single copy genes coordinately regulated by a tumor-promoting phorbol ester. J. Biol. Chem., *261*:12433, 1986.
76. Rahuel, C. et al.: Characterization of cDNA clones for human glycophorin A. Use for gene localization and for analysis of normal and glycophorin-A-deficient (Finnish type) genomic DNA. Eur. J. Biochem., *172*:147, 1988.
77. Siebert, P.D., and Fukuda, M.: Molecular cloning of a human glycophorin B cDNA: Nucleotide sequence and genomic relationship to glycophorin A. Proc. Natl. Acad. Sci. USA, *84*:6735, 1987.
78. Rahuel, C. et al.: Alteration of the genes for glycophorin A and B in glycophorin-A-deficient individuals. Eur. J. Biochem., *177*:605, 1988.
79. Mattei, M.G. et al.: Localization of the gene for human erythrocyte glycophorin C to chromosome 2,q14-21. Hum. Genet., *74*:420, 1986.
80. Colin, Y. et al.: Isolation of cDNA clones and complete amino acid sequence of human glycophorin C. J. Biol. Chem., *261*:229, 1986.
81. Le Van Kim, C. et al.: Gerbich blood group deficiency of the Ge: -1, -2, -3 types. Immunochemical study and genomic analysis with cDNA probes. Eur. J. Biochem., *165*:571, 1987.
82. High, S., and Tanner, J.A.: Human erythrocyte membrane sialoglycoprotein beta. The cDNA sequence suggests the absence of a cleaved N-terminal signal sequence. Biochem. J., *243*:277, 1987.
83. Huang, C-H. et al.: Delta glycophorin (Glycophorin B) gene deletion in two individuals homozygous for the S-s-U- blood group phenotype. Blood, *70*:1830, 1987.
84. Kopito, R.R., Andersson, M., and Lodish, H.F.: Structure and organization of the murine band 3 gene. J. Biol. Chem. *262*:8035, 1987.
85. Kopito, R.R., Andersson, M., and Lodish, H.F.: Multiple tissue-specific sites of transcriptional initiation of the mouse anion antiport gene in erythroid and renal cells. Proc. Natl. Acad. Sci. USA, *84*:7149, 1987.
86. Kim, H-R.C. et al.: Two different mRNAs are transcribed from a single genomic locus encoding the chicken erythrocyte anion transport proteins (band 3). Mol. Cell. Biol., *8*:4416, 1988.
87. Palumbo, A.P. et al.: Chromosomal localization of a human band 3-like gene to region 7q35-7q36. Am. J. Hum. Genet., *39*:307, 1986.

88. Demuth, D.R. et al.: Cloning and structural characterization of a human band 3-like protein. EMBO J., 5:1205, 1986.
89. Tanner, M.J.A., Martin, P.G., and High, S.: The complete amino acid sequence of the human erythrocyte membrane anion-transport protein deduced from the cDNA sequence. Biochem. J., 256:703, 1988.
90. Lux, S.E. et al.: Cloning and characterization of band 3, the human erythrocyte anion transport channel. Proc. Natl. Acad. Sci. USA, 86, in press, 1989.
91. Nudel, U. et al.: The nucleotide sequence of the rat cytoplasmic beta-actin gene. Nucleic Acids Res., 11:1759, 1983.
92. Kost, T.A., Theodorakis, N., and Hughes, S.H.: The nucleotide sequence of the chick cytoplasmic beta-actin gene. Nucleic Acids Res., 11:8287, 1983.
93. Nakajima-Iijima, S., Hamada, H., Reddy, P., and Kakunaga, T.: Molecular structure of the human cytoplasmic beta-actin gene interspecies homology of sequences in the introns. Proc. Natl. Acad. Sci. USA, 82:6133, 1985.
94. Moos, M., and Gallwitz, D.: Human beta-actin-related processed genes. EMBO J., 2:757, 1983.
95. Howard Hughes Medical Institute, New Haven, Human Gene Mapping Library: Chromosome plots. Regional localization of genes and DNA segments on human chromosomes. Number 4, November 1988.
96. Morton, N.E.: The detection and estimation of linkage between the genes for elliptocytosis and the Rh blood type. Am. J. Hum. Genet., 8:80, 1956.
97. Cook, P.J.L., Noodes, J.E., Newton, M.S., and Demey, R.: On the orientation of the Rh E1, linkage group. Ann. Hum. Genet., 41:157, 1977.
98. Keats, B.J.B.: Another elliptocytosis locus on chromosome 17. Hum. Genet., 50:227, 1979.
99. Kimberling, W.J., Taylor, P.A., Chapman, R.G., and Lubs, H.A.: Linkage and gene localization of hereditary spherocytosis (HS). Blood. 52:859, 1978.
100. Bass, E.B., Smith, S.W., Stevenson, R.E., and Rosse, W.F.: Further evidence for location of the spherocytosis gene on chromosome 8. Ann. Intern. Med., 99:192, 1983.
101. Costa, F.F. et al.: Dominant hereditary spherocytosis (HS) is linked to the gene for the erythrocyte membrane protein ankyrin. Blood. 72:38a, 1988.
102. Palek, J.: Hereditary elliptocytosis and related disorders. Clin. Haematol., 14:45, 1985.
103. Bernstein, S.E.: Inherited hemolytic disease in mice: a review and update. Lab. Anim. Sci., 30:197, 1980.
104. Anderson, R., Huestis, R.R., and Motulsky, A.G.: Hereditary spherocytosis in the deer mouse, its similarity to the human disease. Blood, 15:491, 1960.
105. Lux, S.E.: Spectrin-actin membrane skeleton of normal and abnormal red blood cells. Semin. Hematol., 16:21, 1979.
106. Greenquist, A.C., Shohet, S.B., and Bernstein, S.E.: Marked reduction of spectrin in hereditary spherocytosis in the common house mouse. Blood, 51:1149, 1978.
107. Bodine, D.M., Birkenmeier, C.S., and Barker, J.E.: Spectrin deficient inherited hemolytic anemias in the mouse: Characterization by spectrin synthesis and mRNA activity in reticulocytes. Cell, 37:721, 1984.
108. Barker, J.E. et al.: Synthesis of spectrin and its assembly into the red cell membrane cytoskeleton of normal and mutant mice. J. Cell Biochem., 9B:3, 1985.
109. Castle, W.B., and Daland, G.A.: Susceptibility of erythrocytes to hypotonic hemolysis as a function of discoidal form. Am. J. Physiol., 120:371, 1937.
110. Becker, P.S., and Lux, S.E.: Hereditary spherocytosis and related disorders. Clin. Haematol., 14:15, 1985.
111. Agre, P. et al.: Partial deficiency of erythrocyte spectrin in hereditary spherocytosis. Nature, 314:380, 1985.
112. Agre, P., Asimos, A., Casella, J.F., and McMillan, D.: Inheritance pattern and clinical response to splenectomy as a reflection of erythrocyte spectrin deficiency in hereditary spherocytosis. N. Engl. J. Med., 315:1579, 1986.
113. Coetzer, T.L. et al.: Partial ankyrin and spectrin deficiency in severe, atypical hereditary spherocytosis. N. Engl. J. Med., 318:230, 1988.
114. Hanspal, M., Hanspal, J., Prchal, J.T., and Palek, J.: Molecular basis of spectrin and ankyrin deficiencies in atypical hereditary spherocytosis (HS): Synthesis of unstable ankyrin. Blood 72:28a, 1988.
115. Agre, P., Orringer, E.P., and Bennett, V.: Deficient red-cell spectrin in severe, recessively inherited spherocytosis. N. Engl. J. Med., 306:1155, 1982.
116. Goodman, S.R., Shiffer, K.A., Casoria, L.A., and Eyster, M.E.: Identification of the molecular defect in the erythrocyte membrane skeleton of some kindreds with hereditary spherocytosis. Blood, 60:772, 1982.
117. Wolfe, L.C. et al.: A genetic defect in the binding of protein 4.1 to spectrin in a kindred with hereditary spherocytosis. N. Engl. J. Med., 307:1367, 1982.
118. Becker, P.S., Morrow, J.S., and Lux, S.E.: Abnormal oxidant sensitivity and beta chain structure of spectrin in hereditary spherocytosis associated with defective spectrin-protein 4.1 binding. J. Clin. Invest., 80:557, 1987.
119. Zarkowsky, H.S., Mohandas, N., Speaker, C.B., and Shohet, S.B.: A congenital haemolytic anaemia with thermal sensitivity of the erythrocyte membrane. Br. J. Haematol., 29:537, 1975.

120. Mohandas, N. et al.: A technique to detect reduced mechanical stability of red cell membranes: relevance to elliptocytic disorders. Blood, *59*:768, 1982.
121. Liu, S.C., Palek, J., Prchal, J., and Castlebury, R.P.: Altered spectrin dimer–dimer association and instability of erythrocyte membrane skeletons in hereditary pyropoikilocytosis. J. Clin. Invest., *68*:597, 1981.
122. Liu, S.C., Palek, J., and Prchal, J.: Defective spectrin dimer–dimer association in hereditary elliptocytosis. Proc. Natl. Acad. Sci. USA, *79*:2072, 1982.
123. Knowles, W.J. et al.: Molecular and functional changes in spectrin from patients with hereditary pyropoikilocytosis. J. Clin. Invest., *71*:1867, 1983.
124. Lawler, J., Liu, S.C., Palek, J., and Prchal, J.: A molecular defect of spectrin in a subset of patients with hereditary elliptocytosis. Alterations in the alpha-subunit domain involved in spectrin self-association. J. Clin. Invest., *73*:1688, 1984.
125. Dhermy, D. et al.: Spectrin beta-chain variant associated with hereditary elliptocytosis. J. Clin. Invest., *70*:707, 1982.
126. Ohanian, V., Evans, J.P., and Gratzer, W.B.: A case of elliptocytosis associated with a truncated spectrin chain. Br. J. Haematol., *61*:31, 1985.
127. Morle, L. et al.: Spectrin Tunis (alpha I/78): a new alpha I variant that causes asymptomatic hereditary elliptocytosis in the heterozygous state. Blood, *71*:508, 1988.
128. Lawler, J., Liu, S.C., Palek, J., and Prchal, J.: Molecular defect of spectrin in hereditary pyropoikilocytosis. J. Clin. Invest., *70*:1019, 1982.
129. LeComte, M.C. et al.: Hereditary elliptocytosis with spectrin molecular defect in a white patient. Acta Haematol (Basel), *71*:235, 1984.
130. Dhermy, D. et al.: Molecular defect of spectrin in the family of a child with congenital hemolytic poikilocytic anemia. Pediatr. Res., *18*:1005, 1984.
131. Dhermy, D. et al.: Hereditary elliptocytosis: clinical, morphological, and biochemical studies of 38 cases. Nouv. Rev. Fr. Hematol., *28*:129, 1986.
132. Garbarz, M. et al.: Double inheritance of an alpha I/65 spectrin variant in a child with homozygous elliptocytosis. Blood, *67*:1661, 1986.
133. LeComte, M.C. et al.: A new abnormal variant of spectrin in black patients with hereditary elliptocytosis. Blood, *65*:1208, 1985.
134. Lawler, J. et al.: Sp alpha I/65: a new variant of the alpha subunit of spectrin in hereditary elliptocytosis. Blood, *66*:706, 1985.
135. Alloisio, N. et al.: Sp alpha I/65 hereditary elliptocytosis in North Africa. Am. J. Hematol., *23*:113, 1986.
136. Lawler, J. et al.: Molecular heterogeneity of hereditary pyropoikilocytosis: identification of a second variant of the spectrin alpha-subunit. Blood, *62*:1182, 1983.
137. LeComte, M.C. et al.: Pathologic and non-pathologic variants of the spectrin molecule in two black families with hereditary elliptocytosis. Hum. Genet., *71*:351, 1985.
138. Marchesi, S.L. et al.: Abnormal spectrin in hereditary elliptocytosis. Blood, *67*:141, 1986.
139. Marchesi, S.L. et al.: Mutant forms of spectrin alpha-subunits in hereditary elliptocytosis. J. Clin. Invest., *80*:191, 1987.
140. Lambert, S., and Zail, S.: A new variant of the alpha subunit of spectrin in hereditary elliptocytosis. Blood, *69*:473, 1987.
141. Lane, P.A. et al.: Unique alpha-spectrin mutant in a kindred with common hereditary elliptocytosis. J. Clin. Invest., *79*:989, 1987.
142. Coetzer, T.L., and Palek, J.: Partial spectrin deficiency in hereditary pyropoikilocytosis. Blood, *67*:919, 1986.
143. Sahr, K.E. et al.: use of the polymerase chain reaction (PCR) for the detection of mutations causing hereditary elliptocytosis (HE). Blood *72*:34a, 1988.
144. Roux, A.F. et al.: Molecular basis of Sp alpha I/65 hereditary elliptocytosis in North Africa: insertion of a TTG triplet between codons 147 and 149 in the alpha-spectrin gene of five unrelated families. Blood, *72*:33a, 1988.
145. Sahr, K.E. et al.: Sequence and exon-intron organization of the DNA encoding the alpha I domain of human spectrin: applications to the study of mutations in hereditary elliptocytosis. J. Clin. Invest. *84*:1243, 1989.
146. Garbarz, M. et al. The DNA's from HE and HPP related patients with the spectrin alpha I/78 kD variant contain a single base substitution in exon-3 of the alpha spectrin gene. Blood, *72*:41a, 1988.
147. Tchernia, G., Mohandas, N., and Shohet, S.B.: Deficiency of cytoskeletal membrane protein band 4.1 in homozygous hereditary elliptocytosis: implications for erythrocyte membrane stability. J. Clin. Invest., *68*:454, 1981.
148. Feo, C.J., Fischer, S., and Plau, J.P.: Première observation de l'absence d'une proteine de la membrane érythrocytaire (band 4.1) dans un cas d'anemie elliptocytaire familiale. Nouv. Rev. Fr. Hematol., *22*:315, 1981.
149. Alloisio, N., Dorleai, E., Girot, R., and Delaunay, J.: Analysis of the red cell membrane in a family with hereditary elliptocytosis—total or partial—of protein 4.1. Hum. Genet., *59*:68.

150. Alloisio, N. et al.: The heterozygous form of 4.1 (−) hereditary elliptocytosis [the 4.1(−) trait]. Blood, *65*:46, 1985.
151. Lambert, S. et al.: A molecular study of heterozygous protein 4.1 deficiency in hereditary elliptocytosis. Blood, *72*:1926, 1988.
152. Talakuwan, Y. et al.: Restoration of normal membrane stability to unstable protein 4.1-deficient erythrocyte membranes by incorporation of purified protein 4.1. J. Clin. Invest., *78*:80, 1986.
153. Garbarz, M. et al.: A variant of erythrocyte membrane skeletal protein band 4.1 associated with hereditary elliptocytosis. Blood, *64*:1006, 1984.
154. Letsinger, J.T., Agre, P., and Marchesi, S.L.: High molecular weight protein 4.1 in the cytoskeletons of hereditary elliptocytes. Blood, *68*:38a, 1986.
155. Conboy, J., Kan, Y.W., Agre, P., and Mohandas, N.: Molecular characterization of hereditary elliptocytosis due to an elongated protein 4.1. Blood, *68*:34a, 1986.
156. Marchesi, S.L. et al.: Characterization of high and low molecular weight varients of protein 4.1 associated with elliptocytosis. Blood *72*:31a, 1988.
157. Rossi, M.E. et al.: Reconstitution with normal protein 4.1 restores mechanical stability to unstable membranes of hereditary elliptocytes with mutant, dysfunctional protein 4.1. Blood, *72*:48a, 1988.
158. Alloisio, N. et al.: Red cell membrane sialoglycoprotein beta in homozygous and heterozygous 4.1 (−) hereditary elliptocytosis. Biochim. Biophys. Acta, *816*:57, 1985.
159. Mueller, T.J., William, J., Wang, W., and Morrison, M.: Cytoskeletal alterations in hereditary elliptocytosis. Blood, *58*:47a, 1981.
160. Anstee, D.J. et al.: Two individuals with elliptocytic red cells apparently lack three major sialoglycoproteins. Biochem. J., *218*:615, 1984.
161. Zail, S.S., and Coetzer, T.L.: Defective binding of spectrin to ankyrin in a kindred with recessively inherited elliptocytosis. J. Clin. Invest., *74*:753, 1984.
162. Hayashi, S. et al.: Abnormality in a specific protein of the erythrocyte membrane in hereditary spherrocytosis. Biochem. Biophys. Res. Commun., *57*:1038, 1974.
163. Rybicki, A.C. et al.: Deficiency of protein 4.2 in erythrocytes from a patient with a Coombs negative hemolytic anemia. Evidence for a role of protein 4.2 in stabilizing ankyrin on the membrane. J. Clin. Invest., *81*:893, 1988.

Chapter 10

MOLECULAR GENETICS OF THE PLASMA APOLIPOPROTEINS

Lawrence Chan • Eric Boerwinkle • Wen-Hsiung Li

Atherosclerotic cardiovascular disease is a major cause of mortality and morbidity in the United States.[1] The atherosclerotic plaque, the basic underlying lesion in atherosclerosis, contains cholesteryl esters that are derived from the circulating lipids. Plasma lipids are thus essential to the development of the atherosclerosis. Not surprisingly, therefore, the propensity to develop atherosclerotic cardiovascular disease is directly related to the plasma lipid concentration and distribution.

Lipid is insoluble in aqueous solutions. Circulating lipids are transported in plasma associated with special carrier proteins. Such lipid–protein complexes are known as lipoproteins and the carrier proteins, apolipoproteins. The physical properties and chemical composition of the major classes of lipoproteins are listed in Table 10-1, and their metabolism is described in the next section.

PLASMA LIPOPROTEINS

The lipoproteins can be separated by many techniques, including ultracentrifugal flotation, zonal electrophoresis, gradient gel electrophoresis, and gel filtration chromatography. These techniques can be used both analytically and on a preparative scale. In certain situations, other techniques have also been used, such as chemical precipitation and affinity chromatography using monoclonal or polyclonal antisera to particular apolipoproteins. The various methods of lipoprotein isolation have been the subject of many reviews.[2,3] The physicochemical properties of the various lipoprotein classes are described in Table 10-1. Their origin and metabolism are summarized in the following sections.

Chylomicrons

Chylomicrons are triacylglycerol-rich lipoproteins of intestinal origin with Svedbergs of flotation (SF) > 400. They are spherical particles by electron microscopy in negatively stained preparations and range in size from 75 to 1000 nm in diameter. Chylomicrons are secreted by the intestinal epithelial cells into the mesenteric lymph, through which they ultimately reach the intravascular compartment. Redgrave[4] first showed that in functionally hepatectomized rats, chylomicrons are metabolized to chylomicron "remnants," smaller particles depleted of much of the triacylglycerol but still retaining most of the cholesteryl esters. In normal animals, these chylomicron remnants are rapidly removed from the plasma by the liver. The peripheral metabolism of chylomicron to chylomicron remnants depends on lipoprotein lipase,[5] an enzyme anchored on the surface of the capillary endothelium of adipose tissues and muscle. The apolipoprotein (apo) composition of the particles also changes during the metabolism of chylomi-

Table 10–1.   Physical Properties and Chemical Composition of Human Plasma Lipoprotein Families

	Electrophoretic Definition	Particle Size (nm)	Molecular Weight	Density (g/mL)	Surface Components (mol %)			Core Lipids (mol %)	
					Choles-terol*	Phospho-lipids	Apolipo-protein	Trigly-cerides	Cholesteryl esters
Chylomicrons	Remains at origin	75–1000	~400,000,000	0.93	35	63	2	95	5
VLDL	Pre-β-lipoproteins	30–80	10–80,000,000	0.93–1.006	43	55	2	76	24
IDL	Slow pre-β-lipoproteins	20–35	5–10,000,000	1.006–1.019	38	60	2	78	22
LDL	β-Lipoproteins	18–25	2,300,000	1.019–1.063	42	58	0.2	19	81
HDL$_2$	α-Lipoproteins	9–12	360,000	1.063–1.125	22	75	2	18	82
HDL$_3$	α-Lipoproteins	5–9	175,000	1.125–1.210	23	72	5	16	84

* May be distributed between the surface and core, and in the case of large chylomicrons, more cholesterol may be in the core than the surface.
HDL, high-density lipoprotein; IDL, intermediate-density lipoprotein; LDL, low-density lipoprotein; VLDL, very low-density lipoprotein.
Modified from Smith, L. C. et al.: *In Supramolecular Structure and Function.* Edited by G. Pifat and J. N. Herak. New York, Plenum Press, 1983. p. 210.[171]

crons.[6] ApoB-48, synthesized by the small intestine in adult humans, stays with the particles and is a good marker for chylomicron and remnant particles. ApoA-I, synthesized both by the liver and intestine, is largely lost from the chylomicrons when they reach the ciirculation. In contrast, apoE and much of the C-apolipoproteins are acquired by the chylomicrons in the capillaries of the intestine.[6–8] Most of the latter apolipoproteins are lost during remnant conversion, but apoE remains with the remnant particles and seems likely as the ligand for the hepatic chylomicron remnant receptors.[9,10]

Very Low-Density Lipoproteins

Very low-density lipoproteins (VLDL) are produced primarily by the liver. They consist of particles that vary in size from approximately 30 to 80 nm and have flotation rates ranging from SF 20 to SF 400. They are the major carriers of endogenously produced lipids.[11,12] ApoB-100 is an obligatory component of nascent VLDL, which also contain apoE and the C-apolipoproteins. After acquiring additional C-apolipoproteins from high-density lipoproteins (HDL), VLDL are acted on by lipoprotein lipase, producing VLDL remnants.[5] Most of the VLDL remnants (also known as intermediate-density lipoproteins, IDL, which have flotation rates of SF 12 to 20 and sizes of 20 to 35 nm) are further modified to become low-density lipoproteins (LDL).[12–13] During the process, most of the remaining triacylglycerols are removed, and all proteins except apoB-100 leave the lipoprotein particle. In normal individuals, a small amount of the IDL is taken up by the liver by adsorptive endocytosis.

Low-Density Lipoproteins

As discussed in the last section, LDL are largely the product of VLDL metabolism.[13] They comprise lipoprotein particles of approximate size 18 to 25 nm with an SF of 0 to 12. Four major density groups, designated LDL-I through LDL-IV, have been described.[14] These subclasses appear to be familially transmitted,[14,15] and the small, dense LDL particles are significantly associated with increased risk of myocardial infarction.[16] LDL are a major physiologic ligand for the LDL (apoB-E) receptor,[17] and high LDL levels are strongly correlated with the development of atherosclerosis.[18] LDL contain only one apoprotein, apoB-100, as their major protein component.[7]

High-Density Lipoproteins

HLD are isolated from the density range of 1.21 to 1.063 g/mL. HDL particles range in size from approximately 5 to 12 nm. Two major subclasses, HDL_2 and HDL_3, have been identified with mean hydrated densities of 1.090 and 1.145 g/mL, respectively.[19] Other additional subpoplations of HLD have also been described.[20,21]

HDL are secreted in a nascent state in the form of discoidal particles[22] consisting of a bilayer of phospholipids onto which the apolipoproteins (apoA-I, A-II, and E) are embedded.[21] HDL acquire some additional apoproteins during the catabolism of the triacylglycerol-rich lipoproteins, chylomicron and VLDL.[23] Cholesteryl esters present in the core of HDL particles are derived from the action of an enzyme, lecithin-cholesterol acyltransferase (LCAT) associated with HDL.[24] LCAT esterifies cholesterol with a fatty acyl residue derived from the major phospholipid of lipoproteins, lecithin (phosphatidyl choline). The cholesteryl esters are rapidly transferred to other lipoproteins. The cholesterol acted on by LCAT is derived from the surface of lipoproteins or of cells.

HDL are secreted by the liver, possibly the small intestine and other tissues.[11,21,22,25] By transferring the cholesteryl esters to the liver via remnants or other lipoproteins, HDL form an important part of the "reverse cholesterol transport" pathway first proposed by Glomset.[24] The plasma concentration of HDL is inversely related to the propensity for developing atherosclerosis,[26,27] perhaps partly because of their participation in this pathway.

PLASMA APOLIPOPROTEINS

There are ten major apolipoproteins and several additional minor ones. The physical properties of the major apolipoproteins are listed in Table 10-2. The apolipoproteins all share one common property, the ability to bind to lipids. This characteristic allows these proteins to be an essential component of the lipoprotein complex. Therefore, they all play important structural roles in lipid transport. Most of the apolipoproteins have also acquired highly specialized functions. ApoA-I, and possibly apoC-I and apoA-IV, activate LCAT[28-30] and therefore participate actively in reverse cholesterol transport. ApoC-II is an essential co-factor for the action of lipoprotein lipase[31]; a heritable deficiency of apoC-II results in hyperchylomicronemia and a phenotype often indistinguishable from that of lipoprotein lipase deficiency.[32] ApoC-III may have some inhibitory effect on apoC-II activation of lipoprotein lipase[33]; it also impedes the uptake of apoE-containing lipoprotein particles by the liver.[34] ApoB-48 is a marker for intestinal chylomicron and chylomicron remnants,[7] whereas apoB-100 is an important physiologic ligand for the LDL (apoB-E) receptor.[17] ApoE also confers receptor-binding properties to lipoprotein particles for both the LDL (apoB-E) receptor[17,35] and a specific apoE receptor in the liver.[9,36] There is speculation that, as an almost ubiquitous protein, apoE potentially serves many other roles such as immunomodulation, neural degeneration, and even paracrine functions.[36,37] The functions and structure–function relationships of the various apolipoproteins will be discussed further in the section entitled Relative Rates of Evolution and Functional Aspects of Apolipoprotein Structure.

Table 10–2. Physical Properties and Functions of the Human Plasma Apolipoproteins

Protein	Molecular Weight	Distribution in Lipoproteins*				Major Function(s)
		HDL	LDL	IDL	VLDL	
ApoA-I	28,000	+ + + +				Activator of LCAT
ApoA-II	17,000	+ + + +				May inhibit hepatic triacylglycerol lipase at high concentrations
ApoA-IV†	44,500	+ +				Activates LCAT
ApoB-48*	264,000					Structural: obligatory protein for chylomicrons
ApoB-100	550,000		+ + +	+ +	+	Ligand for the LDL (apoB-E) receptor
ApoC-I	6,600	+ + +		trace	+	Activates LCAT
ApoC-II	8,900	+ + +		+	+	Activator of lipoprotein lipase
ApoC-III	8,800	+ +		+	+	Modulates hepatic uptake of apoE-containing lipoproteins
ApoD	22,000	+ + +				May be involved in cholesterol ester transfer
ApoE	34,000	+ +		+ +	+ +	Ligand for the LDL (apoB–E) receptor as well as for the apoE receptor

* Chylomicron, at various stages of its metabolism, contains small amounts of all the apolipoproteins except apoB-100. ApoB-48 is present exclusively in chylomicron and its remnants.
† Plasma apoA-IV is mainly unassociated with lipoproteins in the 1.21 G/mL infranate.
HDL, high density lipoprotein; IDL, intermediate-density lipoprotein; LCAT, lecithin-cholesterol acyltransferase; VLDL, very low-density lipoprotein.
+ to + + + + indicate relative distributions of various apolipoproteins in lipoproteins, with + + + + representing the highest degree of abundance.

The metabolism of the lipoprotein lipid transport system is complex. The interconversion and metabolic fate of the various particles, and the receptor-mediated metabolism of LDL and IDL are beyond the scope of this chapter. The interested reader should consult recent reviews for details.[17,38,39]

THE APOLIPOPROTEIN GENES

All the major human apolipoprotein cDNAs and chromosomal genes have been cloned and sequenced, including those for apoA-I, apoA-II, apoA-IV, apoE, apoC-I, apoC-II, apoC-III, apoB-100, and apoD. In this section, we shall review the chromosomal localization and structural organization of the human apolipoprotein genes.

Chromosomal Localization and Linkage Analysis

The chromosomal locations of some of the apolipoprotein genes have previously been established by classic family studies and linkage analysis. Recently, most of the apolipoprotein genes have been further mapped onto specific human chromosomes by two newly developed techniques: (1) Southern blot analysis of somatic cell hybrids involving human and nonhuman cells and (2) in-situ nucleic acid hybridization on chromosome spreads. In somatic cell hybrid analysis, hybrid cells formed between human cells and nonhuman (usually rodent) cells are established in vitro. In human–rodent hybrids, preferential loss of human chromosomes is observed. Thus, human chromosomes that are retained or lost in the hybrids may be identified and correlated with the presence or absence of a specific apolipoprotein gene detected by Southern blot analysis.[40] In-situ nucleic acid hybridization was first used to detect repetitive sequences on DNA. Subsequent modifications of the original method allowed the localization of single-copy genes (such as apolipoprotein genes) in mitotic chromosome preparations.[41]

Using the two techniques, all the major apolipoprotein genes have been mapped to specific human chromosomes. Their locations are summarized in Table 10-3.

Structure of Apolipoprotein Genes

The genomic structures of the seven major soluble apolipoproteins (apoA-I, A-II, A-IV, C-I, C-II, C-III, and E) (Fig. 10-1) are all very similar to one another, whereas the structure of the apoB gene is quite different from the others (Fig. 10-2). For the soluble apolipoproteins, with the exception of apoA-IV, there are four exons and three introns in each gene. Among these genes, the introns occur at very similar locations: intron-I interrupts the 5′ untranslated region of the gene; intron-II interrupts a coding region of the gene very close to the signal peptidase cleavage site; and intron-III interrupts the part of the gene encoding the mature peptide. ApoA-IV differs from the other genes in that the intron in the 5′ untranslated region is missing. The other two introns in this

Table 10–3. Chromosomal Location of Human Apolipoprotein Genes

ApoA-II	Chromosome 1, 1p21–qter
ApoB	Chromosome 2, 2p23–p24
ApoD	Chromosome 3, 3p14.2–qter
ApoA-I, C-III, A-IV	Chromosome 11, 11q13–qter
ApoE, C-I, C-II	Chromosome 19, 19q

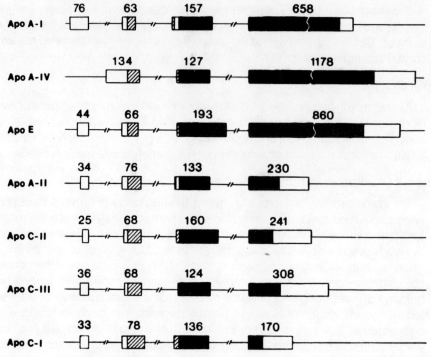

Fig. 10–1. Structural organization of the human apolipoprotein A-I, A-II, A-IV, C-I, C-II, C-III, and E genes. Transcription is from *left* to *right*. The *wide bars* represent the exons, and the *thin line* represents the 5' flanking region, introns, and 3' flank region of the respective genes. The *wide bars* are divided into several regions: the *open bars* at the two ends represent the 5' and 3' untranslated regions; the *hatched bars*, the signal peptide regions; and the *solid bars*, the mature-peptide regions of the respective genes. In apoA-I and apoA-II, the prosegment is represented by a *narrow open bar* between the signal peptide and mature peptide region. The *numbers above* the exons indicate the length (number of nucleotides) of the exons. The lengths of the exons are drawn to scale, except for the last exons in apoA-I, A-IV, and E.

gene, however, are present at the same locations as in the other apolipoprotein genes. The lengths of the first three (two for apoA-IV) exons are very similar among genes, and the differences in the total length of messenger RNAs (mRNAs) are accounted for mainly by differences in the length of exon 4 (exon 3 for apoA-IV). This striking similarity in the genomic structure of the seven soluble apolipoproteins supports the hypothesis

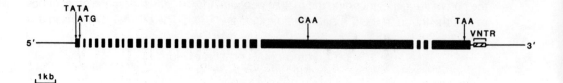

Fig. 10–2. Structure of the human apolipoprotein B gene. This map is based on the study of Blackhart et al.[43] The exons are represented by *black bars* and are drawn to scale. The introns are represented by blank *spaces* in between the exons and are not shown to scale. The 3' polymorphic locus of a variable number of tandem repeats (VNTR) is drawn to scale. Positions of the TATA box, initiation codon, ATG, and termination codon, TAA, are shown. The location of the CAA codon that is edited to UAA in apoB-48 messenger RNA (mRNA) is also shown.

that these agents have arisen from a common ancestor and that the individual apo-lipoprotein genes have evolved through partial and complete gene duplications.[42]

Compared to the other apolipoproteins, the human apoB gene shows many re-markable differences[43] (Fig. 10-2). Although it spans 43 kilobases (kb), the total length of the apoB gene is actually relatively small for the size of its gene product, which is the largest monomeric protein known. The exons together comprise approximately one third of the gene, and the total length of the 28 introns is only about twice that of the exons. Two of the exons (exons 26 and 29) are unusually long: 7572 and 1906 base pairs (bp), respectively. The distribution of introns is extremely asymmetric, with a concentration in the 5'-terminal one-third of the gene. For example, one intron occurs in the 5'-untranslated region, and another 11 introns occur within the first 500 codons, and a total of 19 introns appear within the first 1000 codons. The evidence shows that if the apoB gene is evolutionarily related to the other apolipoprotein genes, it has undergone considerable change with time and no longer has the 3-intron/4-exon structure common to most of the other genes.

The organization of the apoD gene is quite different from those of the apoB or the other soluble apolipoprotein genes.[44] It has at least one intron in the 5' untranslated region. Because the transcription-initiation site has not been defined, additional introns in this region are possible. Unlike the other soluble apolipoprotein genes, the signal peptide region in the apoD gene does not contain any introns. Further, in the mature peptide-coding region, the apoD gene contains three introns compared to only one in the other genes. The drastically different structural organization of the apoD gene suggests that it is not a member of the apolipoprotein multigene family (see following section).

APOLIPOPROTEIN mRNA TRANSLATION AND POST-TRANSLATIONAL MODIFICATION

Co-Translational Processing of Preapolipoproteins

The first model apolipoprotein that was studied by in-vitro translation and radio-sequencing is apoVLDL-II, an estrogen-inducible apolipoprotein in the chicken.[45] The nascent, in-vitro translation product, designated preapoVLDL-II, contains a 24 amino acid NH_2-terminal extension of the mature plasma protein.[46] The properties of this extra sequence are very similar to those previously described for signal peptides found in secretory protein precursors.[47–50] The hydrophobic residues are predominantly re-stricted to the interior segment of the peptide, while the charged and hydroxyl-con-taining residues are located near the ends. The sequence represents the first docu-mentation of a signal peptide in a plasma apolipoprotein precursor. The translocation of preapoVLDL-II represents a unique case of vectorial migration of a protein through the membrane of the rough endoplasmic reticulum.[51] ApoVLDL-II is unique when compared to other secretory proteins in that, being an apolipoprotein, it has a structural component that would normally spontaneously bind to a phospholipid matrix such as the rough endoplasmic reticulum outer surface (see the following section on Lipid-Binding Domains of Apolipoproteins). Translocation of this phospholipid-binding pro-tein requires some mechanism to prevent interaction of the amphipathic region with the outer membrane. Whether control is afforded by a specific topogenic sequence[52] or whether the signal sequence regulates the folding of the nascent polypeptide chain[53] and prevents this interaction is uknown at present.

Subsequent studies indicate the existence of the signal peptide in all apolipoprotein precursors: apoA-I, apoA-II, apoA-IV, apoE, apoC-I, apoC-II, apoC-III, apoB-100, and apoD (refs. to each of these instances are given in Li et al.[42]). The lengths of the peptides vary between 18 and 27 amino acids, well within the usual limits for eukaryotic signal peptides.[49,50] These peptides encompass a positively charged NH_2-terminal region, a central hydrophobic region, and a more polar C-terminal region that seems to define the cleavage site. Thus, they show characteristics present in other eukaryotic signal peptides.

In addition to the structural similarities of the apolipoprotein signal sequences to other eukaryotic signal sequences, additional similarities in structure occur among the individual apolipoprotein sequences, suggesting that these proteins may be related in evolution (see the section entitled Molecular Evolution of Apolipoprotein Genes, p. 197).

ApoB-48 Synthesis in the Small Intestine: A Unique Case of RNA Editing

In adult humans, apoB-100 is synthesized exclusively in the liver, and apoB-48, in the intestine.[7,54] The early fetal intestine synthesizes mainly apoB-100, and the capacity to synthesize apoB-48 is acquired during maturation of the fetus.[55] ApoB-48 shares antigenic determinants with the NH_2-terminal half of apoB-100[56,57] and has a molecular weight approximately 48% of the latter.[7,54] Studies using monoclonal antibodies,[58] genetic mutants,[59] Southern blotting,[60] and genomic cloning and sequencing[43,61] indicate that apoB-48 and apoB-100 are the products of the same gene. However, experiments involving labeled amino acid precursor incorporation and immunochemical identification indicate that apoB-48 is not a post-translational cleavage product of apoB-100.[62]

These puzzling observations can be explained by some recent studies by Powell et al.[63] and Chen et al.[64] They examined apoB-48 synthesis and structure by five different approaches: DNA-excess hybridization of small intestinal mRNA using apoB-100 complementary DNA (cDNA) probes, direct sequencing of tryptic peptides of apoB-48 purified from chylous ascites fluid, sequencing of cloned intestinal apoB cDNAs, direct sequencing of intestinal mRNAs, and oligonucleotide hybridization of intestinal cellular DNA. ApoB-48 was shown to be the product of an intestinal mRNA that has an in-frame UAA stop codon resulting from a C→U change in the codon CAA encoding glutamine (Gln)-2153 in apoB-100 mRNA. In other words, the same apoB gene on chromosome two expresses both apoB-48 in the intestine and apoB-100 in the liver by this organ-specific alteration of the transcribed mRNA. Oligonucleotide hybridization of intestinal cellular DNA indicates that, like leukocyte DNA, enterocyte DNA had a C in this location. The COOH-terminal isoleucine (Ile)-2152 of apoB-48 purified from chylous ascites fluid had apparently been cleaved from the initial translation product, leaving methionine (Met)-2151 as the new COOH-terminus.[64] DNA-excess hybridization and sequencing data indicate that ~85% of the intestinal mRNAs terminate within approximately 110 bases downstream from the in-frame stop codon. The other ~15% of the mRNAs have lengths similar to hepatic apoB-100 mRNA, but they also have the same in-frame stop codon.[64] The organ-specific introduction of a stop codon into intestinal apoB-48 mRNA is an unprecedented observation, and the exact mechanism is currently under active investigation.

Post-Translational Processing of Mammalian Apolipoproteins

ApoA-I

When either human or rat apoA-I mRNA is translated in vitro in the presence of signal peptidase, an 18 amino acid signal peptide is co-translationally removed by the enzyme. Examination of the cleaved product reveals that it has an additional 6-amino acid residue segment attached to the NH_2-terminal of mature plasma apoA-I. This larger protein has been designated proapoA-I.[65] The primary structure of the proseg-ment has been deduced by both in vitro translation and radiosequencing as well as from the nucleotide sequence of apoA-I cDNA clones. In the human, the rat, and the dog,[65–67] the prosegment contains a COOH-terminal Gln–Gln dipeptide, unusual amino acids for protein precursors that are processed proteolytically. In contrast, the rabbit proapoA-I sequence ends with Gln-arginine (Arg),[68] and the chicken with Gln-histidine (His).[69]

Pulse-labeling experiments in the human hepatoma cell line, HepG2, indicate that the hexapeptide is present in greater than 95% of newly secreted apoA-I.[70,71] Obser-vations using rat hepatocytes and enterocytes in culture, as well as perfused liver and intestinal segments, indicate that most of the proapoA-I is secreted intact without significant proteolytic processing.[65,72] Thus, in these experimental systems, the proteo-lytic processing of proapoA-I appears to be an exclusively extracellular event. The pres-ence of proapoA-I in the plasma partly accounts for the heterogeneity of plasma apoA-I isoforms. In contrast, Banerjee et al.[73] found that in cultured chick hepatocytes and in HepG2 cells, the proportion of secreted proapoA-I vs. mature apoA-I showed wide variations, depending on the hormonal environment and on the presence or absence of fetal bovine serum in the medium.

ApoA-II

ApoA-II synthesized in vitro in the presence of signal peptidase also contains a pentapeptide attached to the NH_2-terminus of the mature plasma protein. The se-quence of this peptide, AlaLeuValArgArg, is identical in the human and rat proapoA-II.[74–77] It resembles most prosegments in that it terminates with paired basic amino acid residues. In the human hepatoma cell line, HepG2 proapoA-II processing occurs predominantly extracellularly. The cells appear to secrete an enzyme(s) capable of the cleavage. The cleavage activity, which contains a Cathepsin B-like protein, is not in-hibited by serine protease inhibitors but is blocked by many thiol protease inhibitors.

Post-Translational Modifications of Apolipoproteins

All the major apolipoproteins are polymorphic in plasma. Each exists in two or more isoforms, which differ in charge or mass and are detected by two-dimensional gel electrophoresis. The structural basis for some of these apparently heterogeneous forms has been worked out for some of the apolipoproteins and includes amino acid sub-stitution, glycosylation, deamidation, proteolytic cleavage, acylation, and phosphory-lation. ApoE is an excellent case of protein charge heterogeneity resulting from allelic differences in amino acid sequence (see discussion later on apoE isoforms in the section entitled Genetic Variation and Its Effects). In addition, the following apolipoproteins are known to be glycosylated; apoA-II, B, C-III, and E. ApoC-II appears to be secreted in a glycosylated form and is deglycosylated in the circulation.[78] The different isoforms of plasma apoA-I are, in part, the products of deamidation.[79] Some of the circulating apolipoproteins undergo further proteolytic cleavage, in many cases possibly as inter-mediate steps in degradation. Whether in some cases the proteolytic cleavage products are artifacts of isolation or incubation conditions is unclear. In the case of apoC-II, a minor isoform designated apoC-II$_{1/2}$ by Fojo et al.[78] was found to be a truncated protein missing the first six N-terminal amino acid residues of the major isoform, apoC-II.

Using the human hepatoma cell line HepG2, Hoeg et al.[80] have established that, in addition to proteolytic processing, secreted nascent apoA-I is acylated with palmitate. They also suggest that apoA-II, A-IV, B-100, C-II, and C-III all have some degree of fatty acid acylation. The biologic consequence of the acylation is unknown. Hoeg et al. speculate that the increased hydrophobicity conferred on the protein by the fatty acid might facilitate protein–lipid interaction.

In cultured rat hepatocytes, Davis et al.[81] found that newly secreted apoB-48, but not B-100, was phosphorylated. In radiolabeling experiments when the cells were incubated with ^{32}P-orthophosphate, at least 20% of the ^{32}P associated with apoB-48 was in the form of phosphoserine. In vivo labeling experiments confirmed that phos-phorylation of apoB-48 occurs in vivo in the rat. Furthermore, ^{32}P-labeled apoB-48 loses its radioactivity on incubation in rat serum. The physiologic significance of apoB-48 phosphorylation in rat liver is unknown.

STRUCTURAL FEATURES OF APOLIPOPROTEINS

Lipid-Binding Domains of Apolipoproteins

A basic function of all apolipoproteins is lipid transport. Based on experimental ob-servations, and on model building of apolipoproteins with known sequences, Segrest et al.[82] proposed that these proteins contain amphipathic helical regions that interact with lipid. An important feature of the amphipathic helix is the presence of two clearly defined faces, one hydrophobic, which is inserted between the fatty acyl chains of the phospholipid molecules, and the other hydrophilic, which interacts with the phos-pholipid head groups and the aqueous phase. Another feature of the model is the distribution of the charged amino acid residues on the polar face. The negatively charged residues such as glutamic acid (Glu) and aspartic acid (Asp) tend to occur along the center of the polar face, while the positively charged residues, lysine (Lys) and Arg, are located on the lateral edges of the polar face.

The lipid–protein interaction between lipids and apolipoproteins differs from that between lipids and membrane-spanning proteins. In the latter case, long segments of exclusively nonpolar amino acid residues facilitate the actual insertion of the polypep-tide into the hydrophobic environment. In contrast, the hydrophilic residues on one side of the amphipathic helix are in contact with the surrounding aqueous phase and keep the apolipoproteins at the surface of the lipoprotein particle. Such a location facilitates transfer between lipoprotein particles and interaction with other molecules such as enzymes (e.g., lipases) and specific cell-surface receptors.

ApoB-100 is unique among the apolipoproteins in having a relatively high β-sheet content (~20%), both by predictive analysis of its primary sequence[83] and by circular dichroism and infra-red absorption studies of LDL.[84] Unlike other apolipoproteins, LDL apoB-100 does not transfer among lipoprotein particles. These observations have led to the speculation that apoB-100 binds lipid by its hydrophobic β-sheet regions.[85] The conformation of apoB-100 on LDL particles has been inferred by the ability of trypsin to preferentially release peptides from apoB-100 on LDLs[86]. Such experiments dem-onstrate that certain regions of apoB-100 are consistently released, while other regions remain associated with the partially digested lipoprotein particles. The regions that are preferentially released do not necessarily coincide with apoB-100 sequences that are accessible to monoclonal antibody binding.[87] Being a highly complex protein, the conformation of apoB-100 on different lipoprotein particles must be examined by the simultaneous application of different experimental approaches.

ApoD also differs from the other apolipoproteins in that the cDNA-deduced amino acid sequence contains little, if any, predicted amphipathic helical structure.[88]

Internal Repeats in the Soluble Apolipoproteins

All soluble apolipoprotein primary structures contain multiple repeats of 22 amino acids (22-mer), each of which is a tandem array of two 11-mers (Fig. 10-3). Repeating structures of two 22-mers connected in the middle by a helix breaker, Pro, appear to represent a structural paradigm of lipid-binding domains of the plasma apolipoproteins. Oligopeptides with such a structure mimic closely the surface properties and conformations of apoA-I.[89] The centrally located Pro residue appears to keep the hydrophobic faces in phase, and the resulting concavity of the 44-mer is suited for the adsorption of the peptide to the highly curved surface of human plasma HDL-3 (radius 40 to 50 Å).

All the soluble apolipoproteins share a common block of 33 amino acids upstream from the junction of exon 3 and intron 3 (top part of Fig. 10-3). A very striking feature of this block is that many columns, e.g., columns 8 and 9 within segments A and B and column 6 with segment C in Figure 10-3 (top) are completely or almost completely occupied by hydrophobic residues (green color). In addition, some columns are predominantly occupied by acidic residues (red), some others by indifferent residues (uncolored), and one by basic residues (blue). When the block is divided into three segments (A, B, and C) of 11 columns, the first column of each segment consists predominantly of hydrophobic residues, and that in segments A and B, columns 8 and 9, consists almost completely of hydrophobic residues. Secondary structure analysis of the sequence of the 33 amino acids that make up the common block by the method of Eisenberg et al.[90] indicates that they form a well-defined domain at the N-terminus of each apolipoprotein. The three segments in the common block might have arisen from triplication of an 11-mer (11 amino acids).

The region encoded by exon 4 shows an even more conspicuous pattern of internal repeats (bottom part of Fig. 10-3). The fundamental unit of repeat in this region is not an 11-mer, but a 22-mer that is made up of two 11-mers. The main reason for this is that most 11-mers are more similar to the 11-mer one unit removed than to the adjacent 11-mer.[27–29] This basic structure of 22-amino acid units connected to one another by a proline residue is highly reminiscent of the basic lipid-binding 44-amino acid structures described previously.[89]

Some remarkable features emerge from the alignment shown in the bottom part of Figure 10-3. Most prolines (purple) appear in the first column of group A. Columns 6 and 10 of both groups A and B are mostly occupied by hydrophobic amino acids (green), the rest being indifferent (uncolored) amino acids. In both groups, columns 4 and 5 are predominantly occupied by acidic amino acids (red), and column 9 by basic amino acids (blue), while column 11 does not have a predominant pattern (color).

Under the pattern of internal repeats depicted in Figure 10-3, the mature peptide of each of these proteins is almost completely made up of internal repeats. For example,

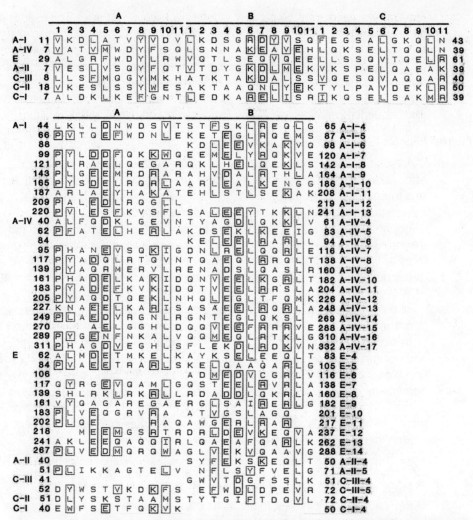

Fig. 10-3. Internal repeats in the soluble human apolipoproteins. Top: The last 33 amino acids of exon 3 can be divided into 3 repeats of 11 amino acids. Colors indicate proline (purple, P); aspartic acid or glutamic acid (red, D and E); arginine or lysine (blue, R and X); and methionine, valine, leucine, isoleucine, phenylalanine, tryrosine, or tryptophan (green, M, V, L, I, F, Y, and W). The remaining amino acids, glycine, alanine, serine, threonine, asparagine, glutamine, histidine, and cysteine (G, A, S, T, N, Q, H, and C) are uncolored and called indifferent. Any column containing four or more amino acids of a single color is said to possess that character, and the amino acids of that color are boxed. Numbers along the left and right margins are residue numbers in the mature peptide. We refer to these three repeats as the first three repeats in each gene. Bottom: Most of the repeats in exon 4 are 22-mers, each of which is made up of two 11-mers, and the other repeats are 11-mers. If more than 16 amino acids in a column are of a single color, the amino acids with that color are boxed. We refer to the first repeat in exon 4 as the fourth repeat in each gene; the repeats are denoted as A-I-4, A-I-5, etc.

in apoA-II, A-I, and C-III, only the first and the last six or seven amino acids of the mature peptide are not included in the repeats shown in Figure 10-3.

The similarities in genomic structure and in the pattern of internal repeats among the genes coding for the soluble apolipoproteins can be incorporated into a common model for the structure of their mRNAs (Fig. 10-4). Based on such a model, one can

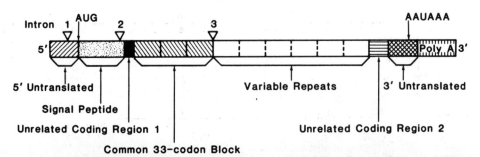

Fig. 10-4. Common messenger RNA (mRNA) structure for the soluble apolipoproteins. The mRNAs that share this common structure include: apoA-I, A-II, C-I, C-II, C-III, and E. ApoA-IV mRNA has a similar structure except that it has lost the intron I represented by the first triangle.

directly compare regions of the gene (and protein) among different apolipoproteins, as well as compare the same apolipoproteins (and their genes) among different species. From such sequence comparisons, one can estimate the rates of evolution of individual apolipoproteins and their subdomains and infer the evolutionary relationships among the members of the apolipoprotein multigene family (see p. 200).

ApoB-100 Structure and Internal Repeats

ApoB-100 is the largest of the apolipoproteins. For many years, investigations into the structure of apoB have been hampered by its large size, its insolubility in aqueous buffers after delipidation, and its susceptibility to proteolytic cleavage during purification. ApoB-48, the form of apoB present in chylomicrons, shares epitopes[56,57] and sequences[63,64] common to the NH$_2$-terminal half of apoB-100. ApoB-100 is a 4536-amino acid protein, the largest polypeptide ever studied.[83,86,91,92] It is the most hydrophobic of all the apolipoproteins. The sequence is puncturated by 25 cysteines, 12 of which are located in the NH$_2$-terminal 500 residues. Inspection of the sequence has led to the identification of two potential LDL receptor-binding domains based on their homology to the receptor-binding region in apoE.[91] A synthetic peptide (residues 3345 to 3381) of one of the two regions was found to bind to the LDL receptor and suppress 3-hydroxy-3-methylglutaryl coenzyme A (HMG Co-A) reductase activities in cultured human fibroblasts.[92]

When human apoB-100 was digested with trypsin, the amino acid composition of the remaining parts of the protein showed little difference from that of the undigested protein.[93,94] This observation suggests that apoB-100 contains internal repeats. This is an interesting possibility because it might explain how such an exceptionally large protein has evolved and because internal repeats have been found in all the apolipoproteins discussed previously. The recent studies by Yang et al.[92] and DeLoof et al.[95] have provided evidence for the existence of many internal repeats in apoB-100. Of particular interest was the discovery of the following two classes of repeats: amphipathic helical regions and hydrophobic proline-rich domains[95]; (Tables 10-4 and 10-5).

The first class of internal repeats, together with the 22-residue consensus, are shown in Table 10-4. Representation of the consensus sequence in an Edmundson-wheel diagram indicates that it has an amphipathic structure: hydrophobic amino acids are located on one side of the helix, whereas the polar residues are located on the other side.[95] This is also true for each of the repeats in Table 10-4. Moreover, calculations of the mean hydrophobicity and the helical hydrophobic moment of each repeat and

Table 10–4. 22 Residue Consensus

First Residue		Score	Mean Hydrophobicity	Mean Helical Hydrophobic Moment
2079	QFVRKYRAALGKLPQQANDYLN	12.54	−.20	0.98
2135	DAKINFNEKLSQLQTYMIQFDQ	11.63	−.08	0.89
2173	NIIDEIIEKLKSLDEHYHIRVN	12.09	−.06	1.02
2384	TFIEDVNKFLDMLIKKLKSFDY	11.59	.06	1.01
2407	QFVDETNDKIREVTQRLNGEIQ	12.31	−.32	1.02
4150	RVTQEFHMKVKHLIDSLIDFLN	12.68	.03	1.00
4237	DVISMYRELLKDLSKEAQEVFK	11.95	−.12	0.99
4397	EYIVSASNFTSQLSSQVEQFLH	11.85	.13	0.69
4463	DYHQQFRYKLQDFSDQLSDYYE	12.90	−.31	0.82
Consensus	**DFIDEFNEKLKDLSDQLNDFLN**		−.13	0.98

22-Residue-long consensus sequence derived by the iterative alignment procedure. Identical residues are printed bold, and related amino acids are underlined. For details on the identifiction of these repeats, the calculations of the comparison scores, mean hydrophobicities, and mean helical hydrophobic moments, see DeLoof et al.[95]

the consensus sequence indicate that all the sequences have a high helical hydrophobic moment consistent with an amphipathic structure. Such structures have also been found in other apolipoproteins and are thought to be important for phospholipid binding (see p. 193). Interestingly, on intact LDL particles, and domains containing these repeats are readily releasable by trypsin,[86,92] further supporting the hypothesis that these are involved in lipid bindings.

In contrast, the proline-rich repeats (Table 10-5) are unique to apoB-100 and are characterized by the preponderance of hydrophobic residues. Their secondary structure is predicted to be composed of predominantly β-sheets and β-turns (as a result of proline residue), which might interact with lipids in a particular way.[95] Computer modeling of these proline-rich sequences in the presence of dipalmitoylphosphatidylcholine suggests that the first part of each segment consists of a β-sheet that might penetrate into the acyl chains. After a turn around a proline residue, the segment can form a second β-sheet parallel to the first one but with a reverse orientation. Such a structure would be able to penetrate more deeply into the LDL than the amphipathic helices. Cooperativity in the lipid binding of these two different classes of subdomains, i.e., amphipathic vs. proline-rich regions, might account for the observation that in contrast

Table 10–5. "Proline-Rich" Consensus Sequences

First Residue		Score
1283	LKMLETVRTPALHFKSVGFHLPSREFQVPTFTIPKLYQLQVP-LLGVLDLSTN	11.88
2574	EVSLQALQKATFQTPDFIVPLTDLRIPSVQINFKDLKNIKIPSRFSTPEFTI-	11.31
2666	LRDLKVEDIPLARITLPDFRLPEIAIPEFIIPTLNLNDFQVP-DLHIPEFQLP	12.28
3245	YVFPKAVSMPSFSILGSDVRVPSYTLILPSLELPVLHVPRNL-KLSLPDFKEL	11.98
3711	NDLNSVLVMPTFHVPFTDLQVPSCKLDFREIQIYKKLRTSSF-ALNLPTLPEV	11.40
3805	SDGIAALDLNAVANKIADFELPTIIVPEQTIEIPSIKFSVPA-GIAIPSFQAL	12.26
Consensus	**LDSLKALDMPTFHIPSSDFRLPSITIPEPTIEIPKLKNSQVP-ALSIPDFQEL**	

52-Residue-long consensus sequence derived by the iterative alignment procedure. Identical residues are printed bold, and related amino acids are underlined. For details on the identification of these repeats, and the calculations of the comparison scores, see DeLoof et al.[95]

to the smaller apolipoprotein, apoB does not exchange between different lipoprotein particles.[7]

ApoD Structure: Lack of Internal Repeats

ApoD is a glycoprotein with an apparent relative molecular mass (M_r) of 33,000 found in HDL.[96,97] Its function is unknown. The protein, however, was found complexed with LCAT, and cholesteryl ester transport is one of its postulated functions.[96,97] As deduced from its cloned cDNA sequence,[88] human apoD is a 169-amino acid protein with a predicted α-helical content of less than 5%. The protein does not have any internal repeats, in contradiction to all the other apolipoproteins sequenced to date.

MOLECULAR EVOLUTION OF APOLIPOPROTEIN GENES

Principles of Molecular Evolution

One important aspect in the study of evolution of genes is to know the rate of nucleotide substitution. Because of the degeneracy of the genetic code, nucleotide substitutions in protein-coding genes can be of two types: synonymous and nonsynonymous. A synonymous substitution causes no amino acid change, while a nonsynonymous substitution results in an amino acid replacement. Treating these two types of substitutions separately can give a better insight into the mechanisms of nucleotide substitution in evolution. Moreover, nucleotide sites can also be classified into synonymous and nonsynonymous. The conventional approach is to count a nucleotide site as a nonsynonymous site if all possible changes at that site are nonsynonymous, two-thirds nonsynonymous and one-third synonymous if two of the three possible changes at that site are nonsynonymous, and so on.[98] Following this classification, one can then calculate the number (K_A) of substitutions per nonsynonymous site and the number (K_S) of substitutions per synonymous site when comparing two homologous genes.[98] If the divergence time between the two genes is known, then one can compute the synonymous and nonsynonymous rates of nucleotide substitution.

Knowing the rate of nucleotide substitution can serve at least two purposes. First, it enables one to infer the dates of certain types of evolutionary events such as gene duplication. By this approach, rough estimates of the duplication dates have been obtained from several pairs of the apolipoprotein genes.[99] Second, it may enable one to infer the stringency of structural requirements in a protein or parts of a protein. Comparative studies of protein and DNA sequences have led to the general conclusion that functionally more important molecules or parts of a molecule evolve more slowly than less important ones.[98,100–102] A good example is the proinsulin polypeptide, which consists of three peptides, A, B, and C. The C peptide connects the A and B chains of the proinsulin and is cleaved off following disulfide linkage and assembly of the latter two chains to form the mature insulin molecule. The A and B chains comprise the bioactive insulin that binds to the insulin receptor, while the C peptide, once cleaved from proinsulin, does not have any known function. The rate of nonsynonymous substitution is found to be much higher in the region coding for the C peptide than in the regions coding for the A and B peptides, suggesting that the former is subject to less stringent structural constraints than the latter.[102] Using this idea, the structure–function relationships of different regions of many individual apolipoproteins can be subject to detailed analysis.

Another important aspect in the study of evolution of genes is to infer the evolutionary relationships among genes. For the apolipoprotein genes, this can be done by considering similarities in DNA or amino acid sequences, pattern of internal repeats, and genomic structure[99] (see also Fig. 10-4).

Rates of Nucleotide Substitution in Apolipoprotein Genes

The primary structures for all the major human apolipoproteins have been determined. In addition, the structures of a number of these proteins in several other species, including rat, mouse, cow, rabbit, chicken, and fish, have been reported. Based on the available information, the rates of nucleotide substitution in each apolipoprotein can be computed. Table 10-6 shows the rate of nucleotide substitution in the genes coding for apoA-I, A-II, A-IV, C-II, C-III, and E. In this discussion, the rate of synonymous substitution is not considered for two reasons. First, this rate is not useful for inferring the stringency of structural requirements at the amino acid level. Second, the degree of divergence between different apolipoprotein genes is very large at synonymous sites, so that it is difficult to obtain reliable estimates of the K_S values. Therefore, the K_S values cannot be used to infer the duplication dates or the evolutionary relationships among the apolipoprotein genes. In the present section, we consider the substitution rates in the signal peptide region and in the mature peptide region as a whole. The substitution rates in individual repeat regions are considered later.

In the signal peptide region, the K_A value for apoC-III is 0.06 between human and dog, 0.15 between dog and rat, and 0.09 between human and rat. Because the three species diverged at about the same time,[103] we can show that the K_A value is $(0.06 + 0.09 - 0.15)/2 = 0.00$ in the human lineage, 0.06 in the dog lineage, and 0.09 in the rat lineage; for the method, see Fitch and Margoliash.[104] This suggests that no nonsynonymous substitution has occurred in the signal peptide region in the human apoC-III gene since the time of mammalian radiation, i.e., about 80 million years ago.[103] The variation in the K_A value among the other genes can be largely attributed to statistical fluctuations, and so the signal peptides in the other proteins have evolved at about the same rate. Actually, in the divergence between dog and rat, the signal peptide in apoC-III has also evolved at a rate similar to those in the other apoproteins in Table 10-6. It is not clear why the signal peptide in human apoC-III has been so well conserved in evolution.

In the mature peptide region of the apoC-III gene, the K_A value is 0.17 between human and dog, 0.33 between dog and rat, and 0.32 between human and rat. Therefore, the K_A value is 0.08 in the human lineage, 0.09 in the dog lineage, and 0.24 in the rat lineage. Thus, apoC-III has evolved about three times faster in the rat lineage than in the human and dog lineages. This difference in rate is considerably larger than Wu and Li's[105] estimate that the nonsynonymous rate is, on the average, 1.3 times higher in the rodent lineage than in the human lineage. It will be interesting to know whether any other apolipoproteins also evolve much faster in the rodent lineage than in the human lineage.

In the mature peptide region, the K_A value between the human and dog apoC-II genes is somewhat larger than that between the human and dog apoC-III genes, suggesting that apoC-II is somewhat less conservative than apoC-III. The K_A value between the human and rat apoC-III genes is close to that between the human and rat apoA-II genes. As shown in Table 10-6, the apoA-II gene evolves two to three times faster than the β-globin gene, which evolves at the average rate for 35 mammalian

Table 10–6. Number of Substitutions Per Nonsynonymous Site Between Human and Dog or Rat Apolipoprotein Genes

Genes	Signal-Peptide Region	Repeats in the Mature-Peptide Region															Total Mature Peptide
		1 + 2 + 3	4	5	6	7	8	9	10	11	12	13	14	15	16	17	
β-Globin	0.13 (0.06)																0.12 (0.02)
B-100	0.06 (0.05)																0.09 (0.01)
C-II	0.15 (0.08)	0.26 (0.07)	0.11 (0.05)														0.22 (0.04)
C-III	0.09 (0.05)	0.10 (0.04)	0.08 (0.06)	0.23 (0.08)													0.17 (0.03)
		0.26 (0.07)	0.28 (0.13)	0.26 (0.08)													0.33 (0.05)
		0.23 (0.06)	0.34 (0.15)	0.31 (0.10)													0.32 (0.05)
A-II	0.19 (0.08)	0.25 (0.06)	0.29 (0.12)	0.51 (0.14)													0.35 (0.05)
A-I	0.15 (0.07)	0.08 (0.03)	0.25 (0.08)	0.13 (0.05)	0.23 (0.10)	0.19 (0.07)	0.49 (0.14)	0.43 (0.12)	0.51 (0.14)	0.21 (0.07)	0.14 (0.08)	0.46 (0.12)					0.25 (0.02)
E	0.22 (0.09)	0.08 (0.03)	0.19 (0.07)	0.07 (0.04)	0.23 (0.11)	0.21 (0.07)	0.07 (0.04)	0.10 (0.05)	0.48 (0.15)	0.27 (0.11)	0.26 (0.09)	0.12 (0.05)	0.14 (0.06)				0.19 (0.02)
A-IV	0.16 (0.08)	0.09 (0.04)	0.15 (0.06)	0.40 (0.13)	0.26 (0.14)	0.25 (0.08)	0.41 (0.12)	0.45 (0.14)	0.23 (0.08)	0.26 (0.09)	0.19 (0.07)	0.39 (0.12)	0.13 (0.06)	0.28 (0.10)	0.23 (0.08)	0.15 (0.06)	0.24 (0.02)

For apoC-II, the comparison is between human and dog; for apoC-III, the three comparisons are between human and dog, dog and rat, and human and rat, respectively; for the other genes, the comparison is between human and rat. the apoB-100 is a partial sequence. The values in parentheses are standard errors. The repeat numbers in different genes do not necessarily correspond to each other. The β-globin gene is given for comparison because it evolves at the average rate for 35 mammalian genes [98].

genes.[98] We may therefore conclude that apoA-II, apoC-II, and apoC-III all evolve very rapidly. Further, we note that apoA-I and apoA-IV have evolved twice as fast as the β-globin and that apoE has also evolved considerably faster than the β-globin. Thus, all of these apolipoproteins are not conservative in evolution. In contrast, the partial apoB-100 sequence corresponding to residues 595→979 seems to be more conservative than the β-globin (Table 10-6). It will be interesting to know whether this is true for the entire apoB-100 molecule. It is also interesting to note that the shorter proteins A-II, C-II, and C-III are less conservative than the longer proteins A-I, E, A-IV, and apoB-100.

Evolutionary Relationships Among the Soluble Apolipoproteins

As discussed previously, the genes coding for apoA-I, A-II, C-I, C-II, C-III, and E each has a total of three introns at the same locations, and the gene encoding apoA-IV also has the same genomic structure except that it has lost the first intron. All of these genes have a similar repeat pattern, and their third exon shares a common block of 33 codons (see Figs. 10-3 and 10-4). These observations lead to the proposal that these genes are evolutionarily related and arose from a common ancestral gene.[99] The evolutionary scenario proposed by Luo et al.[99] for these genes is as follows (Fig. 10-5).

The common ancester of the apolipoprotein genes was very similar to the present day apoC-I in structure and length. It contained three introns at the same locations as observed in the present-day genes. The third exon of this gene contained the common block of 33 codons mentioned previously, and the fourth exon contained the basic repeat unit of 11 codons, as does the present-day apoC-I. This gene was duplicated into two; one of them led to apoC-I, and the other became the common ancestor of the other apolipoprotein genes. In the first lineage, a recent duplication event resulted in the production of the apoC-I pseudogene. In the second lineage, the first 11 codons of exon 4 were duplicated, and then the whole gene was duplicated into two. One of the two genes became the present-day apoC-II, and the other became the common ancestor of the apolipoprotein genes other than apoC-I and apoC-II. In the latter lineage, a duplication of the first 22 codons of exon 4 occurred, and then a duplication of the whole gene followed. In one of the two resultant genes, a deletion of the first 11 codons of exon 4 occurred, and the gene was later duplicated, one part leading to apoA-II and the other to apoC-III. The other gene experienced six or seven duplications of the 22-codon repeat and a duplication or deletion of 11 codons in exon 4 and was then duplicated: One of them gained two duplications of 22 codons and became the present-day apoE; the other gained one duplication of 22 codons in exon 4 and was then duplicated, one leading to apoA-I and the other to apoA-IV after gaining three duplications of 22 codons in exon 4. After the separation of apoA-IV from apoA-I, the gene for apoA-IV lost the first intron.

ApoA-II and apoC-III are more closely related to each other than either of them is to the other genes (Fig. 10-5), because they are very similar in structure and length; their mature regions contain the same number of repeats, and their signal peptide regions encoded by exon 2 differ by only one codon in length. Similarly, apoA-I and apoA-IV are closer to each other than either is to apoE for two reasons. First, apoA-I and apoA-IV seem to share more repeats in common (Fig. 10-3). Actually, their repeat patterns are very similar; e.g., most of their repeats start with proline (see Fig. 10-3 bottom). Second, the degree of sequence divergence is smaller between apoA-I and

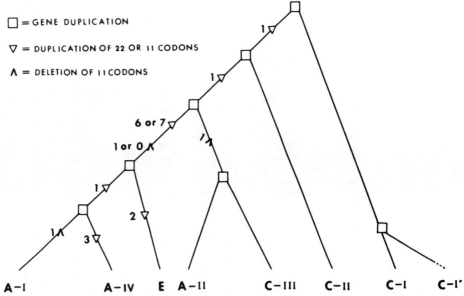

□ = GENE DUPLICATION

▽ = DUPLICATION OF 22 OR 11 CODONS

Λ = DELETION OF 11 CODONS

Fig. 10–5. A hypothetical scheme for the evolution of apolipoprotein genes. The primordial apolipoprotein gene was probably very similar to apoC-I in structure and length. It was duplicated in two; one of them led to apoC-I, and the other became the common ancestor of the apolipoprotein genes. A recent duplication of the apoC-I gene processed the apoC-I pseudogene. In the other lineage, the first 11 codons of exon 4 were duplicated, and the the whole gene was duplicated into two. One of them led to apoC-II, and the other became the common ancestor of the apolipoprotein genes other than apoC-I and C-II. In the latter lineage, the first 22 codons of exon 4 were duplicated, and a duplication of the whole genes followed. In one lineage, a deletion of the first 11 codons of exon 4 occurred, and then the gene was duplicated into two, leading to apoA-II and C-III. In the other lineage, at least six duplications of 22 codons occurred (these six repeats are referred to as A-I-7, A-I-8, A-I-9, A-I-11, A-I-12, and A-I-13 in apoA-I; A-IV-7, A-IV-8, A-IV-9, A-IV-14, A-IV-16, and A-IV-17 in apoA-IV; and E-7, E-8, E-10, E-12, E-13, and E-14 in apoE) and possibly a duplication of 11 codons or a duplication of 22 codons followed by a deletion of 11 codons (this explains the half repeats, A-I-6, A-IV-6, and E-6). A duplication of the whole gene then occurred. One of the two resultant genes led to apoE after gaining two duplications of 22 codons (repeats E-9 and E-11). The other gained one duplication of 22 codons (A-I-7 in apoA-I and A-IV-7 in apoA-IV) and was duplicated into two. One of them became apoA-IV after gaining three duplications of 22 codons (A-IV-12, A-IV-13, and A-IV-15). The other became apoA-I after losing 11 codons in repeat A-I-12. For more details, see Luo et al.[99]

apoA-IV than between either of them and apoE.[99] At any rate, the evolutionary tree shown in Figure 10-5 is the most parsimonious in terms of internal duplications and deletions.

The estimated divergence times between each pair of genes in Figure 10-5 are as follows: ApoA-I and apoA-IV diverged about 270 million years (Myr) ago, and their ancestor and apoE diverged about 420 Myr ago. ApoA-II and apoC-III separated from each other about 285 Myr ago, and their ancestor separated from that of apoA-I, apoA-IV, and apoE about 430 Myr ago. The last estimate is consistent with the observation that apoA-II is present in fish,[106–109] because fish and mammals diverged about 400 Myr ago.[103] ApoC-II branched off about 570 Myr ago, and apoC-I and the other genes in Figure 10-5 shared a common ancestor about 680 Myr ago. Luo et al.[99] cautioned that these are very rough estimates.

Table 10–7.   Percentage of Similarities Between ApoE and ApoB-100 Segments and Between ApoA-IV and ApoB-100 Segments

ApoE vs. ApoB-100				ApoA-IV vs. ApoB-100			
Segment in E	Segment in B-100	L	S	Segment in A-IV	Segment in B-100	L	S
21–67	1011–1057	47	23	77–104	323–350	28	39
42–67	2463–2488	26	35	104–135	381–412	32	31
47–75	2314–2342	29	31	110–133	2431–2454	24	33
48–86	316–354	39	31	148–173	1523–1548	26	35
66–86	2431–2451	21	38	202–226	1718–1742	25	32
121–156	745–780	36	39	217–237	1378–1398	21	33
122–143	464–485	22	36	232–260	1774–1802	29	38
133–155	3350–3372	23	39	224–254	561–591	31	32
156–177	3627–3648	22	36	243–264	1358–1379	22	36
203–223	1571–1591	21	38	269–289	1890–1910	21	38
209–228	2476–2495	20	35	331–363	2512–2544	33	30
231–252	2322–2343	22	41	332–353	1660–1681	22	30
233–261	2415–2443	29	31				
234–254	334–354	21	38				
238–262	4267–4291	25	32				
242–264	1092–1114	23	39				

L, number of amino acid residues; S, similarities.

Are ApoB and ApoD Related to Other Apolipoproteins

We have used the dot-matrix method to search for potential homologous regions between apoB-100 and each of apoA-I, apoA-IV, and apoE. Table 10-7 shows a partial list of regions of high similarity between apoE and apoB-100 and between apoA-IV and apoB-100. With one exception, all the similarities shown in Table 10-7 are equal to or higher than 30%. Although it is difficult to know if any of these segment pairs are true homologs, the results suggest that apoB-100 might be related to apoA-I, A-IV, and E (see also DeLoof et al.[95]).

A dot-matrix comparison between apoD[88] and the other apolipoproteins failed to show any significant homology. This suggests that, evolutionarily, apoD is not a member of the apolipoprotein multigene family that comprises the other apolipoproteins discussed in this review. Furthermore, analysis of RNA blots showed that apoD mRNA is most abundant in the adrenal gland and is present in considerably higher concentrations in the kidney, pancreas, and small intestine than in the liver.[88] Such a tissue distribution is quite distinct from that of all the other apolipoproteins and suggests that apoD might belong to a different class of proteins. The recent availability of apoD genomic structure supports such an interpretation.[44] Unlike the other soluble apolipoproteins, the mature peptide-coding portion of the gene contains three introns instead of one. Furthermore, the signal peptide region is not interrupted by any intron. The structural organization of the apoD gene is thus drastically different from those of the other soluble apolipoproteins. Indeed, the protein has been shown to display a high degree of homology to human plasma retinol-binding protein, α_1-microglobulin, ungulate β-microglobulin, rodent $\alpha_{2\mu}$-globulin, and tobacco hornworm insecticyanin—all members of the $\alpha_{2\mu}$-globulin superfamily.[44,48]

RELATIVE RATES OF EVOLUTION AND FUNCTIONAL ASPECTS OF APOLIPOPROTEIN STRUCTURE

Common Domains

As mentioned previously, all of the apolipoprotein genes shown in Table 10-6 and Figures 10-3 and 10-4 contain a common block of 33 codons in the mature peptide region. This block is located at the end of exon 3 and is referred to as repeats 1, 2, and 3 in Table 10-6. In this block, the K_A value (the number of nucleotide substitutions per nonsynonymous site) is low for apoA-I, A-IV, and E, suggesting that this block of 33 amino acids may be structurally important for the function of these apolipoproteins. This block is also well conserved in human and dog apoC-III, though not in apoC-II, apoA-II, and rat apoC-III.

The relatively high overall rates of nonsynonymous substitution in apoA-I, apoA-IV, and apoE suggest that the structural requirement of lipid binding is not stringent, for all three proteins contain many lipid-binding domains.[82] For example, each of the repeats A-I-8, A-I-9, A-I-10, A-I-13, A-IV-7, A-IV-10, E-11, and E-12 contains an amphiphathic helix; yet, they all exhibit high rates of nonsynonymous substitution.

ApoC-II

In apoC-II, exon 4 (repeat 4) is well conserved, probably because this part is important for the activation of lipoprotein lipase and in chylomicron and VLDL metabolism. Support for the importance of the carboxylterminal part of the molecule in lipoprotein lipase activation comes from the structure of apoC-II_Toronto and apoC-II_St. Michael mutant apoC-II proteins isolated from two different patients with homozygous deficiency.[110,111] The sequence of apoC-II_Toronto is identical to that of normal apoC-II from residues 1→68. Residues 69→79 are missing, being replaced by an unrelated hexapeptide: changes consistent with the deletion of a nucleotide for the codon of either threonine (Thr)-68 or Asp-69 and a translation reading frame shift. ApoC-II_St. Michael contains 96 amino acid residues. Its sequence is the same as normal apoC-II in the first 69 residues. In apoC-II_St. Michael, Gln-70 was replaced by Pro-70, and the sequence terminated with Pro-96 with an unrelated peptide sequence spanning these two prolines. The defect is consistent with a base insertion in the codon for Asp-69 or Gln-70 and a subsequent translation reading frame shift. Both apoC-II_Toronto and apoC-II_St. Michael are totally nonfunctional,[110,111] indicating the importance of residues 69→79 for the normal lipase-activating activity of the protein. This conclusion is consistent with previous experimental observations.[112,113] Studies using proteolytic fragments as well as synthetic peptides of human apoC-II indicate that the site of interaction of apoC-II with lipoprotein lipase is contained within residues 56 to 79, which encompasses repeat 4. Additional residues upstream (50 to 55 and 44 to 55) seem to enhance the enzyme activation in the presence of phospholipid-stabilized triolein emulsion because of the phospholipid-binding activity that these additional residues confer to the peptide. Interestingly, Kinnunen et al.[112,113] found the last three residues (77 to 79) to be absolutely required for lipase activation; deletion of these residues resulted in total inactivation of the protein. In the dog, two of the last three amino acid residues are not conserved (Gly-Asp-Ser instead of Gly-Glu-Glu), suggesting that the structural requirement for the terminal dipeptide -Glu-Glu is not absolute.[114] Perhaps the critical requirement in the carboxyterminal tripeptide is the carboxylic acid side chains at the penultimate residue.

ApoC-III

In apoC-III, repeats 1, 2, and 3, and exon 4 are not well conserved. This observation suggests that the function of apoC-III (e.g., the modulation of hepatic uptake of lipoprotein remnant particles) does not have a very stringent structural requirement.

ApoA-II

The extremely high K_A values in all regions in apoA-II probably are due to weak functional constraints. ApoA-II does not seem to play an important functional role, although its exact function is unknown. In fact, apoA-II is either absent or expressed at a very low level in the dog,[115] pig,[116] cow,[117] or chicken.[118]

ApoA-I

A major function of apoA-I is the activation of LCAT,[24,28] an important enzyme in lipoprotein metabolism, which converts unesterified cholesterol to its ester form in plasma.[24] Soutar et al.[29] have shown that both the amino- and carboxy-terminal cyanogen bromide fragments (residues 1 to 85 and 147 to 243 of apoA-I) activate LCAT; in the latter fragment, residues 145 to 182 seem to be involved in the activation process.[119] Surprisingly, this part of apoA-I sequence, which corresponds to repeats A-I-9 and A-I-10, is less conserved than the other repeats (Table 10-6). Thus, the structural requirements for LCAT activation may not be stringent. This conclusion is supported by other lines of evidence: (1) another region of apoA-I also activates LCAT (see preceding discussion); (2) at least three other apolipoproteins, apoC-I,[29] apoA-IV,[30] and apoE[120,121] can also activate LCAT; and (3) synthetic model peptides that mimic apoA-I surface properties but differ from apoA-I in primary sequences are effective in LCAT activation.[119,122,123] While LCAT activation may be a major function of apoA-I, a high rate of evolution may still occur because the protein can undergo considerable change in its primary structure without impairment of this function.

ApoA-IV

The function of apoA-IV is unknown, though the protein also has LCAT-activating potential.[30] However, like apoA-I, although some of the regions in apoA-IV are relatively well conserved, most of them are not conserved. On the whole, apoA-IV evolves at a rate very similar to that in apoA-I. Thus, the function of apoA-IV also does not seem to have stringent structural constraints.

ApoE

ApoE is an important determinant in the interaction between apoE-containing lipoproteins and cell-surface receptors.[124] Studies with specific proteolytic and chemical cleavage fragments of apoE indicate that the receptor-binding domain exists in the region between residues 129 and 191.[125] The use of monoclonal antibodies has localized the receptor-binding region to the vicinity of residues 140 to 150.[126] Studies of apoE mutants of known structure have further indicated the importance of residues 142, 145, and 146 in mediating receptor binding.[127–129] The positive charge of arginine 158 also appears to be important in maintaining the correct conformation necessary for normal binding.[130] Examination of the degree of conservation of the various repeats in apoE (Table 10-6) indicates that E-8 (residues 139 to 166), which encompasses the receptor-binding region, is indeed one of the most highly conserved sequences in apoE.

ApoB-100

ApoB-100 is an obligatory constituent of VLDL, IDL, and LDL and is the ligand that binds to the LDL receptor. It thus plays a central role in cholesterol homeostasis; in fact, the plasma concentration of apoB-100 is strongly correlated with the development of atherosclerosis.[131,132] The protein is unique among the apolipoproteins in its remarkable size and in its hydrophobicity and high β-sheet content.[83,91,92] As discussed previously, apoB-100 appears to contain numerous internal repeated sequences. With such a structural organization, one would expect a considerable redundancy in apoB-100 structure and a relatively high rate of evolution. Interestingly, our analysis of the rate of nonsynonymous substitution in a substantial portion (about 1.2 kb[133]) of the apoB-100 gene indicates that it is the most conservative of all the apolipoprotein genes. The low rate of nonsynonymous substitution is all the more remarkable in that the sequence analyzed corresponds to residues 595→979, which do not overlap the putative receptor-binding domains.[91,92] This observation suggests that the apparent built-in redundancy in apoB-100 structure might be important in its overall function. It further supports the suggestion that apoB is a very ancient gene. An apoB-like protein, i.e., an LDL protein with an apparent M_r or >250,000 Daltons, and remarkably similar amino acid composition that is insoluble in aqueous media following delipidation, has been found in a number of fish species, including members of the lowest vertebrates, the hagfish, and the shark.[107,108,134] The very similar properties of fish apoB-100 and human apoB-100 are also consistent with a low rate of evolution for this protein. However, to confirm whether apoB-100 is truly a well-conserved protein, we need additional sequences from the rest of the apoB-100 molecule from the rat as well as sequences of the protein from other species.

EPIDEMIOLOGIC GENETIC APPROACHES TO LIPOPROTEIN VARIATION

Background

Researchers have recognized for many years that coronary heart disease (CHD) aggregates in families.[135] Certain families bear an unequal brunt of the disease burden relative to the population prevalence and may be segregating for gene mutations that contribute to their disease liability. Underlying the aggregation of the clinical end point is the familial nature of its contributing factors such as lipid, lipoprotein, and apolipoprotein levels. Human geneticists used the observed familial aggregation of a trait to determine the contribution of genetic factors to their variability in the population. The polymorphic genes contributing to familial aggregation also contribute to interindividual variation in lipid, lipoprotein, and apolipoprotein levels in the general population. It is relevant at this juncture to note the three units of inference in these studies: the individual, the family, and the population.

Genetic epidemiologists have estimated the proportion of interindividual variability of several CHD risk factors attributable to polymorphic genetic effects. Familial aggregation for diseases such as CHD is evident when one or usually more of these genetic mutations segregates among related individuals. Using information gleaned from these population and family studies combined wiith the techniques of molecular biology, one hopes to predict individuals having a genetic liability for premature disease. Table 10-8 summarizes the contribution of polymorphic genetic variability to variation of lipid, lipoprotein, and apolipoprotein levels in the population. The effects of two types of gene mutations are considered in these studies. The first type includes those with rare alleles, e.g., the LDL-receptor defect leading to familial hypercholesterolemia, with very large effects on plasma lipid levels. These *major gene* mutations mandate an increased risk to disease for the individuals carrying them. For many of the traits in Table 10-8, significant evidence indicates a gene with a large effect.[136] The identity and characteristics of these major gene mutations are not addressed in most biometric genetic studies. Most individuals at increased risk to disease, however, are at risk because of the combined effects of numerous genetic and environmental factors with moderate effects. Genes with polymorphic allele frequencies and moderate or small effects on the traits of interest are classically termed *polygenes*. The heritability given in Table 10-8 is a simplified, but useful, summary of the contribution of polygene effects to variation of each of the risk factor phenotypes. For each trait, these studies have determined that a significant fraction of variation is attributable to polymorphic genetic variability. For example, the proportion of variability of plasma apoB levels attributable to genetic factors is 51%.[137] The multifactorial etiology of a complex disease such as CHD is well recognized. An unknown subset of contributing genes and environments interact to impart disease liability to each individual. Family members share many things in addition to genes such as those factors making up their common household environment. Also summarized in Table 10-8 is the contribution of common household effects to the variation in these risk factor levels. Interestingly, the contribution of genetic factors usually exceeds that of shared environmental effects.

Although these statistical genetic studies indicate that genes are contributing to development of CHD, they do not provide information on the identity and role of specific genes or gene mutations, i.e., which genes or gene mutations are contributing to the total polygenetic variance, or the frequency of their polymorphic allelic variation. Perhaps most important, one needs to know the effects of these genes on the levels of the phenotypes of interest and their contribution to the phenotypic variance in the general population. One does not know, for example, whether many genes, each with small effects, are involved or whether only few genes with larger effects are involved. Information about the effects of specific loci is a prerequisite for precisely predicting those individuals carrying specific mutations and for characterizing the genetic factors involved in the atherosclerotic process. Such information will aid in early identification of individuals at increased risk and assigning them to a more efficacious intervention program.

Recent advances in atherosclerosis research and molecular biology have identified several genes and gene products that are likely contributing to cardiovascular disease. A large number of these genes and gene products have been described earlier in this chapter. They include enzymes such as lipoprotein lipase, which is involved in the conversion of one lipoprotein species to another; the lipoprotein receptors such as the LDL (apoB-E) receptor, whose alteration or deficiency produces familial hypercholesterolemia; and the apolipoproteins themselves. Our thesis is that genetic variability at these candidate genes contributes to differences in CHD risk factor levels among individuals. The action of these genes constitutes the statistical genetic effects presented in Table 10-8. Recent genetic studies have begun to use this measured genotype information[138] to investigate the contribution of polymorphic genetic variation in the candidate genes to lipid, lipoprotein, and apolipoprotein variation in the general population. In the sections that follow, we will not catalog the variability in the apolipoprotein genes and review their estimated association with lipid levels. Several such reviews have been published recently (e.g., refs. 139 and 140). Instead, we will discuss

Table 10–8. Polygenetic and Common Environmental Components of Variance Along With Evidence for the Effects of a Major Gene for Several CHD Risk Factors

Trait	Reference	Polygenetic Effects (%)	Shared Environmental Effects (%)	Single Locus
Systolic blood pressure	172	34	11	No
Fasting glucose	173	27	10	No
Triglycerides	137	33	23	No
Total cholesterol	137	64	17	Yes
Low-density lipoprotein	137	67	10	Yes
High-density lipoprotein	137	42	9	Yes
Apolipoprotein AI	137	43	5	Yes
Apolipoprotein AII	137	30	10	No
Apolipoprotein B	137	51	14	Yes

CHD, coronary heart disease.

the characteristics and effects of three types of variation in the apoE and apoB genes. These two examples address basic principles that we feel apply to a wide range of issues in common chronic disease research.

Laboratory methods enable one to detect large amounts of genetic variability in targeted candidate genes. Some of this genetic variability has a direct biochemical and physiologic effect on the gene product and on lipid metabolism. An example of this type of effect is that of the apoE polymorphism on lipid metabolism (see following discussion). Not all variability, however, affects the phenotype of interest; it is "silent" with respect to its phenotypic effects. Marker loci with no effects may be associated in the population with genetic variability with direct physiologic effects. This nonrandom association between alleles at two gene loci is termed linkage or gametic disequilibrium.[141] Therefore, a marker with no direct effect on the phenotype may have a statistical association with lipid levels or a disease end point. An example of this type of effect is the association of the XbaI restriction site polymorphism in the apoB gene involving codon 7464 with CHD risk.[142] The relationship between loci with direct causal effects on a trait with associated marker loci is schematically shown in Figure 10-6. The realized effect of a marker locus is a function of two parameters: the size of the effect of the phenotypically important mutation and the degree of disequilibrium between the two loci. When the disequilibrium is absolute, the realized effect of the marker locus and the real physiologic effects are comparable. As the degree of disequilibrium decreases, the realized effect of the marker locus also decreases until it is zero when the two loci are independent.

Genetic Variation and Its Effects

The molecular structure of the major apolipoprotein genes is known and has been discussed earlier in this chapter. The apolipoprotein genes, like other genes, exhibit considerable DNA sequence variation among individuals in a population. This sequence heterogeneity can be detected using restriction enzyme analysis, allele-specific oligonucleotide hybridization, or direct DNA sequencing. Most of the DNA sequence heterogeneity occurs in nonexpressed or noncoding regions of the gene and is not reflected in the gene product. A small proportion, however, is transcribed and translated,

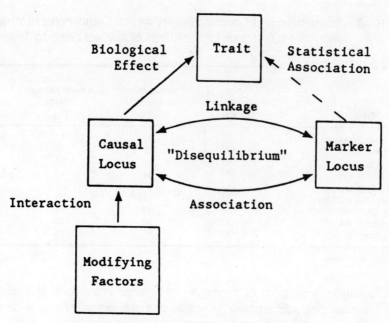

Fig. 10–6. Schematic diagram summarizing the relevant factors in genetic association studies. One assumes that some causal locus has a direct biologic effect on the trait of interest. The traits of interest are, for example, levels of plasma lipid, lipoproteins, or apolipoproteins, or other measures of lipid metabolism. The effect of the causal locus may be modified by other factors such as high-fat diet, smoking, or the action of another gene. Genotypes at the marker locus may be in association or disequilibrium with genotypes at the causal locus. This disequilibrium results in an association between genotypes at the marker locus and lipid, lipoprotein, or apolipoprotein levels.

yielding altered apolipoprotein species in some cases. Protein heterogeneity is often detectable by immunochemical analysis, electrophoresis, or isoelectric focusing.

The best characterized apolipoprotein variant is that in the human apoE gene. ApoE is a structural component of chylomicrons, VLDL, and HDL. The protein is thought to have a major regulatory role in lipid metabolism because of its role as a physiologic ligand for the LDL (apoB-E) receptor and for a specific apoE receptor.[124] Human apoE is polymorphic.[143,144] There are three common alleles, designated ϵ_2, ϵ_3, and ϵ_4. The mutational variation that is responsible for this polymorphism has been well characterized at both the protein and DNA level.[145,146] The mature apoE polypeptide is 299 amino acids long and rich in arginine. Although other variations have been reported, the common apoE polymorphism is attributable to variations at amino acids 112 and 158. The ϵ_2-isoform has cysteine at both positions, and the ϵ_4 has arginine at both positions. The ϵ_3-isoform has cysteine at position 112 and arginine at position 158. These amino acid substitutions result in charge differences that allow the apoE isoforms to be readily separated by isoelectric focusing. The relative frequencies of these three alleles in several populations are presented in Table 10-9. In the combined sample of 5556 Caucasian individuals, the relative frequencies of the ϵ_2, ϵ_3, and ϵ_4-alleles are 0.11, 0.76, and 0.13, respectively. Average total cholesterol levels for each of the six apoE genotypes are shown in Table 10-10. Individuals carrying an ϵ_2 allele tend to have significantly lower total cholesterol levels, and individuals carrying an ϵ_4 allele tend to have significantly higher levels than the population mean. The apoE poly-

Table 10–9. Apolipoprotein E Allele Frequencies

Population	Reference	N	Frequency ϵ_2	ϵ_3	ϵ_4
Canada	174	102	.078	.770	.152
France	175	223	.130	.742	.128
Scotland	176	400	.083	.770	.147
New Zealand	177	426	.120	.720	.160
West Germany	147	563	.063	.793	.144
Finland	178	615	.041	.733	.226
USA	179	1209	.073	.786	.141
Holland	180	2018	.167	.750	.083
Summary		5556	.109	.760	.131

morphism has a large effect on both total cholesterol and plasma apoE levels.[147] The average effects of the apoE alleles on plasma levels of total cholesterol, apoB, and apoE are shown in Table 10-11. The average effect of an allele is the expected deviation from the population mean of an individual carrying that allele. In this and most other populations, the average effect of the ϵ_2-allele is to significantly lower total cholesterol levels, and the average effect of the ϵ_4-allele is to significantly raise total cholesterol levels. The average effects of the apoE alleles on plasma apoE levels are in the opposite direction as those for cholesterol levels. Genetic variation in the apoE protein accounts for 20% of the variability in plasma apoE levels. Using family data, Boerwinkle and Sing[148] directly estimated that the apoE polymorphism accounts for 12.5% of the overall polygenetic variance of total serum cholesterol levels.

The mechanism by which the apoE polymorphism has its effect on serum cholesterol levels in the general population is not well understood. The inverse effect of this locus

Table 10–10. Average Plasma Cholesterol for Each ApoE Genotype

Study	Reference	Number	Genotype Means (mg/dl) ϵ_2/ϵ_2	ϵ_3/ϵ_2	ϵ_3/ϵ_3	ϵ_4/ϵ_2	ϵ_4/ϵ_3	ϵ_4/ϵ_4
Canada	174	102	136	161	174	178	184	180
Finland	178	207	173	202	214	196	223	239
France	175	223	199	207	235	237	241	240
West Germany	147	563	184	164	184	208	189	192
USA	179	1209	184	195	210	214	217	214
Holland	180	2018	209	202	216	207	219	225

Table 10–11. Estimated Average Effects of the ApoE Alleles and Serum Cholesterol

Phenotypic Levels	ϵ_2	ϵ_3	ϵ_4
Apolipoprotein E (mg/dL)	0.95	−0.04	−0.19
Apolipoprotein B (mg/dL)	−9.46	−0.18	4.92
Total cholesterol (mg/dL)	−14.2	−0.16	7.09

on apoE levels as compared with apoB and cholesterol levels may shed light on this mechanism. Lipoprotein particles containing the ϵ_2-isoform appear to have reduced binding affinity for LDL (apoB-E) receptors.[149,150] Using immunoelectrophoresis and apoE specific antibodies, Utermann[151] has shown that in $\epsilon_{2/2}$-homozygotes apoE accumulates in lipoproteins with β motility and slow α motility. Boerwinkle and Uterman[147] report that the ϵ_2-allele raiises average apoE levels in the general population, resulting from the accumulation of apoE-containing lipoprotein particles. Conversely, the average effect of the ϵ_4-allele is to lower plasma apoE levels, a result of increased turnover rates of apoE-4-containing lipoproteins. In support of this hypothesis, Gregg et al.[152] found that the ϵ_4-isoform has a faster rate of clearance from the plasma than does the ϵ_3-isoform. This may be attributable to the failure of the ϵ_4-isoform to form the apoE-A-II complex.[153,154]

From these results, a model has been proposed[147] by which the apoE polymorphism affects plasma LDL levels and, indirectly, total serum cholesterol and apoB levels in the population at large. Cholesterol delivered to the liver regulates several metabolic events determining cholesterol homeostasis. These events include regulating the activity of HMG Co-A reductase, the rate-limiting enzyme in endogenous cholesterol synthesis, and regulating the number of LDL (apoB-E) receptors on the cell surface. In $\epsilon_{2/2}$-homozygotes, chylomicron remnants, VLDL remnants and apoE-containing lipoproteins that depend on functional apoE for uptake by the liver accumulate in the plasma as a result of reduced binding affinity of the ϵ_2-isoform. VLDL remnants containing the apoϵ_2-isoform are also converted to LDL at a lower rate, resulting in a reduced production of LDL and an accumulation of VLDL. We speculate that the lower LDL levels allow less cholesterol to enter the cells of the liver and other tissues. To compensate for reduced intracellular cholesterol levels, LDL (apoB-E) receptors are up-regulated. This up-regulation results in enhanced uptake of LDL into the cells and lowering of LDL in the plasma. In combination with concomitant hyperlipidemia attributable to environmental or other genetic factors, this model may also account for the association between the apoϵ_2-isoform and type-III hyperlipidemia. The converse mechanism probably accounts for elevated LDL levels in individuals with the ϵ_4-allele. In $\epsilon_{4/4}$-individuals, more cholesterol (relative to $\epsilon_{3/3}$-homozygotes) from apoE-containing lipoproteins is internalized by cells of the liver. This relative increase in intracellular cholesterol levels is compensated for by down-regulating the number of LDL receptors. Fewer LDL receptors on the surface of the liver cells results in an increase in plasma LDL levels.

Immunochemical polymorphism of human apoB has been shown in various laboratories.[155–157] Recent studies[158] using monoclonal antibodies against purified LDL containing apoB-100 as its only protein component confirm the existence of such immunochemical polymorphisms. The first evidence for polymorphism at the primary amino acid sequence level was provided by Wei et al.,[159] who found, scattered over the carboxyl-terminal fifth of the protein, differences in single amino acid residues identified by direct sequencing of apoB-100 tryptic peptides and those deduced from the cDNA sequence. Later research determined that such polymorphisms indeed exist throughout the molecule, as revealed by complete cDNA sequences reported by different laboratories.[86] Some of the differences are reflected by restriction fragment length polymorphisms involving the apoB gene,[60,160–162] while the others are confirmed by direct sequence analysis of apoB-100 tryptic peptides.[86] Some of the sequence heterogeneities of apoB-100 may involve crucial parts of the molecule, resulting in differences in function and metabolism of LDL. Indeed, evidence that a specific apoB-

100 allele can result in familial hypercholesterolemia and premature atherosclerosis has been found in a strain of pigs described by Rapacz et al.[163] and Lowe et al.[164] These investigators found that these pigs bear three immunologically defined lipoprotein-associated markers (allotypes), including one that defines a variant apoB in these animals. Recently, variation in a putative LDL (apoB-E) receptor-binding domain of apoB-100 has been reported in humans. This apoB mutation involves an alteration at amino acid 3500 and is associated with low MB19 antibody affinity and premature atherosclerosis.[165,166] Future studies on these genetic markers, especially with respect to apoB structure and function, will be iimportant for understanding apoB-LDL receptor interactions.

Readily detectable protein variation like that in apoE or apoB is a minor component of the total variation in the apolipoprotein genes. A popular method for detecting a small subset of this variation is the use of restriction endonucleases, which cleave DNA at specific sequences. Recently, numerous reports have been made of significant associations between DNA restriction sites and lipid levels. Unlike the apoE protein polymorphism or the apoB 3500 mutation, which have direct physiologic effects, most DNA markers being used in association studies with cardiovascular disease risk factors and end points have no known physiologic effect. Regions of the apolipoprotein B gene exhibit considerable amounts of polymorphic genetic variation. These polymorphic sites may be located outside the transcript region of the gene, in proposed regulatory regions, in introns, or in transcribed sequences where they may or may not alter the amino acid sequence of the gene product. Some are caused by single base changes that alter specific restriction sites, others are the result of insertion or deletion of pieces of DNA, producing restriction fragments of varying lengths. Approximately one in 200 to 500 nucleotide bp is estimated to be heterozygous in an individual.[167] The large amount of DNA sequence variation in the candidate genes, such as apoB, is a valuable tool for studies examining the association between DNA variations and lipid metabolism.

Many studies have investigated the association between restriction site variation in the apoB gene and lipid levels. One such restriction site polymorphism results in a synonymous substitution at codon 2488 of the apoB gene detectable by the enzyme XbaI. IIn most Caucasian populations, approximately one half of the individuals have the DNA sequence 5'-CCC-3' around codon 7464; the other half, however, have the sequence 5-'CTC-3'. Like the apoE polymorphism reviewed previously, considerable research is directed toward investigating the effects of this, and other, genetic variation on lipid levels and CHD risk. Table 10-12 summarizes results from three separate association studies using this XbaI polymorphism. The results summarized in Table 10-12 exemplify several points apparent in these and other studies using detectable DNA variation. First, they typically employ two study designs: one comparing patients with

Table 10–12. Association of the 8.6-kb XbaI Allele of ApoB With Lipids and CHD Risk in Three Studies

	Patient–Control Allele Frequencies	Random Sample Summary
Hegele et al.[142]	.64 vs. .50 (P < .05)	No significant differences
Deeb et al.[181]	.50 vs. .54 (N.S.)	Higher triglycerides (P < .05)
Talmud et al.[182]	.50 vs. .32 (P < .05)	Lower cholesterol (P < .05)

CHD, coronary heart disease; kb, kilobase; NS, not significant.

a matched control group and the other investigating the effects of the gene in a random sample of unrelated individuals. For example, Hegele et al.[142] report that the allele frequencies of this polymorphic restriction site differ between CHD patients and normal individuals; however, no differences in average lipid, lipoprotein, or apolipoprotein levels were found among genotypes in the general population. A comparison of the results between these designs is tenuous. Second, these and numerous other reports are yielding statistically significant associations of DNA variation with risk factor levels and overt disease. These results generally indicate that genetic variation in and around the apolipoprotein genes is affecting lipid, lipoprotein, and apolipoprotein levels and may affect one's risk to disease. Humphries[139] and Cooper and Clayton[140] summarize results from numerous DNA-disease–association studies. Some of these associations are real and are assumed to be attributable to disequilibrium between the marker DNA variation and a nearby gene locus with a direct physiologic effect. Third, the two studies differ with respect to their conclusions. This may be attributable to separate populations of inference, chance sampling variability, and different statistical strategies used. Because of these and other inconsistencies, we believe that restriction site variation is best applied to family studies to estimate the effects of genetic variation in the apolipoprotein genes on lipid levels.

Linkage and Family Studies

Different from case-control studies estimating the effects of polymorphisms in a sample of unrelated individuals is the use of genetic markers in family studies. Here the use of measured genotype information asserts its greatest potential.[138] Large amounts of polymorphic variability in the apolipoprotein genes enable one to follow the segregation of these genes in families. One can therefore assess the characteristics of mutational variability without directly measuring those loci with physiologic effects on the phenotypes of interest. The most widely recognized examples of genetic alterations affecting lipid metabolism and risk to CHD are mutations in the LDL-receptor gene resulting in familial hypercholesterolemia (FH).[17] Individuals with heterozygous FH have plasma levels of LDL twice normal and often have heart attacks as early as age 30. Studies of FH have contributed significantly to the understanding of the cellular metabolism of cholesterol. Using restriction site polymorphisms in the LDL-receptor gene, Leppert et al.[168] report very tight linkage of this gene to factors responsible for hypercholesterolemia and early CHD in a large pedigree. With a frequency of less than 1 in 500, however, FH contributes only modestly to the overall CHD morbidity and mortality in the general population. As previously noted, Boerwinkle and Sing[148] have used family data to estimate directly the contribution of the apoE polymorphism to the familial aggregation of several risk factor phenotypes. They estimate that the proportion of interindividual variability of total cholesterol levels attributable to the apoE polymorphism and to residual polygenic effects is 8 and 56%, respectively.

The utility of genetic markers in the apolipoprotein genes and other candidate genes for the study of CHD risk is determined in large part by the extent of allelic variation that exists at a particular locus. On the 3' end of the apoB gene is a polymorphic region consisting of a variable number of tandemly repeated (VNTR) short DNA sequences. Boerwinkle et al.[169] recently described a general method that uses the polymerase chain reaction to type this apoB VNTR locus rapidly and accurately. The sensitivity of this new method makes it superior to traditional Southern blot analysis. With this new method, the large number of alleles in this locus make it a valuable marker for association studies and family studies with the apoB gene.

COMMENTS

Up to about 15 years ago, molecular biology was a foreign discipline to lipoprotein metabolism and atherosclerosis research. The subsequent 5 years witnessed the beginning of a molecular approach mainly in the area of apolipoprotein synthesis in the avian model system.[45] At about the same time, the LDL-receptor pathway was biochemically characterized. In the last 10 years, the application of recombinant DNA techniques has led to an unpredecented explosion of information. The most notable of these advances is the complete elucidation of the LDL-receptor pathway at the molecular level and the genetic basis of familial hypercholesterolemia.[17] The other major advances have been summarized in this chapter.

In the next decade, we will undoubtedly experience additional breakthroughs in our knowledge in this area. At the clinical level, some of the hypercholesterolemic syndromes not caused by LDL-receptor defects will be defined. Similarly, many of the other dyslipoproteinemic syndromes will be elucidated at the molecular level. The structure–function relationship of apoB will be further characterized. Rapid progress is occurring on the use of molecular genetic and epidemiologic information in the investigation of the common chronic diseases.[170] Pathologic states like atherosclerosis are multifactorial in origin. They are much more difficult to characterize and are all the more challenging. We expect that additional genetic factors will be defined. Such advances, together with an increased understanding of the mechanisms of various environmental factors, will help define future preventive measures as well as therapeutic inventions. At the basic science level, several exciting areas are being explored. The molecular mechanisms underlying the RNA editing responsible for apoB-48 mRNA biogenesis will be characterized. The elucidation of this "reaction" is important in our understanding of the control of gene expression. Finally, the possible evolutionary pathways summarized in this chapter will be better defined and modified as studies on additional species accumulate. We expect that the application of molecular biology will continue to contribute to our knowledge of lipoprotein metabolism and basic and clinical aspects of atherosclerosis for the next decade.

ACKNOWLEDGMENTS

The work in the authors' laboratories described in this chapter was supported by grants from the National Institutes of Health HL-16512 and HL-27341 to L.C., HL-40013 to E.B., and GM-39927 to W.-H.L., and a grant from the March of Dimes Birth Defects Foundation to L.C. We thank Ms. Suzanne Mascola for typing this chapter.

REFERENCES

1. Arteriosclerosis: A report by the National Heart and Lung Institute Task Force on Arteriosclerosis. DHEW Publication No. [NIH] 72-219. Washington, DC: U.S. Government Printing Office, 1971. Vol. 2.
2. Segrest, J.P., and Albers, J.J. (eds.): Plasma Lipoproteins: Part A. Preparation, Structure and Molecular Biology, Methods in Enzymology, Orlando, FL, Academic Press, 1986. Vol. 128.
3. Kane, J.P.: Serum Lipoproteins: Structure and Metabolism. *In* Lipid Metabolism in Mammals. Edited by F. Snyder. New York, Plenum Press, 1977. p. 209.
4. Redgrave, T.: Formation of cholesterol ester-rich particulate lipid during metabolism of chylomicrons. J. Clin. Invest., 49:465, 1970.
5. Nilsson-Ehle, P., Garfinkle, A.S., and Schotz, M.C.: Lipolytic enzymes and liproprotein metabolism. Annu. Rev. Biochem., 49:667, 1980.

6. Mjos, O.D., Faergeman, O., Hamilton, R.L., and Havel, R.J.: Characterization of remnants produced during the metabolism of triglyceride-rich lipoproteins in blood plasma and intestinal lymph in the rat. J. Clin. Invest., *56*:603, 1975.

7. Kane, J.P.: Apoprotein B: structural and metabolic heterogeneity. Annu. Rev. Physiol., *45*:637, 1983.

8. Wu, A.L., and Windemuller, H.G.: Relative contributions by liver and intestine to individual plasma apolipoprotein in the rat. J. Biol. Chem., *254*:7316, 1979.

9. Sherrill, B.C., Innerarity, T.L., and Mahley, R.W.: Rapid hepatic clearance of the canine lipoproteins containing only the E apoprotein by high affinity receptor. J. Biol. Chem., *255*:1804, 1980.

10. Jones, A.L., et al.: Uptake and processing of remnants of chylomicrons and very low density lipoproteins by rat liver. J. Lipid Res., *25*:1151, 1984.

11. Marsh, J.B.: Lipoproteins in a nonrecirculating perfusate of rat liver. J. Lipid Res., *15*:544, 1974.

12. Havel, R. J.: Lipoprotein biosynthesis and metabolism. Ann. N.Y. Acad. Sci., *348*:16, 1980.

13. Havel, R. J.: The formation of LDL: mechanisms and regulation. J. Lipid Res., *25*:1570, 1984.

14. Krauss, R. M., and Burke, D. J.: Identification of multiple subclasses of plasma low density lipoproteins in normal humans. J. Lipid Res., *23*:97, 1982.

15. Fisher, W.R., Hammond, M.G., Mengel, M.C., and Warmke, G.L.: A genetic determinant of the phenotypic variance of the molecular weight of low density lipoprotein. Proc. Natl. Acad. Sci. USA, *72*:2347, 1975.

16. Austin, M.A. et al.: Low density lipoprotein subclass patterns and risk of myocardial infarction. JAMA, *260*:1917, 1988.

17. Brown, M.S., and Goldstein, J.L.: A receptor-mediated pathway for cholesterol homeostasis. Science, *232*:34, 1986.

18. Kannel, W.B., Castelli, W.P., and Gordon, T.: Serum cholesterol, lipoproteins, and the risk of coronary heart disease. Ann. Intern. Med. *74*:1, 1971.

19. Lindgren, F.T., Nichols, A., and Freeman, N.K.: Physical and chemical composition studies on the lipoproteins of fasting and heparinized human sera. J. Phys. Chem., *59*:930, 1955.

20. Krauss, R.M.: Regulation of high density lipoprotein levels. Med. Clin. North Am., *66*:403, 1982.

21. Eisenberg, S.: High density lipoprotein metabolism. J. Lipid Res., *25*:1017, 1984.

22. Hamilton, R.L., Williams, M.C., Fielding, C.J., and Havel, R.J.: Discoidal bilayer structure of nascent high density lipoproteins from perfused rat liver. J. Clin. Invest., *58*:667, 1976.

23. Tall, A.R., Green, P.H.R., Glickman, R.M., and Riley, J.W.: Metabolic fate of chylomicron phospholipids and apoproteins in the rat. J. Clin. Invest. *664*:977, 1979.

24. Glomset, J.A.: The plasma lecithin-cholesterol acyltransferase reaction. J. Lipid Res., *9*:155, 1968.

25. Green, P.H.R., and Glickman, R.M.: Intestinal lipoprotein metabolism. J. Lipid. Res., *22*:1153, 1981.

26. Barr, D.P., Russ, E.M., and Eder, H.A.: Protein–lipid relationship in human plasma: II. In atherosclerosis and related conditions. Am. J. Med., *11*:480, 1951.

27. Miller, G.J., and Miller, N.E.: Plasma high density lipoprotein concentration and development of ischaemic heart disease. Lancet, *1*:16, 1975.

28. Fielding, C.J., Shore, V.G., and Fielding, P.E.: A protein cofactor of lecithin-cholesterol acyltransferase. Biochem. Biophys. Res. Commun., *46*:1493, 1972.

29. Soutar, A.L. et al.: Effect of the human apolipoproteins and phosphatidylcholine acyl donor on the activity of lecithin-cholesterol acyltransferase. Biochemistry, *14*:3057, 1975.

30. Steinmetz, A., and Utermann, G.: Activation of lecithin cholesterol acyltransferase by human apolipoprotein A-IV. J. Biol. Chem., *260*:2258, 1985.

31. LaRosa, J.C. et al.: A specific apoprotein activator for lipoprotein lipase. Biochem. Biophys. Res. Commun., *41*:57, 1970.

32. Breckenridge, W.C. et al.: Hypertriglyceridemia associated with deficiency of apolipoprotein C-II. N. Engl. J. Med., *298*:1265, 1978.

33. Brown, W.-V., and Baginsky, M.L.: Inhibition of lipoprotein lipase by the apolipoprotein of human very low density lipoprotein. Biochem. Biophys. Res. Commun., *46*:375, 1972.

34. Wang, C.S., McConathy, W.J., Kloer, H.U., and Alaupovic, P.: Modulation of lipoprotein lipase activity by apolipoproteins. Effect of apolipoprotein C-III. J. Clin. Invest., *75*:384, 1985.

35. Innerarity, T.L., and Mahley, R.W.: Enhanced binding by cultured human fibroblasts of apoE-containing lipoproteins as compared with low density lipoproteins. Biochemistry, *17*:1440, 1978.

36. Lin, C.-T., Xu, Y., Wu, J.-Y., and Chan, L.: Immunoreactive apolipoprotein E is a widely distributed cellular protein. Immunohistochemical localization of apolipoprotein E in baboon tissues. J. Clin. Invest., *78*:947, 1986.

37. Hui, D.Y., Innerarity, T.L., and Mahley, R.W.: Lipoprotein binding to canine hepatic membranes: metabolically distinct apoE and apoB,E receptors. J. Biol. Chem., *256*:5646, 1981.

38. Havel, R.J., (ed.): Symposium on Lipid Disorders. Med. Clin. North Am., *66*: 1982.

39. Mahley, R.W., Innerarity, T.L., Rall, S.C., Jr., and Weisgraber, K.H.: Plasma lipoproteins: apolipoprotein structure and function. J. Lipid Res., *25*:1277, 1984.

40. Kao, F.T., and Chan, L.: Genetic mapping of apolipoprotein genes in the human genome by somatic cell hybrids. Methods Enzymol., *128*:851, 1986.

41. Harper, M.E., and Chan, L.: Chromosomal fine mapping of apolipoprotein genes by *in situ* nucleic acid hybridization to metaphase chromosomes. Methods Enzymol., *128*:863, 1986.
42. Li, W.-H. et al.: The apolipoprotein multigene family: biosynthesis, structure, structure–function relationships and evolution. J. Lipid Res., *29*:245, 1988.
43. Blackhart, B.D. et al.: Structure of the human apolipoprotein B gene. J. Biol. Chem., *261*:15364, 1986.
44. Drayna, D.T. et al.: Human apolipoprotein D gene: gene sequence, chromosome localization, and homology to the α_2-globulin superfamily. DNA, *6*:199, 1987.
45. Chan, L.: Hormonal control of apolipoprotein synthesis. Annu. Rev. Physiol., *45*:615, 1983.
46. Chan, L., Bradley, W.A., and Means, A.R.: Amino acid sequence of the signal peptide of apoVLDL-II, a major apoprotein in avian very low density lipoproteins. J. Biol. Chem., *255*:10060, 1980.
47. Blobel, G., and Dobberstein, B.: Transfer of proteins across membranes: I. Presence of proteolytically processed and unprocessed nascent immunoglobulin light chains on membrane-bound ribosomes of murine myeloma. J. Cell Biol., *67*:835, 1975.
48. Blobel, G., and Dobberstein, B.: Transfer of proteins across membranes: II. Reconstitution of functional rough microsomes from heterologous components. J. Cell Biol., *67*:852, 1975.
49. Chan, L., and Bradley, W.A.: Signal peptides: properties and interactions. *In* Cellular Regulation of Secretion and Release. Edited by P.M. Conn. Orlando, FL: Academic Press, 1982. p. 301.
50. von Heijne, G.: Signal sequences: the limits of variation. J. Mol. Biol., *184*:99, 1985.
51. Walter, P., and Lingappa, V.R.: Mechanism of protein translocation across the endoplasmic reticulum membrane. Annu. Rev. Cell Biol., *2*:499, 1986.
52. Blobel, G.: Intracellular protein topogenesis. Proc. Natl. Acad. Sci. USA, *77*:1496, 1980.
53. Wickner, W.: The assembly of proteins into biological membranes: the membrane trigger hypothesis. Annu. Rev. Biochem., *48*:23, 1979.
54. Kane, J.P., Hardman, D.A., and Paulus, H.E.: Heterogeneity of apolipoprotein B: isolation of a new species from human chylomicron. Proc. Natl. Acad. Sci. USA, *77*:2465, 1980.
55. Glickman, R.M., Rogers, M., and Glickman, J.N.: Apolipoprotein B synthesis by human liver and intestine in vitro. Proc. Natl. Acad. Sci. USA, *83*:5296, 1986.
56. Marcel, Y.L., Hogue, M., Theolis, R., Jr., and Milne, R.W.: Mapping of antigenic determinants of human apolipoprotein B using monoclonal antibodies against low density lipoproteins. J. Biol. Chem., *257*:13165, 1982.
57. Cardin, A.D. et al.: Structural organization of apolipoprotein B-100 of human plasma low density lipoproteins. Comparison to B-48 of chylomicrons and very low density lipoproteins. J. Biol. Chem., *261*:16744, 1986.
58. Young, S.G. et al.: Parallel expression of the MB19 genetic polymorphism in apoprotein B-100 and apoprotein B-48. Evidence that both apoproteins are products of the same genes. J. Biol. Chem., *261*:2995, 1986.
59. Young, S.G. et al.: Genetic analysis of a kindred with familial hypobetalipoproteinemia. Evidence for two separate gene defects: one associated with an abnormal apolipoprotein B species, apolipoprotein B-37; and a second associated with low plasma concentration of apolipoprotein B-100. J. Clin. Invest. *79*:1842, 1987.
60. Chan, L. et al.: The human apolipoprotein B-100 gene: a highly polymorphic gene that maps to the short arm of chromosome 2. Biochem. Biophys. Res. Commun., *133*:248, 1985.
61. Ludwig, E. H. et al.: DNA sequence of the human apolipoprotein B gene. DNA, *6*:363, 1987.
62. Reuben, M.A. et al.: Biosynthetic relationships between three rat apolipoprotein B peptides. J. Lipid Res., *29*:1337, 1988.
63. Powell, L.M. et al.: A novel form of tissue-specific RNA processing produces apolipoprotein B-48 in intestine. Cell, *50*:831, 1987.
64. Chen, S.H. et al.: Apolipoprotein B-48 is the product of an mRNA with an organ-specific in-frame stop codon. Science, *238*:363, 1987.
65. Gordon, J.I. et al.: The primary translation product of rat intestinal apolipoprotein A-I mRNA is an unusual preproprotein. J. Biol. Chem., *257*:971, 1982.
66. Stoffel, W., Kruger, E., and Deutzmann, R.: Cell-free translation of human liver apolipoprotein A-I and A-II mRNA. Processing of primary translation products. Hoppe-Seyler's Z. Physiol. Chem., *364*:227, 1983.
67. Cheung, P., and Chan, L.: Nucleotide sequence of cloned cDNA of human apolipoprotein A-I. Nucleic Acids Res., *11*:3703, 1983.
68. Pan, T., Hao, Q., Yamin, T.-T., Dai, B., Chen, S., Kroon, P.A., and Chao, Y.: Rabbit apolipoprotein A-I mRNA and gene. Eur. J. Biochem., *170*:99, 1987.
69. Yang, C.-Y. et al.: The complete amino acid sequence of proapolipoprotein A-I of chicken high density lipoproteiins. FEBS Lett., *224*:261, 1987.
70. Gordon, J.I. et al.: Proteolytic processing of human preproapolipoprotein A-I. A proposed defect in the conversion of proA-I to A-I in Tangier's disease. J. Biol. Chem., *258*:4037, 1983.
71. Zannis, V.I. et al.: Intracellular and extracellular processing of human apolipoprotein A-I: secreted apolipoprotein A-I isoprotein 2 is a propeptide. Proc. Natl. Acad. Sci. USA, *80*:2574, 1983.

72. Stoffel, W., Bode, C., and Knyrim, K.: Serum apolipoprotein A-I synthesis in rat hepatocytes and its secretion as a proform. Hoppe Seyler's Z. Physiol. Chem., *364*:439, 1983.
73. Banerjee, D. et al.: Regulation of apoA-I processing in cultured hepatocytes. J. Biol. Chem., *261*:9844, 1986.
74. Gordon, J.I. et al.: Biosynthesis of human preproapolipoprotein A-II. J. Biol. Chem., *258*:14054, 1983.
75. Moore, M.N., Kao, F.T., Tsao, Y.K., and Chan, L.: Human apolipoprotein A-II: nucleotide sequence of a cloned cDNA, and localization of its structural gene on human chromosome. 1. Biochem. Biophys. Res. Commun., *123*:1, 1984.
76. Gordon, J.I. et al.: Extracellular processing of proapolipoprotein A-II in HepG2 cell cultures is mediated by a 54-kDa protease immunologically related to cathepsin B. J. Biol. Chem., *260*:14824, 1986.
77. Gordon, J.I. et al.: Proteolytic processing and compartmentalization of the primary translation products of mammalian apolipoprotein mRNAs. CRC Crit. Rev. Biochem., *20*:37, 1986.
78. Fojo, S.S. et al.: Human preproapolipoprotein C-II. Analysis of major plasma isoforms. J. Biol. Chem., *261*:9591, 1986.
79. Ghiselli, G. et al.: Origin of apolipoprotein A-I polymorphism in plasma. J. Biol. Chem., *260*:15662, 1985.
80. Hoeg, J.M. et al.: Human apolipoprotein A-I, post-translational modification by fatty acid acylation. J. Biol. Chem., *261*:3911, 1986.
81. Davis, R.A. et al.: Intrahepatic assembly of very low density lipoproteins. Phosphorylation of small molecular weight apolipoprotein B. J. Biol. Chem., *259*:3383, 1984.
82. Segrest, J.P., Jackson, R.L., Morrisett, J.D., and Gotto, A.M., Jr.: A molecular theory of lipid–protein interactions in the plasma lipoproteins. FEBS Lett. *38*:247, 1974.
83. Chen, S.H. et al.: The complete cDNA and amino acid sequence of human apolipoprotein B-100. J. Biol. Chem., *261*:12918, 1986.
84. Gotto, A.M., Jr., Levy, R.I., and Frederickson, D.S.: Observations on the conformation of human beta-lipoprotein: evidence of the occurrence of beta structure. Proc. Natl. Acad. Sci. USA, *60*:1436, 1968.
85. Osterman, D. et al.: A synthetic amphiphilic beta-strand tridecapeptide: a model for apolipoprotein B. J. Am. Chem. Soc., *106*:6845, 1984.
86. Yang, C.-Y. et al.: Structure of apolipoprotein B-100 of human low density lipoproteins. Arteriosclerosis, *9*:96, 1989.
87. Chen, P.-F. et al.: Primary sequence mapping of human apolipoprotein B-100 epitopes. Comparisons of trypsin accessibility and immunoreactivity and implication for apoB conformation. Eur. J. Biochem., *175*:111, 1988.
88. Drayna, D. et al.: Cloning and expression of human apolipoprotein D cDNA. J. Biol. Chem., *261*:16535, 1986.
89. Nakagawa, S.H., Lau, H.S.H., Kezdy, F.J., and Kaiser, E.T.: The use of polymer-bound oximes for the synthesis of large peptides usable in segment condensation: synthesis of a 44 amino acid amphipathic peptide model of apolipoprotein A-I. J. Am. Chem. Soc., *107*:7087, 1985.
90. Eisenberg, D., Weiss, R.M., and Tergwillager, T.C.: The helical hydrophobic moment: a measure of the amphiphicity of a helix. Nature, *299*:371, 1982.
91. Knott, J. et al.: Complete protein sequence and identification of structural domains of human apolipoprotein B. Nature, *323*:734, 1986.
92. Yang, C.-Y. et al.: Sequence, structure, receptor binding domains and internal repeats of human apolipoprotein B-100. Nature, *323*:738, 1986.
93. Margolis, S., and Langdon, R.G.: Studies on human serum, beta₁-lipoprotein. III. Enzymatic modifications. J. Biol. Chem., *241*:485, 1966.
94. Chapman, M.J., Goldstein, S., and Mills, G.L.: Limited tryptic digestion of human serum low density lipoprotein. Isolation and characterization of the protein-deficient particles and of its apoproteins. Eur. J. Biochem., *87*:475, 1978.
95. DeLoof, H. et al.: Human apolipoprotein B: analysis of internal repeats and homology with other apolipoproteins. J. Lipid Res., *28*:145, 1987.
96. Fielding, P.E., and Fielding, C.J.: A cholesterol ester transfer complex in human plasma. Proc. Natl. Acad. Sci. USA, *77*:3327, 1980.
97. Albers, J.J., Cheung, M.C., Ewens, S.L., and Tollefson, J.H.: Characterization and immunoassay of apolipoprotein D. Atherosclerosis, *39*:395, 1981.
98. Li, W.-H., Wu, C.-I., and Luo, C.-C.: A new method for estimating synonymous and nonsynonymous rates of nucleotide substitution considering the relative likelihood of nucleotide and codon changes. Mol. Biol. Evol. *2*:150, 1985.
99. Luo, C.-C., Li, W.-H., Moore, M.N., and Chan, L.: Structure and evolution of the apolipoprotein multigene family. J. Mol. Biol., *187*:325, 1986.
100. Dayhoff, M.O.: Atlas of Protein Sequence and Structure. Silver Spring, MD, National Biochemical Research Foundation, 1972.
101. Kimura, M.: The Neutral Theory of Molecular Evolution. Cambridge, England, Cambridge University Press, 1983.

102. Li, W.-H., Luo, C.-C., and Wu, C.-I.: Evolution of DNA sequences. *In* Molecular Evolutionary Genetics. Edited by R.J. MacIntyre. New York, Plenum Press, 1985. p. 1.
103. Romer, A.S.: Vertebrate Palentology. Chicago, University of Chicago Press, 1966.
104. Fitch, W.M., and Margoliash, E.: Construction of phylogenetic trees. A method based on mutation distances, as estimated from cytochrome C sequences is of general applicability. Science, *155*:279, 1967.
105. Wu, C.-I., and Li, W.-H.: Evidence for higher rates of nucleotide substitution in rodents than in man. Proc. Natl. Acad. Sci. USA, *82*:1741, 1985.
106. Pontes, M. et al.: cDNA sequences of two apolipoproteins from lamprey. Biochemistry, *26*:1611, 1987.
107. Chapman, M.J., Goldstein, S., Mills, G.L., and Leger, C.: Distribution and characterization of the serum lipoproteins and their apoproteins in the rainbow trout (*Salmo gairdnerii*). Biochemistry, *17*:4455, 1978.
108. Chapman, M.J.: Animal lipoproteins: chemistry, structure, and comparative aspects. J. Lipid Res., *21*:789, 1980.
109. Babin, P.J.: Apolipoproteins and the association of egg yolk proteins with plasma high density lipoproteins after ovulation and follicular atresia in the rainbow trout (*Salmo gairdneri*). J. Biol. Chem., *262*:4290, 1987.
110. Connelly, P.W., Hoffman, G.F.T., and Little, J.A.: Structure of apolipoprotein C-II$_{Toronto}$: a non-functional human apolipoprotein. Proc. Natl. Acad. Sci. USA, *84*:270, 1987.
111. Connelly, P.W., Maguire, G.F., and Little, J.A.: Apolipoprotein CII$_{St. Michael}$. Familial apolipoprotein C-II deficiency associated with premature vascular disease. J. Clin. Invest., *80*:1597, 1988.
112. Kinnunen, P.K.J. et al.: Activation of lipoprotein lipase by native and synthetic fragments of human plasma apoC-II. Proc. Natl. Acad. Sci. USA, *74*:4848, 1977.
113. Smith, L.C. et al.: Activation of lipoprotein lipase by native and synthetic peptide fragments of apolipoprotein C-II. *In* Proceedings of the International Conference on Atherosclerosis. Edited by L.A. Carlson, R. Paoletti, and G. Weber. New York, Raven Press, 1978. p. 269.
114. Datta, S. et al.: Structure and expression of dog apolipoprotein C-II and C-III mRNAs: implications for the evolution and functional constraints of apolipoprotein structure. J. Biol. Chem., *262*:10588, 1987.
115. Edelstein, C. et al.: Isolation and characterization of a dog serum lipoprotein having apolipoprotein A-I as its predominant protein constituent. Biochemistry, *15*:1934, 1976.
116. Jackson, R.L. et al.: A comparison of the major apolipoprotein from pig and human high density lipoproteins. J. Biol. Chem., *248*:2639, 1973.
117. Jonas, A.: Physicochemical properties of bovine serum high density lipoprotein. J. Biol. Chem., *247*:7767, 1972.
118. Hillyard, L.A., White, H.M., and Pangburn, S.A.: Characterization of apolipoproteins in chicken serum and egg yolk. *Biochemistry, 11*:511, 1972.
119. Sparrow, J.T., and Gotto, A.M., Jr.: Phospholipid binding studies with synthetic apolipoprotein fragments. Ann. N.Y. Acad. Sci., *348*:187, 1980.
120. Zorich, N., Jonas, A., and Pownall, H.L.: Activation of lecithin cholesterol acyltransferase by human apolipoprotein E in discoidal complex with lipids. J. Biol. Chem., *260*:8831, 1985.
121. Steinmetz, A., Kaffarnik, H., and Utermann, G.: Activation of phosphatidylcholine-sterol acyltransferase by human apolipoprotein E isoforms. Eur. J. Biochem. *152*:747, 1985.
122. Pownall, H.J. et al.: Activation of lecithin-cholesterol acyltransferase by a synthetic model lipid-associating peptide. Proc. Natl. Acad. Sci. USA, *77*:3154, 1980.
123. Yokoyama, S. et al.: The mechanism of activation of lecithin: cholesterol acyltransferase by apolipoprotein A-I and an amphipathic peptide. J. Biol. Chem., *255*:7333, 1980.
124. Mahley, R. W., and Innerarity, T.L.: Lipoprotein receptors and cholesterol homeostasis. Biochem. Biophys. Acta, *737*:197, 1983.
125. Innerarity, T.L. et al.: The receptor binding domain of human apolipoprotein E: binding of apolipoprotein E fragments. J. Biol. Chem., *258*:12341, 1983.
126. Weisgraber, K.H. et al.: The receptor binding domain of human apolipoprotein E. Monoclonal antibody inhibition of binding. J. Biol. Chem. *258*:12348, 1983.
127. Weisgraber, K.H., Innerarity, T.L., and Mahley, R.W.: Abnormal lipoprotein receptor binding activity of the human apoprotein due to cysteine–arginine interchange at a single site. J. Biol. Chem., *257*:2518, 1982.
128. Weisgraber, K.H. et al.: A novel electrophoretic variant of human apolipoprotein E: identification and characterization of apolipoprotein E1. J. Clin. Invest., *73*:1024, 1984.
129. Rall, S.C., Jr., Weisgraber, K.H., Innerarity, T.L., and Mahley, R.W.: Structural basis for receptor binding heterogeneity of apolipoprotein E from the type III hyperlipoproteinemic subjects. Proc. Natl. Acad. Sci. USA, *79*:4696, 1982.
130. Innerarity, T.L. et al.: Normalization of receptor binding of apolipoprotein E2. J. Biol. Chem., *259*:7261, 1984.
131. Brunzell, J.D., Sniderman, A.D., Albers, J.J., and Kwiterovich, P.O., Jr.: Apoproteins B and A-I and coronary artery disease in humans. Atherosclerosis, *4*:79, 1984.
132. Sniderman, A. et al.: Association of coronary atherosclerosis with hyperbetalipoproteinemia (increased

protein but normal cholesterol levels in human plasma low density lipoproteins). Proc. Natl. Acad. Sci. USA, *77*:604, 1980.

133. Matsumoto, A. et al.: Cloning and regulation of rat apolipoprotein B mRNA. Biochem. Biophys. Res. Commun., *142*:92, 1987.

134. Goldstein, S., Chapman, M.J., and Mills, G.L.: Biochemical and immunological evidence for the presence of an apolipoprotein B-like component in the serum low density lipoproteins of several animal species. Atherosclerosis, *28*:93, 1977.

135. Muller, L.: Angina pectoris in hereditary xanthomatosis. Arch. Intern. Med., *64*:675, 1939.

136. Morton, N.E. et al.: Determinants of blood pressure in Japanese-American families. Hum. Genet., *53*:216, 1980.

137. Hamsten, A., Iselius, L., Dahlen, G., and de Faire, U.: Genetic and cultural inheritance of serum lipids, low and high density lipoprotein cholesterol and serum apolipoproteins A-I, A-II, and B. Atherosclerosis, *60*:199, 1986.

138. Boerwinkle, E., Chakraborty, R., and Sing, C.F.: The use of measured genotype information in the analysis of quantitative phenotypes in man: I. Models and analytical methods. Ann. Hum. Genet., *50*:181, 1986.

139. Humphries, S.E.: DNA polymorphisms of the apolipoprotein gene—their use in the investigation of the genetic component of hyperlipidemia and atherosclerosis. Atherosclerosis, *72*:89, 1988.

140. Cooper, D.N., and Clayton, J.F.: DNA polymorphism and the study of disease association. Hum. Genet., *78*:299, 1988.

141. Hendrick, P.W.: Gametic disequilibrium measures: proceed with caution. Genetics *117*:331, 1987.

142. Hegele, R.A. et al.: Apolipoprotein B-gene DNA polymorphisms associated with myocardial infarction. N. Engl. J. Med., *315*:1509, 1986.

143. Utermann, G., Hees, M., and Steinmetz, A.: Polymorphism of apolipoprotein E and occurrence of dysbetalipoproteinemia in man. Nature, *269*:604, 1977.

144. Bouthillier, D., Sing, C.F., and Davignon, J.: Apolipoprotein E phenotyping with a single gel method—application to the study of informative matings. J. Lipid Res., *24*:1060, 1983.

145. Rall, S.C., Weisgraber, K.H., and Mahley, R.W.: Human apolipoprotein E: the complete amino acid sequence. J. Biol. Chem., *257*:4171, 1982.

146. McLean, J.W. et al.: Human apolipoprotein E mRNA: cDNA cloning and nucleotide sequencing of a new variant. J. Biol. Chem., *259*:6498, 1984.

147. Boerwinkle, E., and Utermann, G.: Simultaneous effects of the apolipoprotein E polymorphism on apolipoprotein E, apolipoprotein B, and cholesterol metabolism. Am. J. Hum. Genet., *42*:104, 1988.

148. Boerwinkle, E., and Sing, C.F.: The use of measured genotype information in the analysis of quantitative phenotypes in man: III. Simultaneous estimation of the frequencies and effects of the apolipoprotein E polymorphism and residual polygenetic effects on cholesterol, beta-lipoproteins and triglyceride levels. Ann. Hum. Genet., *51*:211, 1987.

149. Schneider, W.J. et al.: Familial dysbetalipoproteinemia: abnormal binding of mutant apolipoprotein E to low density lipoprotein receptors of human fibroblasts and membranes from liver and adrenal of rats, rabbits, and cows. J. Clin. Invest., *68*:1075, 1981.

150. Hui, D.Y., Innerarity, T.L., and Mahley, R.W.: Defective hepatic lipoprotein receptor binding of B-very low density lipoproteins from type III hyperlipoproteinemic patients. J. Biol. Chem., *259*:860, 1984.

151. Utermann, G.: Genetic polymorphism of apolipoprotein E—impact on plasma lipoprotein metabolism. *In* Diabetes, Obesity and Hyperlipidemias. Edited by G. Crepaldi. New York, Elsevier, 1985. p. 1.

152. Gregg, R.E. et al.: Abnormal *in vivo* metabolism of apolipoprotein E4 in humans. J. Clin. Invest., *78*:815, 1986.

153. Weisgraber, K.H., and Mahley, R.W.: Apolipoprotein (E-A-II) complex of human plasma lipoproteins: I. Characterization of this mixed disulfide and its identification in a high density lipoprotein subfraction. J. Biol. Chem., *253*:6281, 1978.

154. Innerarity, T.L., Mahley, R.W., Weisgraber, K.H., and Bersot, T.P.: Apolipoprotein (E-A-II) complex of human plasma lipoproteins: II. Receptor binding activity of a high density lipoprotein subfraction modulated by the apo (E-A-II) complex. J. Biol. Chem., *253*:6289, 1978.

155. Blumberg, B.S., Dray, S., and Robinson, J.C.: Antigen polymorphism of a low density beta-lipoprotein allotype in human serum. Nature, *194*:656, 1962.

156. Butler, R., Butler, E., Scherz, R., and Pflugshaupt, R.: The Ag-system of low density lipoprotein—an updating. *In* Proceedings of the 25th Colloquium Protides of the Biological Fluids. Edited by H. Peeters. New York, Pergamon Press, 1978. p. 255.

157. Berg, K.: Inherited Lipoprotein Variation and Atherosclerotic Disease. *In* The Biochemistry of Atherosclerosis Edited by A.M. Scanu, R.W., Wissler, and G.S. Getz. New York, Marcel Dekker. 1979. p. 419.

158. Schumaker, V.N. et al.: Anti-apoprotein B monoclonal antibodies detect human low density lipoprotein polymorphism. J. Biol. Chem., *259*:6423, 1984.

159. Wei, C.-F. et al.: Molecular cloning and expression of partial cDNAs and deduced amino acid sequence of a carboxyl-terminal fragment of human apolipoprotein B-100. Proc. Natl. Acad. Sci. USA, *82*:7265, 1985.

160. Huang, L.S., Miller, D.A., Bruns, G.A.P., and Breslow, J.L.: Mapping of the human *APOB* gene to chro-

mosome 2p and demonstration of a two-allele restriction fragment length polymorphism. Proc. Natl. Acad. Sci. USA, *83*:644, 1986.

161. Mehrabian, M. et al.: Human apolipoprotein B: chromsomal mapping and DNA polymorphisms of hepatic and intestinal species. Somatic Cell Mol. Genet., *12*:245, 1986.

162. Wang, X. et al.: Apolipoprotein B: the Ag(a_1/d) immunogenetic polymorphism coincides with a T-to-C substitution at nucleotide 1981, creating an Alu restriction site. Arteriosclerosis, *8*:429, 1988.

163. Rapacz, J. et al.: Lipoprotein mutations in pigs are associated with elevated plasma cholesterol and atherosclerosis. Science, *234*:1573, 1986.

164. Lowe, S.W., Checovich, W.J., Rapacz, J., and Attie, A.D.: Defective receptor binding of low density lipoprotein from pigs possessing mutant apolipoprotein B alleles. J. Biol. Chem., *263*:15467, 1988.

165. Weisgraber, K.H. et al.: Familial defective apolipoprotein B-100: Enhanced binding of monoclonal antibody MB47 to abnormal low density lipoproteins. Proc. Natl. Acad. Sci. USA, *85*:9758, 1988.

166. Soria, L.F. et al.: Association between a specific apolipoprotein B mutation and familial defective apolipoprotein B-100. Proc. Natl. Acad. Sci. USA, *86*:587, 1989.

167. Neel, J.V.: A revised estimate of the amount of genetic variation in human proteins: implications for the distribution of DNA polymorphisms. Am. J. Hum. Genet., *36*:1135, 1984.

168. Leppert, M.F. et al.: A DNA probe for the LDL receptor gene is tightly linked to hypercholesterolemia in a pedigree with early coronary disease. Am. J. Hum. Genet., *39*:300, 1986.

169. Boerwinkle, E., Xiong, W., Fourest, E., and Chan, L.: Rapid typing of tandemly repeated hypervariable loci by the polymerase chain reaction: application to the apolipoprotein B 3' hypervariable region. Proc. Natl. Acad. Sci. USA, *86*:212, 1988.

170. Bock, G. and Collins, G.M. (editors): Molecular Approaches to Human Polygenic Disease. New York, John R. Wiley & Sons, 1987.

171. Smith, L.C. et al.: *In* Supramolecular Structure and Function. Edited by G. Pifat and J.N. Herrak. New York, Plenum Press, 1983. p. 210.

172. Annest, J.L., Sing, C.F., Biron, P., and Mongeau, J.: Familial aggregation of blood pressure and weight in adoptive families: II. Estimation of the relative contribution of genetic and common environmental factors to blood pressure correlations between family members. Am. J. Edpidemiol. *110*:492, 1976.

173. Boehnke, M., Moll, P.P., Kottke, B.A., and Weidman, W.H.: Partitioning the variability of fasting plasma glucose levels in pedigrees. Am. J. Epidemiol., *125*:679, 1987.

174. Sing, C.F., and Davignon, J.: Role of the apolipoprotein E polymorphism in determining normal plasma lipid and lipoprotein variation. Am. J. Hum. Genet. *37*:268, 1985.

175. Boerwinkle, E. et al.: The use of measured genotype information in the analysis of quantitative phenotypes in man: II. The role of the apolipoprotein E polymorphism in determining levels, variability, and covariability of cholesterol, beta-lipoprotein and triglycerides in a sample of unrelated individuals. Am. J. Med. Genet., *27*:567, 1987.

176. Cumming, A.M., and Robertson, F.: Polymorphism at the apoE locus in relation to risk of coronary disease. Clin. Genet., *25*:310, 1984.

177. Wardell, M.R., Suckling, P.A., and Janus, E.D.: Genetic variation in human apolipoprotein E. J. Lipid Res. *23*:1174, 1982.

178. Ehnholm, C. et al.: Apolipoprotein E polymorphism in the Finnish population: gene frequencies and relation to lipoprotein concentrations. J. Lipid Res., *27*:227, 1986.

179. Ordovas, J.M. et al.: Apolipoprotein E isoform phenotyping methodology and population frequency with identification of apoE1 and apoE5 isoforms. J. Lipid Res, *28*:371, 1987.

180. Smit, M. et al.: Apolipoprotein E polymorphism in the Netherlands and its effect on plasma lipid and apolipoprotein levels. Hum. Genet., *80*:287, 1988.

181. Deeb, S. et al.: Molecular genetics of apolipoproteins and coronary heart disease. *In* Molecular Biology of Homo Sapiens. New York, Cold Spring Harbor, 1986. p. 403.

182. Talmud, P.J. et al.: Apolipoprotein B gene variants are involved in the determination of serum cholesterol levels: a study of normo- and hyper-lipidaemic individuals. Arteriosclerosis, *67*:81, 1987.

Chapter 11

CARDIOVASCULAR MOLECULAR BIOLOGY IN HEALTH AND DISEASE: SUMMARY AND CONCLUSIONS

Shu Chien

Applications of the principles and techniques of molecular biology to cardiovascular research have contributed significantly to our knowledge of circulatory physiology and pathophysiology. Such applications have made it possible to probe the functions of cells and tissues at a level of molecular detail not feasible previously. From the result of complementary DNA (cDNA) sequencing, one can determine the primary structure of the protein encoded and can deduce the secondary structure by computer analysis, thus identifying the various functional domains and sites of interactions with other proteins. From molecular probing of the gene and the cDNA encoding the protein, one can gain insights into the transcription and translation processes leading to the expression of various functionally important proteins, as well as the molecular mechanisms regulating their gene expression. The use of a cDNA probe obtained from a particular protein of one tissue for the cloning and sequencing of cDNAs from other tissues often shows that a protein thought to be specifically present in one tissue actually has a relatively wide distribution and that the analogous proteins in different tissues often have minor molecular variations that account for their different functional manifestations.

Recent molecular biologic studies have also elucidated the interactions among different elements in the circulatory system, e.g., circulating blood elements, vascular endothelial cells, and vascular smooth muscle cells, in normal and disease states. Furthermore, investigations on the molecular mechanisms of regulation of gene expression in the cardiovascular system have shown a striking similarity with those found in tumor cells, viruses, and nervous and endocrine systems. Therefore, molecular biologic studies have led to a remarkable convergence of research fields, including such seemingly divergent areas as cancer, virus, neuroscience, endocrine, and cardiovascular research. These results indicate that molecular biology, which is usually considered to be a science using a reductionist's approach, can offer a global picture of the roles of related proteins in different organs and systems.

The purpose of this chapter is to summarize the materials presented in the preceding chapters, with special emphasis on the common concepts and principles and the application of the molecular biologic approach to physiologic and pathophysiologic investigations.

The functions of the cardiovascular system are effected mainly by the proteins in the system. The mechanical functions of the circulation include the pumping action of the heart, the resistance and capacitance of the vasculature, and the deformation of blood cells and vascular cells. The effector proteins that perform these functions are the contractile proteins in cardiac and vascular smooth muscle cells and the cytoskeletal proteins in blood cells, endothelial cells, and other cardiovascular cells. The functions of the effector cells and effector proteins are mediated and regulated by the interaction of receptors on the cell surface with ligands from extracellular sources. The receptor and ligand proteins may be referred to as mediator proteins. Examples of these mediator proteins are the atrial natriuretic factor (ANF), angiogenin, platelet-derived growth factor (PDGF), and lipoproteins, and their receptor proteins. The paradigm of the control of cardiovascular functions, like that in other physiologic systems, is that the specific interaction between the ligand and its receptor (mediator proteins) causes a signal transduction through a second messenger system to modulate the performance of the effector proteins. In the following sections, a summary is given on the information presented in this book for these two types of proteins, and this is followed by discussions on the physiologic and pathophysiologic relevance of the molecular biology of these proteins.

The purpose of separating the proteins in the cardiovascular system into two categories is to facilitate the organization of the materials, particularly because the contractile and cytoskeletal proteins overlap in structure and function and the mediator proteins have some common characteristics. These two types of proteins, however, also share common features. In many physiologic and developmental conditions, coordinate changes occur in the expression of effector and mediator proteins; e.g., the genes for actin, myosin, and ANF in the heart are regulated in parallel following birth and during myocardial hypertrophy. In addition to the proteins covered in this book, other proteins also play important roles in cardiovascular functions. Examples are the ion channels (e.g., sodium and calcium channels), membrane enzymes (e.g., ATPases), and various neurohumoral mediator proteins and their cardiovascular receptors. A recent publication edited by the author[1] includes discussions on the molecular biology of sodium channel, Na^+-K^+ ATPase, and autonomic receptors in relation to their physiologic functions.

EFFECTOR PROTEINS: CONTRACTILE AND CYTOSKELETAL PROTEINS

The major functioning cells in the cardiovascular system are cardiac muscle and vascular smooth muscle cells, endothelial cells, and blood cells. The muscle cells have contractile proteins as the major functional component; endothelial and blood cells have cytoskeletal proteins that serve to maintain the structural integrity of the cell and mediate functions such as deformation and motility. Each of these proteins exists in isoforms, and many of the isoforms are key components in both the contractile and cytoskeletal proteins.

Tissue Isoforms of Effector Proteins: Relation Between Protein Structure and Cellular Function

As a result of the precise determination of molecular structures of nucleic acids and proteins, many proteins are found to exist in isoforms, varying in their detailed amino acid sequence as a function of (1) the tissues in which the protein is expressed, (2) the developmental stages, (3) the physicochemical stimuli acting on the system, and/or (4) species evolution. An understanding of the roles of the isoforms helps to elucidate the structure–function relationship of the protein and the mechanisms of tissue-specific, growth-dependent, function-adaptive, and evolution-induced differentiations.

The protein isoforms expressed in various regions of the body may subserve similar functions with minor variations or may carry out different functions with considerable tissue specificities. Region-specific protein isoforms are encoded by region-specific messenger RNA (mRNA) isoforms. The results of molecular analysis have shown that mRNA isoforms can arise from (1) the transcription of distinct isoform-specific genes that form a multigene or supergene family, e.g., actins and the myosin regulatory light chains (p. 55, p. 96, p. 103); (2) alternative splicing of a single gene, e.g., band 4.1, β-spectrins (p. 161, 164), and the A chain of PDGF (p. 126); or (3) multiple sites of initiation of transcription (multiple promoters) from a single gene, e.g., the anion exchanger membrane protein band 3 (p. 167). For some proteins, isoforms are produced by a combination of these mechanisms; e.g., tropomyosin isoform diversity is generated by both the first and second conditions mentioned (p. 99), and the isoforms of myosin alkali light chains are transcribed from a single gene by using both the second and the third condition (p. 105).

ACTINS

Actins are a major component of the cytoskeleton of all cells and comprise an essential part of the contractile apparatus of all types of muscle (p. 96). Mammalian cells contain at least six actin isoforms, four of them are muscle actins: α-skeletal muscle, α-cardiac muscle, α-smooth muscle (vascular), and γ-smooth muscle (genital and digestive tracts) and two are nonmuscle actins: β-nonmuscle and γ-nonmuscle. The actin in the erythrocyte membrane skeleton is of the β-nonmuscle type. These actin isoforms are encoded by different genes in a multigene family (p. 55, p. 96). Actin isoforms have highly conserved primary structures, with >90% amino acid homology among them; the sequence conservation is particularly strong for the four muscle actins. The various actin isoforms, though minor in their structural variations, have some functional differences. For example, a switch from nomuscle to α-smooth muscle actin in aortic smooth muscles correlates with an increase in force-generating capacity (p. 98). The differential turnover of actin isoforms in smooth muscles suggests the existence of differences in their intracellular processing.

MYOSINS

Among the various types of muscles, the skeletal and cardiac muscles are more similar, and the smooth muscle differs significantly from these sarcomeric muscles not only in its lack of an organized sarcomere but also in the mechanism and kinetics of contraction. Myosin plays some specific roles in smooth muscle contraction that are different from those fround in sarcomeric muscles. Myosin is also found in nonmuscle cells such as erythrocytes; erythrocyte myosin is much less in quantity and probably less important in function in comparison to muscle myosin.

The amino acid sequences of the isoforms of contractile proteins deduced by using molecular genetic techniques have provided insights into the molecular domains of these proteins that may be important for the functional differences between smooth and sarcomeric muscles. Myosin is composed of myosin heavy chain (MHC), regulatory light chain (RLC), and alkali light chain (ALC). The MHC isoforms of smooth muscles have significant sequence differences from those of sarcomeric muscles, in contrast to the high level of conservation of actins in different types of muscles (p. 101) or the high degree of homology between the α- and β-MHC isoforms in the cardiac muscle (p. 45). An important mechanism that operates specifically in the smooth muscle to generate the energy of contraction involves the phosphorylation of serine-19 on RLC as a result of the activation of myosin light chain kinase (MLCK) by Ca^{2+}–calmodulin complex (p. 93). The RLC isoforms in rat sarcomeric and smooth muscles have amino acid sequences that are similar in the regions flanking the serine-19 but are more different at the two termini that may be involved in the activation of RLC in smooth muscle by MLCK. The MLCK isoforms of smooth muscle and skeletal muscle have 54% homology in the catalytic domain but have much larger differences in the calmodulin-binding domain, which is important in the activation of MLCK in smooth muscle cells to cause RLC phosphorylation.

TROPOMYOSINS

The α-tropomyosin (α-TM) in muscles forms a coiled coil dimer that induces polymerization of the α-muscle actin and lies in the major groove of the F-actin thin filament (p. 98). In the erythrocyte membrane, the nonmuscle TM has a similar relation to the F-actin formed by 12 to 16 monomers of β-nonmuscle actin. α-TM plays a fundamental role in Ca^{2+}-activated contraction of sarcomeric muscles through its interaction with actin and troponins, but it does not appear to be as important for smooth muscle contraction. The TM proteins encoded by different mRNA isoforms in sarcomeric and smooth muscles have the same number of 284 amino acids, and all contain the sequences involved in head-to-tail polymerization of actin. These TM isoforms, however, have minor differences in their primary structures, which may be responsible for the functional differences between sarcomeric and smooth muscles, but the exact structure–function relationship remains to be established.

ERYTHROID AND NONERYTHROID CYTOSKELETAL PROTEINS

The mammalian erythrocyte membrane skeleton comprises many types of proteins. Some of the cytoskeletal proteins are isoforms of the contractile proteins found in muscle, e.g., actin, tropomyosin and myosin, which in the preceding sections are referred to as nonmuscle isoforms. Other membrane proteins in the erythrocyte were first thought to be specific for this cell type, e.g., spectrin, band 3, band 4.1, etc. Subsequent studies by immunologic and molecular biologic techniques, however, have shown that isoforms of these proteins are present in most nonerythroid cells and tissues (pp. 157–160). Examples are the nonerythroid isoforms of spectrin (e.g., fodrin in the brain and TW260/240 in the intestinal brush border), band 3 (e.g., in the kidney), and band 4.1 (e.g., in lymphocytes). The spectrins in various nonerythroid tissues share a common α-subunit and differ in their β-subunit (p. 161). In the chicken, erythoid and muscle cells have a single species of α-spectrin transcript, but these cells have distinct forms of β-spectrin. All α- and β-spectrins have internal repeats of 106-amino acid sequences, which are linked together to form the long-chain molecules (p. 161).

Comparative analysis of human erythroid and lymphoid band 4.1 sequences has revealed complete homology except for five nucleotide sequence "motifs" (or domains), which appear to be inserted or deleted by alternative splicing (p. 164). For example, motif I, which encodes the 21 amino acid segment of the 10-kilodalton (kD) spectrin-actin-binding domain, is present in erythroid and absent in lymphoid DNA, whereas the reverse is the case for motif II, which encodes 34 amino acids near the carboxy terminus of the 22/24-kD domain of band 4.1. The variations in amino acid sequences of these isoforms probably form the basis for functional differences of the 4.1 proteins between erythroid and lymphoid cells.

Regulation of Isoform Expression of Effector Proteins

The expression of muscle protein isoforms is regulated not only in a tissue-specific manner but also during developmental differentiation and in response to various physiologic stimuli. The regulated switching of isoforms in a given type of cell under these conditions apparently serves to adapt its protein expression to meet changing functional demands. Some examples of these adaptive changes follow.

ACTINS

Proliferating skeletal myoblasts contain predominantly the nonmuscle β- and γ-actins, which decline rapidly on cessation of growth and switch to the α-skeletal muscle isoforms (p. 97). The expression of actin isoforms in cardiac and smooth muscles is similarly regulated so that the tissue-specific isoforms become expressed with growth and development. Thus, the fetal rat heart contains α-skeletal muscle actin, and this is switched to α-cardiac muscle actin during development (p. 56). Aortas from newborn rats and gizzards from chick embryos initially contain almost exclusively β-nonmuscle actin, but in the adult, these tissues contain mostly α- and γ-smooth muscle actins, respectively (p. 97). Such developmental expressions of muscle actin genes appear to be regulated at the transcriptional level involving the *cis*-acting elements on the genomic DNA as well as the *trans*-acting factors, which bind to the *cis* elements (p. 97).

MYOSINS

The cardiac MHC isoforms in humans and rats are the products of α- and β-genes; they have distinct patterns of developmental regulation and are differentially affected by cardiac work and hormone status (p. 45 and p. 55). In the rat ventricle, three MHC isoforms have been identified. V_1 myosin, which contains two α-MHC molecules, has a higher adenosine triphosphatase (ATPase) activity than does V_3 myosin, which contains two β-MHC molecules; V_2 myosin contains one α- and one β-MHC molecule and is intermediate in ATPase activity. In the developmental hypertrophy of the normal rat heart, the β-MHC isoform in the fetal heart is switched to α-MHC in the adult, concurrent with the actin isoform switch mentioned previously. Different types of hypertrophy of the adult rat ventricle are associated with different patterns of isoform switching, as discussed later in the section entitled Mechanism of Myocardial Hypertrophy (p. 233).

Human α-sarcomeric muscle MHC gene has a 17-base pair (bp) *cis*-acting sequence identical to that found in genes encoding the muscle isoforms of creatine kinase and α-actin (p. 47). This *cis*-acting sequence has been referred to as a "muscle-specific" element and may be involved in the coordinate control of a subset of muscle genes by *trans*-acting factors, e.g., the thyroid hormone (p. 46) or the myogenic proteins produced by the myogenic regulatory genes (p. 39).

In cultured vascular smooth muscle cells, the expression of MHC isoforms can change with the growth state (p. 101). Subconfluent, proliferating cultures contain mainly the nonmuscle MHC isoform. The postconfluent, growth-arrested cells express increased levels of smooth muscle MHCs, although with the continued expression of large amounts of the nonmuscle MHC. These changes in MHC isoforms during growth arrest of cultured smooth muscle cells are accompanied by the previously mentioned switching of actin from β-nonmuscle to α-smooth muscle isoform.

The tropomyosin in undifferentiated skeletal myoblasts is the nonmuscle isoform, which switches to the striated muscle α-TM isoform as a result of differentiation. A similar developmental switch is found in the cardiac muscle. These developmental changes in TM expression are generated by alternative splicing, probably as a result of the action of developmentally specific *trans*-acting factors that direct α-TM splicing (p. 99).

Sequence Homology Between Different Genes of Effector Proteins: Gene Families

The analysis of nucleotide sequence has allowed the comparison of genes and cDNAs encoding various proteins. Such studies have demonstrated homology not only among protein isoforms, as discussed in the preceding sections, but also among proteins with similar or even seemingly different functions. This suggests that nature uses relatively few basic themes in protein synthesis and that variations from these themes make possible the formation of a great number of diverse types of proteins that can perform functions that are different in some respects but similar in others. The genes that evolve from a common ancestral gene and code for a number of proteins with sequence homologies constitute a multigene or supergene family. Some genes, however, exist as single-copy genes without closely related members, based on tests such as genomic Southern blot analysis. Examples of such single-copy genes are the chicken gizzard smooth muscle and nonmuscle MHCs (p. 102), chicken band 3, mouse band 3, human glycophorin A, and human glycophorin B (pp. 166–168).

MYOSINS

The sarcomeric MHC multigene family includes many isoforms of skeletal and cardiac muscle MHCs, which are expressed in tissue- and developmentally specific manners (see p. 223 and p. 225). Considerable homologies exist between cardiac and skeletal muscle MHC genes; thus, the human cardiac α-MHC/chloramphenicol acetyltransferase (CAT) constructs are expressed also in a skeletal muscle cell line but not in nonmuscle cell lines such as 3T3 or HeLa cells (p. 48). The smooth muscle and nonmuscle MHC genes are far more closely related than they are to the sarcomeric MHC genes (p. 101). In contrast to the large sarcomeric MHC multigene family, smooth muscle and nonmuscle MHCs have few isoforms in their families. Rat α-MHC and growth hormone genes have similar elements for thyroid hormone control (p. 47).

SPECTRINS

α-Spectrin may be part of a large gene family of similar proteins, e.g., α-actinin, a cytoplasmic actin-binding protein, and dystrophin, the protein encoded by the gene found to be deficient in several forms of muscular dystrophy (p. 163). All these proteins have rod-like morphology, with an actin-binding domain and a series of α-helical repeats. One of the internal units in α-spectrin (α-10) has a different structure from the other units, which have the 106-amino acid repeats; this α-10 unit shows homology to the *src* family of proteins, e.g., protein kinases, and phospholipase C, which also associate with the inner membrane surface (p. 163).

MUSCLE-SPECIFIC REGULATORY GENES AND *myc*-ONCOGENE FAMILY

The regulation of gene expression in cardiac myocytes (Chaps. 3 and 4), vascular smooth muscle cells (Chap. 6), endothelial cells (Chap. 7), and blood cells (Chap. 9) in growth and differentiation has been discussed in various chapters. The available results suggest the existence of a paradigm for the mechanism of regulation of gene expression in the cardiovascular system, involving the participation of proto-oncogenes (p. 56).

Several muscle-specific regulatory genes (myogenic genes) are members of a large *myc*-oncogene family and share remarkable homology (p. 39). These myogenic proteins also share homology with proteins encoded by developmentally important genes involved in processes such as sex determination and neurogenesis. These homologies suggest that the regulation of expression of muscle genes involves a mechanism similar to that used in the regulation of growth and development of other types of cells, including neurons, gonads, and tumor cells. Thus, molecular biologic investigations have brought together seemingly diverse fields, such as neuroscience, endocrines, cancer, and cardiovascular research.

Species Comparisons of Effector Proteins: Gene Evolution

Sequence comparison of a given type of gene among different species allows an estimation of the rates of evolution of the gene and its subdomains; such analysis also can provide insights into the structural requirement of protein function. The genes encoding some proteins show particularly remarkable sequence conservation among different species, as well as in various tissues. Examples are the actins (p. 96), smooth muscle MHCs (p. 101), smooth muscle RLCs (p. 103), and nonerythroid spectrins (p. 162). Such conserved genes probably indicate that the proteins they encode have relatively little tolerance to structural modifications for the rather universal functions subserved.

CONTRACTILE PROTEINS

Rat and human cardiac α-MHC genes have remarkable homology in the 5' flanking region several hundred nucleotides upstream of the transcription initiation site but not farther downstream. This sequence conservation may indicate a similarity in the regulatory mechanisms mediated by the region of homology.

A cDNA clone of smooth muscle MHC isolated from the rabbit uterus shows a high degree of amino acid sequence homology (>95%) with the chicken gizzard smooth muscle MHC but shows a much lower level of homology with rabbit or chicken striated muscles. These results suggest that the smooth muscle MHC genes diverged from the ancestral sacromeric MHC genes before the gene duplications that generated the sarcomeric multigene family and probably before the emergence of vertebrates (p. 102).

CYTOSKELETAL PROTEINS

While the nonerythroid spectrins are remarkably conserved, the erythroid spectrins show more structural divergence in different species (pp. 161, 162). These results suggest that the functional requirements placed on the nonerythroid spectrins are relatively uniform in different species, whereas the erythroid spectrins undergo more evolutional alterations to adapt to changing functional needs, e.g., between nucleated and nonnucleated erythrocytes. α-Spectrin is better conserved among species than β-spectrin; restriction enzyme analysis of total mouse and human genomic DNAs has demonstrated the existence of a single erythrocyte α-spectrin (p. 161). The cDNA for human band 3 shows homology in the coding region to that for chicken band 3, and it shows homology to the cDNA for mouse band 3 in both the coding and noncoding regions (p. 167). Comparisons of human and chicken actin genes have demonstrated greater than 85% homology in nucleotide sequences and identical amino acid sequences; similar comparisons between human and rat have shown greater than 90% nucleotide homology and a difference in only one amino acid (p. 168).

MEDIATOR PROTEINS: HUMORAL AND RECEPTOR PROTEINS

Tissue Isoforms of Mediator Proteins

PLATELET-DERIVED GROWTH FACTOR AND RECEPTOR

PDGF is a dimer formed by different combinations of two subunits (A and B chains), which have a high degree of homology (p. 126). All three dimeric isotypes of PDGF (PDGF-AA, -AB and -BB) are biologically active. The PDGF from human platelets is an AB heterodimer, that from porcine platelets is a BB homodimer, and those from smooth muscle cells and some tumor-derived cell lines are putative AA homodimers. The PDGF isotypes have different functional properties (p. 127). The AA homodimer has a lower mitogenic activity, a weaker chemotactic activity, and a lesser ability to cause vasoconstriction and membrane ruffling than the BB form. The AA homodimer, however, has a higher efficacy for activating protein kinase C.

Known effects of PDGF are mediated via its interaction with specific receptors (p. 128). Two types of PDGF receptors have been cloned; one type binds preferentially the BB form of PDGF, whereas the other binds all three isoforms of PDGF. These differences in ligand specificity between the two types of receptors may account for the different functional activities of the PDGF isoforms. Although PDGF A and B chains are homologous, the ligand-binding domains of the two types of PDGF receptors have only about 30% homology (p. 129). The cytoplasmic region of both receptors contains a conserved split tyrosine kinase domain, which functions in signal transduction. It is possible that these two types of receptors may transmit different intracellular signals following PDGF binding and mediate the different functions for the PDGF isoforms.

ANGIOGENIN

The angiogenin gene appears to be expressed in all cells and tissues examined and also in normal plasma (p. 139), and there are no known isoforms of angiogenin. Apparently, there is a single angiogenin gene.

APOLIPOPROTEINS

Each of the major apolipoproteins exists in two or more isoforms that differ in charge or mass and are detected by two-dimensional electrophoresis (p. 186). Apolipoprotein mRNAs undergo co-translational and post-translational modifications (p. 189). The polymorphism of the various major apolipoproteins in plasma reflects post-translational modification processes, including amino acid substitution, glycosylation, deamidation, proteolytic cleavage, acylation, and phosphorylation.

Regulation of Isoform Expression of Mediator Proteins

REGULATION OF ATRIAL NATRIURETIC FACTOR

The expression of ANF is cardiac chamber-specific and developmentally regulated (p. 83). In the adult mammal, ANF mRNA is specifically abundant in the atria; it is present in low level in ventricles and very small amounts in the lung, aorta, and central nervous system. Before birth, the level of ventricular ANF transcripts is comparable to atrial level, but during the first week of life, ventricular level falls abruptly while atrial expression increases. The developmental regulation of ANF in the ventricular cells seems to be an on–off switch or a quantitative change, rather than the qualitative change involving isoform switching described previously for the contractile proteins (see section entitled Regulation of Isoform Expression of Effector Proteins, p. 225). The changes of myocardial contractile proteins and ANF with development are such that the isoforms become more tissue-specific in their molecular structure or more chamber-specific in their distribution.

REGULATION OF PDGF

Cultured vascular smooth muscle cells not only bind and respond to PDGF but also are capable of secreting the mitogen themselves (p. 128). This autocrine production of PDGF may be responsible, in part, for the expansion in size of the growing aorta. Smooth muscle cells from newborn rats express both the A and B chains of PDGF and secrete substantial amounts of PDGF. Smooth muscle cells cultured from adult rats express only the A chain of PDGF and secrete only a small fraction as much of the mitogen. The changes in PDGF isoform types may contribute to the developmental regulation of PDGF expression.

Gene Families of Mediator Proteins

ADHESIVE PROTEINS (See Chap. 7)

Blood cells and endothelial cells can express on their membrane surfaces protein molecules that mediate cell adhesions by ligand–receptor interactions.

The integrin family (p. 121). The integrin family of adhesion proteins includes three molecules found on lymphoid and myeloid leukocytes: lymphocyte function-associated antigen-1 (LFA-1), complement receptor CR3, and p150,95. These three molecules have a common β-subunit but different α-subunits. Molecules in the integrin family interact with the adhesion molecules that belong to the immunoglobulin supergene family, which are described in the next section.

The immunoglobulin supergene family (p. 121). The intracellular adhesion molecule-1 (ICAM-1), which is a ligand for the LFA-1 on lymphocyte surface, is expressed on endothelial cells as well as monocytes, lymphocytes, epithelial cells, and fibroblasts. cDNA sequence of ICAM-1 shows that it is composed of five immunoglobulin-like domains, indicating that this glycoprotein is a member of the immunoglobulin supergene family. ICAM-1 has extensive sequence similarity to other cell-adhesion molecules such as the nerve cell-adhesion molecule (N-CAM) and the myelin-associated glycoprotein. cDNA analysis of ICAM-2, a second LFA-1 ligand, reveals that it is a transmembrane protein with two immunoglobulin-like domains. These ICAMs, which bind the same integrin receptor, are more closely related to each other than to other members of the immunoglobulin superfamily and may be considered as part of a subfamily.

Studies on surface molecules on lymphocytes mediating cell adhesion and killing show that CD2, CD3, CD4, and CD8 glycoproteins, and the T-cell receptor are also members of the immunoglobulin supergene family (p. 120). LFA-3 is a ligand for CD2 in lymphocyte adhesion to a number of cells on which LFA-3 is expressed, including endothelial cells, epithelial cells, fibroblasts, erythrocytes, and most cells of hematopoietic origin. LFA-3 is homologous to its ligand CD2, and it too is a member of the immunoglobulin superfamily. Analysis of cDNA of PDGF receptor reveals a ligand-binding region consisting of five immunoglobulin-like domains.

Proteins mediating leukocyte adhesion to endothelial cells (p. 124). Analysis of the cDNA of Endothelial-leukocyte adhesion molecule-1 (ELAM-1) and granule membrane protein-140 (GMP-140), two types of inducible molecules on endothelial cell surface that mediate the adhesion of neutrophils and related cell lines, shows structural similarities. Both have an N-terminal lectin-like domain, an Epidermal growth factor (EGF) domain, and complement-receptor repeats (six for ELAM-1 and nine for GMP-140). One of the "homing" receptors (Mel-14) expressed by the memory T lymphocytes that direct them to lymph nodes for adhesion to the high endothelial venules has a similar structure as well, i.e., an N-terminal lectin-like domain, an EGF repeat, and two complement-receptor repeats. Another homing receptor, CD44, has an N-terminus with a significant homology to cartilage link proteins, and this provide a structural basis for interactions with extracellular matrix.

PLATELET-DERIVED GROWTH FACTOR

Protein sequence analysis of purified PDGF revealed a striking (90%) homology between the B chain of PDGF and the transforming gene product of the v-*sis* oncogene (p. 126). The v-*sis* oncogene is the transforming gene of an acutely transforming retrovirus, the simian sarcoma virus (SSV). Inappropriate expression of the protein product of the v-*sis* gene is responsible for the induction and maintenance of tumor growth by SSV. The cellular protooncogene c-*sis* is the gene that encodes the B chain of PDGF and is the gene from which the viral v-*sis* oncogene is thought to be originally derived. The other chain of PDGF is encoded by the A-chain gene. The PDGF A- and B-chain genes show striking architectural similarities (p. 126). Although located on different chromosomes (7 and 22 for A and B, respectively), both genes contain seven exons, which correlate with the functional subdivisions of the growth factor. In fact, in regions of substantial amino acid homology, the sequences and splice junctions can be precisely aligned. The amino acid sequences of the PDGF gene are conserved across species.

The PDGF A-chain transcriptional unit is more complex than the B chain (p. 126). Three transcripts are generated from the PDGF A-chain gene, which appear to arise by selection of alternative poly(A) addition sites. Additionally, the A-chain transcript undergoes alternative mRNA splicing, but the events regulating this alternative splicing are not known. Additionally, it is unclear whether the different forms of the A chain generated by the splicing process are functionally different.

ANGIOGENIN

Angiogenin has 35% sequence identity to ribonuclease A and belongs to the same gene family (p. 139). This homology has helped in the design of experiments of site-directed and regional mutagenesis to elucidate the structural basis of angiogenic function. Angiogenesis by angiogenin requires both its cell binding and ribonucleolytic activity.

APOLIPOPROTEINS

The seven major soluble human apolipoproteins (apoA-I, A-II, A-IV, C-I, C-II, C-III, and E) have similar genomic structures, patterns of internal repeats, and sequences of signal peptides, and they constitute an apolipoprotein multigene family (p. 187). The nascent translation products of apolipoproteins contain a hydrophobic signal peptide sequence, 18 to 27 amino acids in length (p. 190). The soluble apolipoproteins in the multigene family are composed predominantly of multiple repeats of 22 amino acids, each of which is a tandem array of two 11-mers, and two adjacent 22-mers are connected by a helix-breaker proline (p. 193). Two classes of repeats are found in apoB: the amphipathic helical regions, which are also found in all apolipoproteins in the multigene family, and the hydrophobic proline-rich domains, which are found only in apoB-100. The amphipathic helical regions are organized to form a hydrophobic face in contact with the lipid core and a hydrophilic face in contact with the surrounding aqueous phase, thus facilitating material transfer between the two phases and the interaction of lipoprotein with enzymes, receptors, and other molecules (p. 192). The hydrophobic proline-rich repeats are composed of predominantly β-sheets and β-turns, and the cooperativity in the lipid-binding of the two types of repeat domains in apoB-100 may account for its lack of exchange between different lipoprotein particles (p. 196).

ApoD, which is a 169-amino acid glycoprotein found in high-density lipoprotein (HDL), does not have any internal repeats (p. 197). The ApoD gene has a different structure and may belong to the α_{2u}-globulin superfamily, which includes human plasma retinol-binding protein, α_1-microglobulin, etc. (p. 202).

The similarities and differences among the members of the apolipoprotein multigene family provide insights into their evolutionary development, as discussed later in the section entitled Apolipoproteins on this page.

Species Comparisons of Mediator Proteins: Gene Evolution

ANF

ANF genes are well conserved, with relatively little differences among species.

ANGIOGENIN

Site-directed mutagenesis studies on angiogenin have led to the postulation that the active site of ribonuclease A has evolved to allow the effective binding of angiogenin to a ligand or cell-surface receptor (p. 144).

APOLIPOPROTEINS (pp. 197–206)

The striking similarity in the genomic structure and molecular organization of the seven soluble apolipoproteins indicates that these genes have arisen from a common ancester, and that the individual apolipoprotein genes have evolved through partial and complete gene duplications. The structures of the mRNAs of members of the apolipoprotein multigene family can be incorporated into a common model for the comparison of molecular structure among different apolipoproteins (and their genes), as well as the comparison of the same apolipoproteins (and their genes) among different species. Such comparisons allow the estimation of the rates of evolution of individual apolipoproteins and their subdomains and the inference of the relationships among members of the multigene family.

A comparison of the number of substitutions per nucleotide site between two homologous genes with known divergence time allows the computation of the rate of nucleotide substitution and an estimation of the date of evolutionary events such as gene duplication. Such studies have been performed for the sites of nucleotides in apolipoproteins where nucleotide substitution has led to a change in amino acids. The results on the signal-peptide region indicate that the signal peptide is exceptionally well conserved in the human apoC-III, as compared to the signal peptides of other apolipoproteins or to those of dog and rat apoC-III. The results on the mature peptide region as a whole indicate that all apolipoproteins studied, except for a partial sequence of apoB-100, evolve more rapidly than the average gene, e.g., β-globin. The shorter proteins A-II, C-II, and C-III are less conservative than the longer proteins A-I, E, A-IV, and B-100. Analysis of the structure and organization of various apolipoprotein genes has led to the proposal of an evolutionary tree in which apoC-I represents the earliest evolutionary offspring from a common ancestral gene about 680 million years ago. In the ensuing 400 million years, first apoC-II, then A-II and C-III, followed by E, and finally A-IV and A-I diverged from the main stem. Detailed analysis of potential homologous regions suggests that apoB-100 might be related to apoA-I, A-IV, and E.

In general, functionally more important molecules or parts of a molecule have more stringent structural constraints and evolve more slowly than the less important ones. The most conservative of all apolipoprotein genes is apoB-100, which is an obligatory constituent of very low density lipoprotein (VLDL), intermediate density lipoprotein (IDL), and low density lipoprotein (LDL), the ligand that binds to the LDL receptor, and an important risk factor of atherogenesis. A remarkable degree of homology exists between the human apoB-100 and the apoB-100 of the lowest vertebrates, the hag fish. ApoE, which is another ligand for the LDL receptor, has one of its most highly conserved sequences in the segment encompassing the receptor-binding region (residues 139 to 166).

PHYSIOLOGIC SIGNIFICANCE

The use of molecular biologic techniques has led to the establishment of molecular structures of various proteins of interest to physiologists, e.g., contractile proteins, membrane proteins, hormones, receptors, etc. The structure of the molecule can be correlated with its function, thus elucidating the molecular basis of physiologic processes. Knowledge of the DNA sequence of the genes encoding these proteins allows an analysis of their control and regulation. By genetic engineering techniques, specific mutations can be created at predetermined site(s) of a DNA molecule (in-vitro mutagenesis); the mutant gene can then be reintroduced into the organism for the analysis of the effect of the specific mutation on functions (transgenic mice). In this manner, physiologic functions of proteins can be assessed with precision to the level of a single amino acid or a single base of the gene.

In the preceding sections on effector and mediator proteins, the relevance of molecular biologic studies to physiologic functions has been summarized in terms of the structure–function relationship of protein isoforms, the isoform switching with development and in response to physiologic stimuli, and the mechanisms of regulation of gene expression. For example, studies on contractile proteins have allowed an analysis of the molecular basis of the functional differences between sarcomeric and smooth muscle cells and point to the direction of future research. In the first part of this section, the mechanism of myocardial hypertrophy is used as an example of the application of molecular biologic approaches to study a physiologic process.

Mechanism of Myocardial Hypertrophy

Myocardial hypertrophy, which is a subject of study in Chapters 3 to 5, occurs in physiologic states as well as pathologic conditions. Myocardial hypertrophy is a part of the physiologic processes of normal growth and exercise training, and it also occurs in response to neurohumoral signals, e.g., sympathetico-adrenergic stimulation and thyroid hormone. The use of molecular biologic techniques has provided insights into the molecular mechanisms for cardiac myocyte hypertrophy, which involves the response of the *cis*-acting elements of myogenic genes to the *trans*-acting factors modulated by various types of stimuli. In the proto-oncogene concept, these stimuli may be considered as growth factors, which act on growth-factor receptors to trigger intracellular transducers for the activation of transcription factors (p. 56). The growth factors can be either humoral, e.g., the catecholamines that bind with α_1-adrenergic receptor, or physical, e.g., the mechanical stretch that can alter cell-surface ion channel serving as a mechanoreceptor. In the myocardial hypertrophy induced by adrenergic stimulation, the intracellular transducer mediating the activation of the transcription factor may be protein kinase C (p. 75).

The 5' flanking region of the rat α-MHC gene, just as that of the rat growth hormone gene, has a thyroid hormone-responsive element (TRE) containing a binding site for a nuclear T_3 receptor, which is the product of the proto-oncogene c-*erb*-A (p. 45). The actions of T_3 on the expression of contractile proteins are mediated by this receptor and the TRE. While catecholamines may act on the α_1-adrenergic receptor on the cell surface to trigger the secondary activation of a transcription factor to react with a *cis*-acting sequence of the β-MHC gene (p. 74), the receptor for thyroid hormone T_3 appears to interact directly with the *cis*-acting TRE (p. 46). The lack of an induction of α-MHC mRNA synthesis by T_3 in the presence of inhibitors or protein synthesis suggests the importance of multiple *trans*-acting protein factors in the regulation of α-MHC transcription (p. 49).

Myocardial hypertrophy can involve different patterns of expression of the contractile protein isoforms (p. 55). In the hypertrophy of cardiac myocytes accompanying normal growth and development, specific inductions of cardiac α-actin and α-MHC mRNAs occur. The hypertrophy produced by exercise training or thyroid hormone is also accompanied by an up-regulation of the adult α-MHC isoforms (p. 45). In contrast, the hypertrophy resulting from α_1-adrenergic stimulation or pressure overload of the adult ventricle is accompanied by induction of skeletal α-actin, β-MHC, and c-*myc* mRNAs (p. 69). Thus, while the hypertrophy of ventricles during normal growth and development and following exercise training or thyroid hormone administration involves the down-regulation of a set of fetal genes and replacement by their adult isoforms, the hypertrophy of ventricles resulting from adrenergic stimulation or pressure loading involves the up-regulation and re-expression of these fetal genes. Such inductions of early developmental isogenes reflect a fundamental change in the transcriptional program of the cardiac myocyte nucleus and may have physiologic significance. The re-expression of the fetal β-MHC isoform would decrease the myosin ATPase activity and myofibril shortening velocity; thus, the burden of increased myocardial function under

[handwritten annotation at top: true in systemic circulation where the vascular wall is exposed to high blood pressure but in microcirculation, where the pressure is not high → no stimulation → no hypertrophy]

chronic adrenergic stimulation or hypertension seems to be borne by an increased number of contractile units rather than the enhanced performance of each unit. The re-expression of the ANF gene in the ventricle during cardiac hypertrophy (p. 84) may be an attempt to increase ANF secretion, which would reduce the after-load and facilitate cardiac emptying. Therefore, some of these isoform switchings may be viewed as teleologically beneficial, but further experimental studies are needed to test the validity of this postulation.

Studies on the time course of the transcriptional induction following α_1-stimulation have shown transient increases of the transcription of cardiac α-actin and skeletal α-actin, with that of skeletal α-actin being much longer lasting (p. 72). In contrast, the α_1-induced increases in transcriptions of ribosomal and transfer RNA (rRNA and tRNA) are not transient. It would be interesting to also determine the temporal sequence of the changes in various types of isoforms in the development of myocardial hypertrophy in vivo; this would provide insights into the mechanism for changing the transcriptional program of these genes.

The developmental and regulatory changes in the transcription of the genes encoding cardiac muscle proteins (Chaps. 2 to 4), vascular smooth muscle proteins (Chap. 6), and ANF (Chap. 5) result from the direct or indirect actions of humoral and other *trans*-acting factors on the *cis*-acting elements.

Functions of Cytoskeletal Proteins

The cytoskeletal proteins of the erythrocyte membrane play major roles in maintaining the cell shape under resting state and effecting cell deformation in response to shear stresses imparted by blood flow. These proteins form a network, which is connected with the transmembrane glycoproteins and interacts with the lipids in the membrane bilayer (p. 155). These glycoproteins serve a variety of functions such as ion transport and ligand bindings, and their interactions with the cytoskeletal network affect each other's functions. Molecular biologic investigations have established the nucleotide and amino acid sequences of most of the cytoskeletal and transmembrane proteins in the erythrocyte membrane, thus allowing the elucidation of the molecular sites of protein interactions and the structural basis of membrane functions.

The rather ubiquitous existence of the cytoskeletal proteins, many of them first found in the erythrocyte, as isoforms in various tissues suggests that these proteins may also play a role in maintaining cell shape and mechanical strength in other cell types, as well as mediating other physiologic functions. It has been known for some time that actin undergoes dynamic reorganization during mitosis; recent studies on band 4.1 indicate that this and other cytoskeletal proteins may also play a significant role in cell division (p. 164).

Model of Cell–Cell Adhesion

Molecular clonng has led to new structural data characterizing several molecules expressed on the surface of endothelial cells that appear to play important active roles in mediating interactions with circulating lymphocytes and other blood leukocytes (p. 123). An instructive model has been generated for the binding of T lymphocytes to either antigen-presenting cells (e.g., endothelial cells) or target cells (p. 123). In this model, the association consists of a ligand pair conferring specificity and a series of accessory interactions that stabilize the specific association. This model may have a more general application to other types of cell–cell associations in the cardiovascular and other systems.

Regulation of Circulating Proteins

Proteins or peptides such as ANF, angiogenin, and apolipoproteins are secreted or released into the circulating fluid from the cells in which they are synthesized. Such secreted proteins generally are produced as a precursor (designated with a prefix "pre") with a hydrophobic signal peptide sequence, which is removed intracellularly (co-translational modification) before the release of the protein. In some cases, after the removal of the signal peptide, an additional length of amino acid residues at the amino terminus is removed post-translationally to form the mature protein released; the protein before the final removal of this peptide is referred to with a prefix "pro", e.g., proANF, and the nascent precursors are designated as preproANF, etc. Therefore, the regulation of expression of these proteins involves, in addition to transcription, co-translational and post-translational processes.

Atrial Natriuretic Factor

ANF has many important physiologic functions in the cardiovascular system, including diuresis, natriuresis, and vasodilation, and it plays a significant role in volume and pressure regulation. The developmental regulation of ANF expression has been discussed in the section entitled Regulation of Atrial Natriuretic Factor (p. 229); there is an up-regulation in atrial cells and down-regulation in ventricular cells after birth.

The *cis*-acting sequences, which are 5′ to the ANF gene, have been cloned. By transfecting cultured atrial and ventricular cells with different types of ANF–CAT hybrid genes, it has been shown that a region of 2.5 kilobases (kb) flanking the 5′ end of the ANF gene promotes expression of the marker gene in embryonic and neonatal atrial cells as well as embryonic, but not neonatal, ventricular cells (p. 85). The results indicate that the *cis*-acting regulatory region directs atria-specific expression of the ANF gene after birth and developmental control of ventricular ANF transcription. In addition to the *cis*-acting sequences, developmentally regulated alterations in the *trans*-acting factors may occur in the cardiac cell nuclei. ANF gene transcription and circulating ANF level are increased by adrenergic stimulation, thyroid hormone, and glucocorticoids. These circulating hormones may interact through their receptors with specific *cis*-acting sequences to augment ANF gene expression. There seems to be a coordinate regulation of the ANF gene with cardiac contractile protein genes during development, cardiac hypertrophy, and neurohumoral stimulations (see Regulation of Atrial Natriuretic Factor, p. 229). In view of the faster response of ANF than muscle hypertrophy, however, the time rates of the activation and/or response of these genes probably differ.

ANF mRNA is translated into the precursor hormone preproANF (151 amino acids in humans and 152 in rodents), which is processed intracellularly for the removal of the signal peptide to yield proANF (126 amino acids in all species). In the atria, perinuclear granules store proANF, which can be released in response to stimuli such as atrial stretch and cardiac glycosides. At the time of secretion, the proANF is cleaved by seryl protease to remove 98 amino acids and generate the active hormone ANF, which is only 28 amino acids in length. Ventricular cells do not have the capacity to store ANF, probably not even when ventricular transcription is up-regulated during re-expression of fetal genes. Therefore, the synthesized hormone would be constitutively released to augment the circulating level.

Angiogenin

The presence of angiogenin cDNA in a normal liver library in addition to tumor cell libraries, the expression of angiogenin gene in all cells and tissues examined, and the existence of angiogenin in normal plasma indicate that this protein probably has a physiologic role in addition to its possible function in neoplasia (p. 140). The precursor of angiogenin has a 22 to 24 amino acid signal peptide, which is removed to form the mature protein (p. 140). Site-specific mutagenesis of the angiogenin gene has generated data in support of the concept that the induced production and release of diacylglycerol from membrane inositol phospholipids is one of the steps leading to neovascularization (p. 144).

Lipoproteins

All the major human apolipoprotein cDNAs and chromosomal genes have been cloned and sequenced, and chromosomal localization and structural organization of the human apolipoprotein genes have been established (p. 187). These findings have made possible the structure–function correlation of the various types of apolipoproteins.

An interesting molecular mechanism for the control of organ-specific expression of apoBs of markedly different sizes has been elucidated through the study of biogenesis of apoB-48 mRNA (p. 190). In the mid-section of the 14-kb apoB mRNA, there is a codon CAA (encoding glutamine) in the liver, but it is UAA (stop codon) in the intestine. As a result, the apoB-100 expressed in the liver is a 550-kD apolipoprotein, but the apoB-48 in the intestine is composed of only the first 264 kD of the apoB-100. This type of differential RNA editing by the modification of a single nucleotide allows the generation of two apolipoproteins that are markedly different in size and subserve different functions. ApoB-48 is a marker for intestinal chylomicron and chylomicron remnants, whereas apoB-100 is an important physiologic ligand for the LDL receptor.

Sequence analysis has led to the identification of two putative LDL-receptor-binding domains on apoB-100, which show a high degree of homology to the receptor-binding domain of apoE (p. 195). A synthetic peptide of one of the two regions has been found to bind to the LDL receptor and to elicit the expected functional changes in cultured human fibroblasts.

Extrapolation From In-Vitro Results to In-Vivo Systems

Most of the molecular biologic studies start with a purified protein or nucleic acid coding for the protein. To attain such purified substances, the research is usually conducted on cultured or isolated cell systems, which allow the isolation of one particular cell type from a mixed population of different cell types. Even in the heart, only 25% of the cells are cardiac myocytes (p. 54). Therefore, the availability of cultured systems (e.g., those of cardiac myocytes, vascular smooth muscle cells or endothelial cells) has greatly facilitated molecular biologic studies on these cells. The cultured cells possess many of the characteristics of the corresponding cells in vivo. For example, cultured myocytes, similar to the hearts of intact animals, undergo hypertrophy in response to α_1-adrenergic stimulation with the selective induction of skeletal α-actin, β-MHC, and c-*myc* mRNAs (p. 70) and increase their MHC V_1 isoform and ANF secretion in response to thyroid hormone (p. 45, p. 87); cultured smooth muscle cells can display both "synthetic" and "contractile" phenotypes and undergo phenotypic modulation analogous to that seen in vivo (p. 94); and genetically engineered angiogenin can alter neovascularization in the chick embryo in directions to be predicted from its characteristic ribonuclease activity pattern in vitro (p. 146). Sometimes, however, differences occur between the behaviors of the cultured cells and the in-vivo system, especially when the experimental variables are not strictly comparable. For example, a switch from β-actin to α-actin is observed in the aorta of rats during development from newborn to adult and also in cultured rat aortic smooth muscle cells during growth from proliferative phase to growth arrest. The switching in the aorta during development in vivo appears to be regulated at the level of transcription, whereas the switching in cultured aortic amooth muscle cells during growth appears to be regulated at the level of translation (p. 97). One must be cautious in extrapolating the in-vitro results to in-vivo systems. The best approach is to establish first the basic principles in the culture system and then assess their physiologic relevance by using in-vivo systems, which should provide the ultimate test.

PATHOPHYSIOLOGIC AND CLINICAL IMPLICATIONS

Genetic Disorders in the Cardiovascular System

It is obvious that molecular biologic approaches are valuable in the elucidation of etiology and pathogenesis of genetic disorders and in the advancement of methods of diagnosis and treatment. One of the best examples is provided by the pioneering studies by Brown and Goldstein[2] on the LDL receptor in relation to hypercholesterolemia and atherosclerosis. They have established molecular abnormalities in each of the four steps controlling LDL-receptor metabolism: synthesis, intracellular transport, LDL binding, and receptor internalization. These results serve to elucidate the molecular basis of the genetic disorders and to provide insights into the fundamental mechanisms of regulating LDL metabolism. The following gives a brief summary of several types of genetic diseases discussed in the various chapters of this book.

GENETIC DISORDERS IN LDL

Abnormalities in LDL can cause some of the hypercholesterolemic syndromes (p. 206), in addition to the well-established roles of hereditary abnormalities in the LDL receptors mentioned previously. Studies on genetic disorders have helped to elucidate the structure–function relationship of lipoproteins. An example is the mutant apoC-II proteins from two patients with homozygous deficiency in the activation of lipoprotein lipase (p. 203). ApoC-II$_{Toronto}$ has a base deletion in the codon of either Thr-68 or Asp-69, resulting in a reading frame shift and the consequent replacement of amino acids 69 to 79 by an unrelated hexapeptide. ApoC-II$_{St. Michael}$ has a base insertion into the codon of either Asp-69 or Gin-70, resulting in a reading frame shift and the consequent replacement of amino acids 70 to 79 by an unrelated peptide sequence of 70 to 96, beginning and ending with proline. These cases point to the importance of amino acids 70 to 79 of apoC-II in lipase activation. With the use of synthetic apoC-II peptides, it has been shown that the carboxy-terminal tripeptide 77 to 79 in this protein is critically required in lipase activation. Interspecies comparison of the amino acids in this tripeptide between dog and human has suggested that the critical element may be the carboxylic acid side chain at residue 78.

LEUKOCYTE ADHESION DEFICIENCY

In congenital leukocyte adhesion deficiency, all three members of the integrin family of cell-surface adhesion proteins are not expressed because of a mutation of the common β-subunit; hence, the monocytes and neutrophils from these patients cannot bind to and cross the endothelium at sites of infection, leading to recurrent life-threatening bacterial infection (p. 121).

HEREDITARY HEMOLYTIC DISORDERS (pp. 168–177)

Many hereditary hemolytic disorders involve abnormalities in erythrocyte membrane proteins, e.g., hereditary spherocytosis (HS), hereditary elliptocytosis (HE), and hereditary pyropoikilocytosis (HPP). HS has been shown to result from defects in the β-spectrin and/or ankyrin gene. Because a deficiency in ankyrin reduces the stability of spectrin attachment to the membrane, it also results in a deficiency in membrane spectrin. This may explain the correlation observed between the degree of spectrin deficiency and the degree of spherocytosis or clinical severity in HS. Approximately 30% of HE patients and all HPP patients demonstrate defective spectrin self-association to form oligomers as a result of molecular defects in the amino terminus of α-spectrin or the carboxy terminus of β-spectrin. Another molecular defect that can lead to HE is a deficiency in band 4.1. The identification of various molecular defects in association with different types of hereditary hemolytic disorders has led to the following hypothesis: Defective vertical interactions, i.e., spectrin attachment to the membrane, lead to HS, whereas defective horizontal interactions in the cytoskeletal network, i.e., spectrin self-association within the plane of the membrane, lead to HE or HPP.

Several mutant mice strains have hereditary hemolytic anemias involving the α-spectrin, β-spectrin, or ankyrin genes; these could serve as excellent animal models for studying the human hereditary hemolytic anemias. Experimental procedures have been developed in which the mutant gene or the corrected gene can be introduced into animals to produce transgenic animals for the investigation of the influence of such gene transfers on the induction or correction of deranged physiologic functions.

Hereditary disorders in human patients and animals provide naturally occurring genetic mutants. Molecular biologic studies in these cases can generate important information on the correlation of physiologic and pathophysiologic functions to molecular structure. For example, studies on hereditary hemolytic disorders have made possible the identification of the molecular domains of skeletal proteins responsible for specific functions such as interactions with other proteins and have allowed the assignment of the chromosomal location of the genes coding for these proteins.

Pathophysiology of Cardiovascular Diseases

Myocardial hypertrophy, which is discussed in Chapters 3 to 5, occurs in many types of cardiovascular disorders, including hypertension, valvular diseases, coronary artery disease, and hypertrophic cardiomyopathies. Myocardial hypertrophy is often maladaptive in these clinical conditions, being either inadequate or excessive relative to hemodynamic loading conditions. Based on the proto-oncogene concept, such maladaptive hypertrophy could be considered as representing abnormalities of myocyte proto-oncogenes (p. 56), just as cancer can be produced by over-expression or mutation of proto-oncogene proteins. The proto-oncogene products important for hypertrophy are beginning to be identified and may include a variety of receptors (e.g., α_1-adrenergic, T_3, stretch) and transducers (e.g., protein kinase C). Myocardial contractile and secretory proteins undergo regulatory changes in the hypertrophies associated with cardiac disease, with a re-expression of the fetal genes.

Aortic smooth muscles undergo an actin isoform switch in atherosclerosis (p. 97).

The aortic smooth muscle cells obtained from human atheromatous plaques or rat aorta with experimentally induced intimal thickening contain almost exclusively the β- and γ-nonmuscle actins characteristic of the "synthetic" phenotype, as contrasted to the normal aorta, which contains mainly the "contractile" phenotype of α-smooth muscle actin. Therefore, the atheromatous lesion might also be considered as an abnormal growth with the re-expression of fetal isoforms, just as in the case of myocardial hypertrophy. In both myocardial hypertrophy and aortic thickening, the isoform switches are regulated mainly at the transcriptional level.

The increase in circulating ANF in cardiac patients may result from an increase in atrial stretch, which is a potent stimulus for ANF release (p. 88). The diuresis induced by cardiac glycosides in patients with heart failure may be related to the increased atrial secretion of ANF in response to the drug. In severe cardiac patients with failing hearts, proANF appears in plasma in addition to the 28-amino acid ANF, suggesting an altered post-translational processing in these patients (p. 89).

The functional discrepancies between PDGF isoforms suggest that differences may occur in the mitogen produced in various pathophysiologic processes involving vascular smooth muscle cell replication, e.g. the media thickening in hypertension vs. the focal intimal proliferation in early atherosclerosis, and also in comparison to the physiologic process of normal development (p. 128). An increase in PDGF A-chain expression and an attendant smooth muscle cell proliferation have been found in the intimal hyperplastic lesion developed in the rat carotid artery after balloon injury and also in vascular cells isolated from human atheromatous lesions (p. 128). Further investigations in this area may elucidate the roles of the molecular nature of the mitogen in smooth muscle proliferation in various pathophysiologic states.

Epidemiology of Cardiovascular Diseases (See Chap. 10)

The combination of molecular genetic and epidemiologic studies promises to provide a way of predicting individuals having a genetic liability for premature occurrence of common cardiovascular diseases, as exemplified by the investigations of apolipoproteins in relation to coronary heart disease. Two types of polymorphic genetic variability contribute to variations of lipid, lipoprotein, and apolipoprotein levels in the population. One type of gene mutation is the rare alleles, e.g., the LDL-receptor defects leading to severe familial hypercholesterolemia and increased risk for coronary heart disease.[2] The more common type is an increased risk that is due to the presence of a number of polygenes, which are genes with polymorphic allele frequencies and moderate or small effects on the disease trait of interest, in combination with environmental factors. Statistical analysis on the latter type indicates that a significant fraction of variations in plasma apoB and other lipid levels is attributable to genetic factors and that the contribution of genetic factors usually exceeds that of environmental factors shared by family members.

To characterize the genetic factors involved in the atherosclerotic process, it is necessary to ascertain the effects of specific gene loci on the phenotypes of interest. Like other genes, the apolipoprotein genes exhibit considerable DNA sequence variations among individuals in a population. Most of the DNA sequence heterogeneity occurs in noncoding regions of the genes, but a small proportion is transcribed and translated to yield apolipoprotein heterogeneity. The best-characterized apolipoprotein variant is the polymorphism found in the human apoE gene, which codes for a 299-amino acid rich in arginine. The ϵ_2-isoform has cysteine at amino acids 112 and 158, and ϵ_4 has cysteine at 112 and arginine at 158. In ϵ_2/ϵ_2 homozygotes, the low binding affinity of the apoE reduces the hepatic uptake of chylomicron remnants, VLDL remnants and apoE-containing lipoproteins and causes their accumulation in plasma. The ϵ_2-containing VLDL remnants are converted to LDL at a slower rate, leading to an up-regulation of LDL receptors and a decrease in plasma LDL. In ϵ_4/ϵ_4 homozygotes, the increased internalization of cholesterol from apoE-containing lipoproteins is compensated by a down-regulation of LDL receptors and an increase in plasma LDL.

Restriction fragment-length polymorphisms (RFLPs) and direct-sequence analysis of apoB-100 tryptic peptides have demonstrated polymorphism throughout the molecule, but most polymorphic sites have no known physiologic effect. A specific apoB-100 allele has been found in a strain of pigs to result in familial hypercholesterolemia and premature atherosclerosis. An apoB mutation at amino acid 3,500 close to the putative LDL-receptor-binding domain of apoB-100 in humans has been found to be associated with premature atherosclerosis.

Diagnosis of Cardiovascular Diseases

With the availability of cDNA probes, genetic disorders can be diagnosed by the application of molecular biologic techniques such as RFLP. This approach has been developed in prenatal diagnosis of hereditary hematologic disorders, e.g. sickle cell anemia and thalassemia.[3] The development of similar applications can be anticipated for various cardiovascular diseases. RFLP has been used to identify an abnormal ankyrin gene in HS patients (p. 170), an abnormal 4.1 gene in HE patients (p. 174), and an abnormal glycophorin B gene in En(a-) patients who lack S, s, and U phenotypes (p. 167), as well as the role of apolipoprotein polymorphism in hypercholesterolemia (p. 212). Such approaches can be used for the molecular diagnosis of various hereditary disorders.

Treatment of Cardiovascular Diseases

Molecular biologic research on the cardiovascular system may generate new approaches for therapy of cardiovascular diseases in the future. For example, identification of the molecular signals controlling myocardial hypertrophy might permit manipulation of hypertrophy in disease states (p. 77, p. 89). Considerable clinical benefits might result if adequate hypertrophy could be stimulated, excessive growth could be prevented, or functional derangements resulting from growth abnormality could be corrected. Angiogenin and other angiogenic molecules may prove to be useful for the improvement of wound healing, whereas inhibitors of angiogenic molecules (e.g., the placenta ribonuclease inhibitor) can provide new therapeutic approaches for diseases associated with vascular proliferation, e.g., solid tumor growth and neovascular glaucoma (p. 152). Because genetically altered autologous endothelial cells can be introduced into the vascular wall and survive, they might be used as a drug-delivery system to treat diseases requiring secretion of gene products directly into the circulation. Furthermore, vascular grafts could be made to contain endothelial cells which are capable of secreting vasodilatory, thrombolytic or angiogenic factors to facilitate the recovery of ischemic myocardium, or capable of producing antineoplastic agents for the treatment of primary or metastatic tumors (p. 132).

CONCLUSIONS

The materials covered in this book indicate that applications of molecular biology have made possible the probing of the functions of cells and tissues to molecular details and contributed significantly to our knowledge of cardiovascular physiology and pathophysiology. Rapid increases in knowledge have been made on the molecular structures and dynamics of the proteins that play important roles in cardiovascular functions, including the identification of their functional domains and sites of interactions with other proteins, and the transcription and translation processes regulating their expression. Proteins thought to be specifically present in one tissue often have a relatively wide distribution, and the minor molecular variations among analogous proteins in different tissues provide the structural basis of their functional diversity. Molecular biologic studies have also elucidated the interactions among various components in the circulatory system in normal and disease states. Furthermore, the molecular mechanisms of regulation of gene expression in the cardiovascular system have striking similarity with those found in other systems. Thus, a remarkable convergence of such seemingly divergent areas as cancer, virus, neuroscience, endocrine, and cardiovascular research has resulted from molecular biologic studies, which can offer a global picture of the roles of related proteins in different organs and systems, while probing the details of biologic structure and function down to the molecular level.

The presentations in this book indicate that molecular biology has allowed the elucidation of physiologic functions transcending space and time. In terms of space, it has made possible the comparison of functional manifestations of different isoforms in various tissues of a given species (see sections entitled Tissue Isoforms, p. 222 and p. 228), as well as across species (see sections entitled Species Comparisons, p. 227 and p. 231). In terms of time, molecular biologic approaches allow us to unravel the mechanisms regulating the rapid responses to neurohumoral stimuli (see sections entitled Regulation of Isoform Expression, p. 225 and p. 229), the mechanisms controlling the slower processes of development and growth (see sections on Regulation of Isoform Expression), and the very slow process of the evolution of various genes and the proteins encoded (see sections on Species Comparisons). In the evolutional studies, molecular biology has allowed the use of knowledge derived from different species at a given time (the present) to deduce the process of evolution, which spans millions of years.

Although rapid advances have occurred in the application of molecular biologic approaches to solve important cardiovascular problems of significance in normal and pathophysiologic conditions, this field is still in its infancy, and an exciting future lies ahead. Availability of Genes and cDNA clones encoding the various proteins that subserve important cardiovascular functions is increasing, and techniques for introducing these genes into cells as well as inactivating native genes are improving rapidly. Therefore, the stage has been set for performing in-vivo studies to relate molecular biologic research to physiologic investigations at the organ-system level. Such in-vivo studies on whole organisms should be the ultimate aim for cardiovascular researchers in quest of answers to the intricate mechanisms of cardiovascular regulation in health and disease. Interdisciplinary research spanning molecular biology to organ-system studies will contribute to not only the elucidation of normal physiology but also the understanding of pathophysiology of cardiovascular disorders and the development of new methods for diagnosis and treatment.

ACKNOWLEDGMENTS

I would like to acknowledge the support by the National Heart, Lung and Blood Institute (HL-19454 and HL-44147). I wish to express my sincere thanks to the authors of Chapters 2 to 9 for their valuable inputs.

REFERENCES

1. Chien, S. (ed.): Molecular Biology in Physiology. New York, Raven Press, 1989, pp. 1–17.
2. Brown, M.S., and Goldstein, J.L.: A receptor-mediated pathway for cholesterol homeostasis. Science, *232:*34, 1986.
3. Chang, J.C., and Kan, Y.W.: A sensitive new prenatal test for sickle cell anemia. N. Engl. J. Med., *307:*30, 1982.

GLOSSARY

Allele

One of several alternative forms of a gene occupying a given locus on the chromosome.

Alternative splicing

Splicing of the primary transcript at different locations, i.e., the removal of different exons and the joining of different introns to produce isoforms of mRNA and hence protein isoforms.

***Alu* sequence**

A set of dispersed and related sequences, each ~300 base pairs long, of which ~500,000 copies are in the human genome. These sequences are transposable and create target-site duplication when they are inserted.

Aminoacyl-tRNA synthetase

Enzyme responsible for covalently linking a specific amino acid to the 2'- or 3'-OH position of tRNA.

Amplification

The production of additional copies of a chromosomal sequence, found as either intrachromosomal or extrachromosomal DNA.

Anticodon

A triplet of nucleotides in the structure of tRNA that is complementary to the codon in mRNA to which the tRNA responds.

Antisense RNA

The RNA transcribed from the DNA strand opposite to that which is normally transcribed, as a result of the engineering of a cloned gene. It has a sequence complementary to the normal "sense" RNA transcript and will hybridize with the sense RNA, thereby inhibiting the translational synthesis of the corresponding protein.

Bacteriophage

A virus that infects bacterial cells and causes their lysis. Also called "phage."

Bacteriophage lambda (λ)

A type of bacteriophage that can be used as a cloning vector; some are expression vectors (e.g., λgt11), whereas others are nonexpression vectors (e.g., λgt10).

Base

A base in a nucleotide is the nitrogen-containing ring compound, either a purine or a pyrimidine. The bases in DNA are adenine (A), guanine (G), cytosine (C), and thymine (T). Base pairing occurs between A-T and C-G. In RNA, thymine is replaced by uracil (U).

Base pair (bp)

A partnership of A with T or of C with G in a DNA double helix.

β-galactosidase

An enzyme capablle of digesting the chromogenic sugar 5-bromo-4-chloro-3-indolyl-β-D-gal-actoside (X-gal) to yield a product with blue color.

CCAAT box

Part of a conserved sequence often found about 75 bp upstream from the start point of eukaryotic transcription; it might play a role in regulating the transcription of a gene.

cDNA clone

A duplex DNA molecule representing an mRNA sequence, carried in a cloning vector.

Centromere

The point at which a pair of chromatids is attached in a chromosome.

Chloramphenicol acetyltransferase (CAT)

A marker enzyme commonly used to determine the parts of a eukaryotic gene responsible for the regulation of its expression. The prokaryotic CAT gene is joined to the DNA sequence of the gene to be studied. The recombinant DNA molecule is inserted into a eukaryotic cell, and the regulatory function of the sequence is tested by determining the expression of CAT activity.

Chromatid

One of the two identical DNA molecules formed after replication that remained joined at the centromere and that make up a metaphase chromosome. The pair of chromatids will separate into the two daughter cells at cell division.

Chromatin

A network of nuclear fibrils formed by DNA in association with histones. Each chromatin fiber (approximately 30 nm in diameter) is composed of a series of nucleosomes.

Chromosome

A structure in the cell that contains the compactly coiled DNA that codes for genetic information. In eukaryotes, it exists in the nucleus and contains DNA in association with histones. Each organism of a species is normally characterized by the same number of chromosomes in its somatic cells.

Cis-acting sequence

A nucleotide sequence on genomic DNA that can modulate the transcription of that DNA in response to transcription factors (or *trans*-acting factors).

Cis-configuration

The existence of two sites on the same molecule of DNA.

Cistron

The genetic unit defined by the *cis/trans* test; equivalent to a gene in comprising the smallest unit of DNA serving a single function, as in coding a polypeptide.

Clone

A genetically pure strain of cells descended in culture from a single cell. In recombinant DNA technology, a clone refers to a specific DNA molecule grown or produced in bacterial cells. A genomic clone is a cultured host cell containing a fragment of genomic DNA; a cDNA clone is a cultured host cell containing a molecule of cDNA.

Cloning

The process by which a single cell is grown into a colony of identical cells. In recombinant DNA technology, cloning refers to the procedure by which a particular DNA sequence can be obtained in large quantity: the DNA is inserted into a vector to produce recombinant DNA, which is then introduced into a host cell; those cells bearing the recombinant DNA vector are then selected for growth.

Codon

A series of three adjacent nucleotides (triplet) in one polynucleotide chain of a DNA or RNA molecule coding for a specific amino acid.

Colony hybridization

A procedure to detect specific recombinant clones in bacterial colonies by in situ hybridization using nitrocellulose blotting and a ^{32}P-labeled probe.

Complementary DNA (cDNA)

The DNA that has been copied from mRNA in a reaction catalyzed by the enzyme reverse transcriptase.

Concatemer

A linear series of DNA consisting of unit genomes repeated in tandem arrays as a result of the pairing of the complementary nucleotide sticky ends on the monomers.

Consensus sequence

An idealized sequence in which each position represents the base most often found when many actual sequences are compared.

Core DNA

The 146 bp of DNA contained on a core particle of histone.

COS cell

A monkey cell line that has been transformed by a plasmid carrying the early region of the simian virus SV40 DNA with a defective replication origin.

Cos site

The paired sticky ends and adjacent sequences of the λ genomes in a concatemer. It serves as a recognition site for cleaving the long concatemer into a unit-length monomer for packaging into each λ phage head.

Cosmid

A plasmid into which λ phage cos sites have been inserted; cosmid cloning vector can incorporate a long segment of DNA.

Degeneracy

The coding of the same amino acid by different base combinations in the mRNA triplet codon; this common coding and lack of specificity occur mainly with variations in the third base.

Deletion

The removal of a sequence of DNA, the regions on either side being joined together.

Deoxyribonuclease (DNase)

An enzyme that degrades DNA.

Diploid cell

A cell that contains two copies of each chromosome.

DNA ligase

An enzyme that joins together the cut ends of two fragments of DNA.

DNA polymerase

An enzyme that catalyzes the replication of DNA by building a complementary strand on a DNA template. It requires a primer to which bases are added.

Endonuclease

An enzyme that cuts DNA or RNA within a chain.

Enhancer

A DNA sequence that increases the amount of RNA synthesized from a transcription unit. Enhancers can be several thousand base pairs away from, and can appear in either orientation with respect to the transcription unit.

Eukaryote

An organism whose cells have true nuclei bounded by nuclear membranes and that exhibit mitosis.

Exon

A region of a primary transcript (or the DNA encoding it) that exits the nucleus and reaches the cytoplasm as part of a mature RNA molecule.

Exonuclease

An enzyme that removes nucleotides one at a time from the ends of RNA or DNA.

Expression vector

A vector containing an inserted gene and capable of causing the synthesis of a large amount of the protein coded by the gene when introduced into an appropriate host cell.

Footprinting

A technique for identifying the site on DNA to which some protein binds by virtue of the protection of bonds in this region against attack by nucleases.

Frameshift mutations

Deletions or insertions that are not a multiple of 3 bp and that therefore change the frame in which triplets are read during translation into protein.

Fusion protein

A protein produced from a hybrid gene generated by fusing part of the coding sequence of a gene to the coding sequence of a different gene.

Gene

A sequence(s) of DNA that carries the information that encodes a polypeptide. It is the biological unit of heredity, and it is located at a definite locus on a particular chromosome.

Genome

The complete set of hereditary information in a species. It is composed of all of the genes contained in the chromosomes.

Genomic DNA clones

Sequences of the genome carried by a cloning vector.

Gyrase

A type II topoisomerase of E. coli with the ability to introduce negative supercoils into DNA.

Haploid cell

A cell that contains only one copy of each chromosome.

Heteroduplex DNA (hybrid DNA)

DNA generated by base pairing between single strands that are derived from different parental duplex molecules and are complementary over a region.

Heterogeneous nuclear (hn) RNA

RNA transcripts of genes in the nucleus; hn RNA has a wide size distribution and low stability.

Histones

Conserved DNA binding proteins of eukaryotes that form the nucleosome, the basic subunit of chromatin.

Hybridization

The pairing of complementary DNA strands to give a duplex DNA hybrid or the pairing of complementary RNA and DNA strands to give an RNA-DNA hybrid.

Hybridoma

The cell line produced by fusing a myeloma with a lymphocyte; it may continue to express the parental immunoglobulins.

Hydropathy index

An index of hydrophobicity of a segment (e.g., 10 amino acids) of a polypeptide chain computed from the free energy required to transfer it from nonpolar solvent to water based on the amino acid composition of the segment.

In situ hybridization

A procedure in which the DNA of cells is denatured in situ so that reaction is possible with an added single-stranded DNA or RNA; the added preparation is labeled either by radioactivity or by enzyme or protein marker, and its hybridization is followed by autoradiography, histochemistry, or immunohistochemistry. Studying the cells in metaphase allows the genes to be located on specific chromosomes.

In vitro mutagenesis

A procedure to modify the sequence of a piece of DNA.

Inducer

A small molecule that triggers gene transcription by binding to a regulator protein.

Intron

A part of a primary transcript (or the DNA encoding it) that is not included in a mature cytoplasmic mRNA molecule.

Isoform

A member of a group of molecules (e.g., proteins) with homologous structure and function. Isoforms of a given protein can be found in different tissues and in different species.

Library of DNA

A statistical representation of all of the DNA sequences present in a cell, either those found in a genome (a genomic library) or those found in mRNA (a cDNA library).

Linkage

The tendency of genes to be inherited together as a result of their location on the same chromosome. Linkage is measured by percent recombination between loci.

Linker fragment

A short synthetic duplex oligonucleotide containing the target site for some restriction enzyme; it may be added to the end of a DNA fragment prepared during construction of recombinant DNA.

Meiosis

The form of cell division that produces haploid cells from a diploid cell. The chromosomes are randomly segregated into four nonidentical daughters, and each carries a 1 n chromosome number.

Messenger RNA (mRNA)

The RNA that carries genetic information copied from DNA to the cytoplasm to direct protein synthesis by specifying a sequence of amino acids.

Metallothionein

A metal binding protein that functions in heavy metal detoxification. The transcription of the mouse metallothionein I gene is induced by heavy metals.

Mitosis

The division of a eukaryotic somatic cell.

Mutation

A change in the sequence of genomic DNA.

Nick

The removal of a phosphodiester bond between two adjacent nucleotides on one strand of duplex DNA.

Nick translation

A method for producing radiolabeled DNA in vitro in which E. coli DNA polymerase I uses a nick in DNA to start degradation of one strand of a duplex DNA and synthetic replacement with radioactively labeled nucleotides.

Nonsense codon

Any one of three triplets (UAG, UAA, UGA) that cause termination of protein synthesis (UAG is known as amber, UAA as ochre, and UGA as opal).

Nonsynonymous substitution

A nucleotide substitution in DNA with a resulting change in the amino acid expressed.

Northern blotting

A technique for transferring RNA from an agarose gel to a nitrocellulose filter on which it can be hybridized with complementary probe DNA.

Nuclease

An enzyme that degrades nucleic acids.

Nucleic acid

A high molecular weight polymer formed by joining nucleotides together by phosphodiester linkage. DNA and RNA are both nucleic acids.

Nucleoside

A combination of 5-carbon sugar with a purine or pyrimidine base. Examples are adenosine, cystidine, guanosine, thymidine, and uridine.

Nucleosome

> The fundamental packing unit in chromatin. It consists of a nucleosome bead, which is composed of DNA in association with histones, and the attached linker DNA, which is not in association with histones.

Nucleotide

> A nucleotide consists of a nitrogen-containing base, a 5-carbon sugar, and one or more phosphate groups. Example: adenylate.

Oligonucleotide

> A polymer made up of a small number of nucleotides.

Oncogene

> Viral genetic material that, when introduced into a normal cell, can transform the host cell into a tumor cell. Each oncogene has a counterpart (proto-oncogene) in the normal eukaryotic genome with a closely similar sequence.

Open reading frame

> A sequence of DNA that can be read as a series of triplets coding for amino acids without any termination codons; the sequence is (potentially) translatable into protein.

Operator

> The site on prokaryotic DNA at which a repressor protein binds to prevent transcription from initiating at the adjacent promoter.

Operon

> A complete unit of bacterial gene expression and regulation, including structural genes, regulator gene(s), and *cis* control elements that serve as binding sites for regulator gene product(s).

Origin (ori)

> A sequence of DNA at which replication is initiated.

Palindrome

> A sequence of DNA that consists of inverted repeats, i.e., the nucleotide sequence on one strand from left to right is the same as that on the other from right to left.

Phenotype

> The entire physical, biochemical, and physiological makeup of an organism resulting from the interaction of its genetic constitution with the environment.

Plasmid

> A small circular DNA molecule (1,000–30,000 base pairs) found in several species of prokaryotes. It generally encodes proteins that are required for resistance to antibiotics or other toxic materials, rather than the proteins that are essential for the growth of the cell. Some plasmids are used as vectors, providing an essential tool of recombinant DNA technology.

Point mutation

> A replacement of a single base of a gene with another.

Poly (A) tail

> A string of 50 to 250 adenylate residues at the 3′ end of most eukaryotic mRNA.

Polyadenylation

> The addition of a sequence of polyadenylic acid to the 3′ end of a eukaryotic RNA molecule.

Polymerase chain reaction

A technique to replicate a segment in a double-stranded DNA by the cyclic repetition of denaturation at high temperature, annealing of synthetic oligonucleotide primers at a lower temperature, and extension of the primer-template complex with a DNA polymerase.

Polymorphism

The occurrence of genomes showing allelic variations in the population; the variations can be demonstrated by the different phenotypes produced or the different DNA fragment lengths generated by restriction enzymes. A polymorphic gene is a gene for which the most common allele has a frequency of less than 0.95.

Polysome (polyribosome)

An mRNA associated with a series of ribosomes engaged in translation.

Pribnow box

The consensus sequence of TATAATG centered about 10 bp before the start point of bacterial genes. It is a part of the promoter especially important in binding RNA polymerase.

Primary transcript

The original unspliced RNA product formed by RNA polymerase II in eukaryotes. It is spliced to produce mRNA.

Primer

A short sequence of nucleotides that is paired with one strand of DNA or RNA and provides a free 3'-OH end at which a DNA polymerase starts synthesis of a deoxyribonucleotide chain.

Primer extension

A method for the mapping of mRNA in which a radiolabeled complementary primer oligonucleotide is annealed to the mRNA, and the annealed primer is extended with reverse transcriptase to the 5' end of the mRNA. The DNA-RNA hybrid is denatured and electrophoresed for identification of the site of annealing and derivation of the mRNA sequence from the extended DNA.

Prokaryote

An organism, e.g., a bacterium, that does not have a true nucleus.

Promoter

A specific site on a DNA molecule that controls the transcription of RNA from adjacent sequences; a *cis* control element.

Proto-oncogene

A gene present in the normal vertebrate cell genome and has a close similarity to a viral oncogene as its counterpart. Many proto-oncogenes code for growth factors, growth factor receptors, or intracellular mediators.

Reading frame

One of three possible ways to group triplet codons in a sequence.

Recombinant DNA

A DNA molecule produced by the insertion of a foreign DNA into an autonomously replicating DNA molecule, such as a virus or plasmid.

Recombinant DNA technology

That body of techniques that allow one to clone a particular DNA sequence by forming a recombinant DNA molecule.

Regional mutagenesis

    A procedure to modify the sequence of a stretch of DNA involving a relatively large number of nucleotides.

Regulatory gene

    A gene that codes for an RNA or protein product whose function is to control the expression of other genes.

Replication

    The process of synthesizing a new strand of DNA by making a complementary copy from a DNA template.

Replication fork

    The point at which strands of parental duplex DNA are separated so that replication can proceed.

Restriction enzyme (restriction endonuclease)

    An endonuclease enzyme that recognizes and cleaves unique, specific short sequences of duplex DNA.

Restriction fragment length polymorphism (RFLP)

    Variations between individuals in the lengths of fragments obtained from their genomic DNAs when cut with a restriction endonuclease, reflecting differences in the restriction enzyme cutting sites between the genomic DNAs.

Restriction map

    A linear array of sites on DNA cleaved by various restriction enzymes.

Retrovirus

    An RNA virus that propagates via conversion into duplex DNA.

Reverse transcriptase

    An enzyme found in retroviruses that can catalyze the synthesis of cDNA from an mRNA template. It is used to produce cDNA from mRNA. Its action requires the preexistence of a short primer strand hybridized to the nucleotides near the 3' end of the mRNA.

Rho factor

    A protein involved in assisting E. coli RNA polymerase to terminate transcription at certain (rho-dependent) sites.

Ribonuclease (RNase)

    An enzyme of the transferase class that catalyzes the hydrolysis of phosphodiester bonds of RNA.

Ribonuclease protection assay

    A method to quantify a desired mRNA in cells. A radiolabeled, complementary RNA probe is synthesized and hybridized to the RNA extracted from cells. The probe fragment hybridized to the mRNA of interest is protected from digestion with ribonuclease and is separated on a sequencing gel for identification and quantification.

Ribosomal RNA (rRNA)

    The RNA that combines with several proteins to form ribosomes, structures that have binding sites for all the interacting molecules necessary for protein synthesis.

RNA polymerase (RPase)

    The enzyme that catalyzes the transcription of RNA from DNA. RPase can bind to the DNA at specific sites and catalyze the synthesis of RNA.

S₁ nuclease

An endonuclease that has preference for single-stranded DNA or for RNA molecules over double-stranded molecules.

S₁ nuclease protection assay

A method for transcript mapping in which a hybrid formed between the RNA and a labeled DNA fragment is subjected to S₁ nuclease digestion, which removes the unhybridized portion of the DNA. The protected probe fragments are separated by gel electrophoresis and detected by autoradiography, revealing the number and sizes of the protected fragments.

Sigma (σ) factors

Accessory proteins of RNA polymerase that aid in the recognition of promoter sites on DNA.

Signal peptide

A sequence of 15 to 60 hydrophobic amino acids typically present in the N-terminal region of the precursor of secretory proteins. It provides a sorting signal for the protein and is often removed from the protein proper once the sorting has been accomplished.

Site-directed mutagenesis

A procedure to specifically modify the coding sequence of a piece of DNA involving either one or a few nucleotides.

Site-specific recombination

The recombination that occurs between two specific (not necessarily homologous) sequences, as in phage integration/excision or resolution of cointegrate structures during transposition.

Southern blotting

The procedure for transferring denatured DNA from an agarose gel to a nitrocellulose filter, where it can be hybridized with a complementary probe nucleic acid.

Splicing

The removal of introns and the adjoining of exons involved in the maturation of mRNA; thus introns are spliced out while exons are spliced together.

Stem cell

Undifferentiated cell capable of giving rise to different types of specialized daughter cells.

Sticky ends

Complementary single strands of DNA that protrude from opposite ends of a duplex or from ends of different duplex molecules; can be generated by restriction enzymes, which produce staggered cuts in duplex DNA.

Structural gene

A gene that codes for any protein product other than a regulator.

Synonymous substitution

A nucleotide substitution in DNA without a resulting change in the amino acid expressed because of the coding for the same amino acid by the different codons.

TATA box (TATAA box)

A conserved A-T-rich septamer found about 25 bp before the start point of many eukaryotic RNA polymerase II transcription units; may be involved in positioning the enzyme for correct initiation.

Topoisomerase

An enzyme that can change the number of times the two strands of a DNA duplex cross over each other.

***Trans* configuration**

The presence of two sites on two different molecules of DNA (chromosomes).

***Trans*-acting factor (transcription factor)**

Humoral or nuclear factors that can bind the *cis*-acting sequence on genomic DNA to modulate the level of gene transcription.

Transcription

The process of synthesizing RNA on a DNA template.

Transcription unit

That region between the sites of initiation and termination by RNA polymerase in prokaryotes; may include more than one gene.

Transfection

The transfer of new genetic markers to a cell by the addition of DNA.

Transfer RNA (tRNA)

The RNA that decodes (translates) the base sequence of mRNA into the amino acid sequence of the protein. The activation of an amino acid following its binding to a specific tRNA allows formation of peptide bond in the polypeptide chain during protein synthesis.

Transformation

The acquisition of new genetic markers by incorporation of exogenous DNA; also the conversion of eukaryotic cells to a state of unrestrained growth in culture, resembling the tumorigenic condition.

Transgenic animal

An animal that contains an altered gene that is passed on to its progeny. It is produced by experimental introduction of the altered gene into a chromosome of the germ line cells or by mating animals bred from cultured embryo-derived stem cells transfected with the altered gene.

Translation

The process of synthesizing protein on the mRNA template.

Vector

A molecule capable of autonomous replication, usually either a bacteriophage or a plasmid that can grow in a particular host. It is used to receive, replicate, and allow the selection of the genomic DNA or cDNA of interest following introduction into the host.

Zinc finger

A molecular domain in a family of sequence-specific DNA-binding proteins with amino acids folded around a Zn atom that links two cysteines and two histidines.

INDEX

Note: Page numbers in *italics* refer to illustrations; page numbers followed by t refer to tables.

255